Biologic Effects of Atmospheric Pollutants

PARTICULATE POLYCYCLIC ORGANIC MATTER

Committee on
Biologic Effects of
Atmospheric Pollutants

DIVISON OF MEDICAL SCIENCES
NATIONAL RESEARCH COUNCIL

NATIONAL ACADEMY
OF SCIENCES
WASHINGTON, D.C. 1972

NOTICE: The study reported herein was undertaken under the aegis of the National Research Council with the express approval of the Governing Board of the NRC. Such approval indicated that the Board considered that the problem is of national significance, that elucidation of the problem required scientific or technical competence, and that the resources of NRC were particularly suitable to the conduct of the project. The institutional responsibilities of NRC were then discharged in the following manner:

The members of the study committee were selected for their individual scholarly competence and judgment with due consideration for the balance and breadth of disciplines. Responsibility for all aspects of this report rests with the study committee, to whom sincere appreciation is expressed.

Although the reports of our study committees are not submitted for approval to the Academy membership nor to the Council, each report is reviewed by a second group of scientists according to procedures established and monitored by the Academy's Report Review Committee. Such reviews are intended to determine, inter alia, whether the major questions and relevant points of view have been addressed and whether the reported findings, conclusions, and recommendations arose from the available data and information. Distribution of the report is approved, by the President, only after satisfactory completion of this review process.

The work on which this publication is based was performed pursuant to Contract No. CPA 70–42 with the Environmental Protection Agency.

Available from

Printing and Publishing Office
National Academy of Sciences
2101 Constitution Avenue, N.W.
Washington, D.C. 20418

ISBN 0-309-02027-1

Library of Congress Catalog Card Number 72-76309

Printed in the United States of America

PANEL ON POLYCYCLIC ORGANIC MATTER

JAMES N. PITTS, JR., Statewide Air Pollution Control Center, University of California, Riverside, *Chairman*

ROY E. ALBERT, Institute of Environmental Medicine, New York University Medical Center, New York

RAYMOND J. CAMPION, Products Research Division, Esso Research & Engineering Company, Linden, New Jersey

BERTRAM W. CARNOW, Department of Preventive Medicine, University of Illinois College of Medicine, Chicago

ERNEST H. Y. CHU, Biology Division, Oak Ridge National Laboratory, Oak Ridge, Tennessee

JOHN C. CRAIG, Department of Pharmaceutical Chemistry, University of California, San Francisco

T. TIMOTHY CROCKER, Department of Community and Environmental Medicine, University of California Medical School, San Francisco

CHARLES HEIDELBERGER, McArdle Laboratory, University of Wisconsin, Madison

GEORGE M. HIDY, Science Center, North American Rockwell Corporation, Thousand Oaks, California

AVERILL A. LIEBOW, Department of Pathology, University of California Medical School, La Jolla

EDWARD D. PALMES, Institute of Environmental Medicine, New York University Medical Center, New York

RAYMOND R. SUSKIND, Department of Environmental Health, University of Cincinnati College of Medicine, Cincinnati, Ohio

BENJAMIN L. VAN DUUREN, Institute of Environmental Medicine, New York University Medical Center, New York

ELIZABETH E. FORCE, Division of Medical Sciences, National Research Council, Washington, D.C., *Staff Officer*

CONSULTANTS

EDWARD J. BAUM, Oregon Graduate Center, Portland
SAMUEL S. EPSTEIN, Case Western Reserve University School of Medicine, Cleveland, Ohio
CHRISTOPHER S. FOOTE, Department of Chemistry, University of California, Los Angeles
SHELDON K. FRIEDLANDER, Department of Environmental Engineering, California Institute of Technology, Pasadena
HARRY V. GELBOIN, Chemical Branch, National Cancer Institute, National Institutes of Health, Bethesda, Maryland
JAMES E. GILL, Biomedical Division, Lawrence Radiation Laboratory, University of California, Livermore
DIETRICH HOFFMANN, Division of Environmental Toxicology, American Health Foundation, New York
WILLIAM D. MacLEOD, Jr., Department of Pharmaceutical Chemistry, University of California, San Francisco
PAUL MEIER, Department of Statistics, University of Chicago, Chicago, Illinois
HERBERT L. RATCLIFFE, Penrose Research Laboratory, Zoological Society of Philadelphia, Philadelphia
O. CLIFTON TAYLOR, Statewide Air Pollution Control Center, University of California, Riverside

CONTRIBUTORS

ROSWELL K. BOUTWELL, Department of Oncology and Biochemistry, University of Wisconsin, Madison
DAVID L. COFFIN, Experimental Pathology Section, Biology Research Branch, Division of Health Effects Research, Air Pollution Control Office, Durham, North Carolina
IAN T. HIGGINS, School of Public Health, University of Michigan, Ann Arbor
MORRIS KATZ, Department of Chemistry, York University, Toronto, Ontario, Canada

JOHN B. LITTLE, Department of Physiology, Harvard School of Public Health, Boston, Massachusetts
NATHAN MANTEL, Biometry Branch, National Cancer Institute, National Institutes of Health, Bethesda, Maryland
PAUL NETTESHEIM, Carcinogenesis Program, Biology Division, Oak Ridge National Laboratory, Oak Ridge, Tennessee
LEO ORRIS, Institute of Environmental Medicine, New York University Medical Center, New York
RICHMOND T. PREHN, Institute for Cancer Research and Department of Pathology, School of Medicine, University of Pennsylvania, Philadelphia

COMMITTEE ON BIOLOGIC EFFECTS OF ATMOSPHERIC POLLUTANTS

ARTHUR B. DuBOIS, Department of Physiology, School of Medicine, University of Pennsylvania, Philadelphia, *Chairman*
VINTON W. BACON, College of Applied Science and Engineering, University of Wisconsin, Milwaukee
ANNA M. BAETJER, Department of Environmental Medicine, School of Hygiene and Public Health, The Johns Hopkins University, Baltimore, Maryland
W. CLARK COOPER, School of Public Health, University of California, Berkeley
MORTON CORN, Graduate School of Public Health, University of Pittsburgh, Pittsburgh, Pennsylvania
BERTRAM D. DINMAN, School of Public Health, University of Michigan, Ann Arbor
LEON GOLBERG, Institute of Experimental Pathology and Toxicology, Albany Medical College, Albany, New York
PAUL B. HAMMOND, Department of Physiology and Pharmacology, College of Veterinary Medicine, University of Minnesota, St. Paul
SAMUEL P. HICKS, Department of Pathology, University of Michigan, Ann Arbor
VICTOR G. LATIES, Department of Radiation Biology and Biophysics, University of Rochester Medical Center, Rochester, New York

ABRAHAM M. LILIENFELD, Department of Chronic Diseases, School of Hygiene and Public Health, The Johns Hopkins University, Baltimore, Maryland

PAUL MEIER, Department of Statistics, University of Chicago, Chicago, Illinois

JAMES N. PITTS, JR., Statewide Air Pollution Control Center, University of California, Riverside

GORDON J. STOPPS, The Environmental Health Branch, Health Studies Service, Ontario Department of Health, Toronto, Ontario, Canada

O. CLIFTON TAYLOR, Statewide Air Pollution Control Center, University of California, Riverside

JAROSLAV J. VOSTAL, Department of Pharmacology and Toxicology, University of Rochester Medical Center, Rochester, New York

T. D. BOAZ, JR., Division of Medical Sciences, National Research Council, Washington, D.C., *Executive Director*

Acknowledgments

The preparation of this document, which deals with a very complex subject, in so short a time was possible only because of the exceptional and dedicated efforts of those involved—particularly, the members and consultants of the Panel on Polycyclic Organic Matter, all of whom contributed their time generously. The report represents a team effort and has been critically evaluated *in toto* by the entire Panel.

Responsibility for generating drafts was divided among three task forces—on environmental appraisal; studies of POM in animals, mammalian cells, and vegetation; and human effects—to facilitate the exchange of information and ideas among smaller groups of researchers working in common areas of experience. The contributions of the chairmen of these task forces—Drs. John C. Craig, Charles Heidelberger, and Raymond R. Suskind, respectively—to the development of this document were noteworthy.

Credit must be given to the persons who drafted the chapters. Those who contributed to the chapters on structure and nomenclature of POM, sources, atmospheric physics, chemical reactivity, collection, separation, detection, identification, and quantitation include: Drs. Edward J. Baum, Raymond J. Campion, John C. Craig, Christopher

S. Foote, James E. Gill, George M. Hidy, Dietrich Hoffmann, Morris Katz, and William D. MacLeod, Jr.; the chapters on the effects of POM in animals, mammalian cells, and vegetation: Drs. Roswell K. Boutwell, Ernest H. Y. Chu, David L. Coffin, T. Timothy Crocker, Samuel S. Epstein, Harry V. Gelboin, Charles Heidelberger, Dietrich Hoffmann, Averill A. Liebow, John B. Little, Nathan Mantel, Paul Nettesheim, Leo Orris, Richmond T. Prehn, Herbert L. Ratcliffe, O. Clifton Taylor, and Benjamin L. Van Duuren; and the chapters on the evaluation of effects of particulate POM in humans: Drs. Roy E. Albert, Bertram W. Carnow, Ian T. Higgins, Averill A. Liebow, Paul Meier, and Raymond R. Suskind.

A significant factor in expediting the development of this document was the support given to the Panel by Dr. Robert J. M. Horton, Senior Research Adviser, Dr. Eugene Sawicki, and Dr. Francis G. Hueter of the Environmental Protection Agency.

The document was reviewed by the parent Committee on the Biologic Effects of Atmospheric Pollutants, by several anonymous editors selected by the National Academy of Sciences, by the Academy's Report Review Committee, by the Academy's Advisory Center on Toxicology, and by the National Research Council's Divisions of Biology and Agriculture, Chemistry and Chemical Technology, and Physical Sciences. The report was edited by Mr. Norman Grossblatt, Editor for the Division of Medical Sciences. Their substantial contributions to this document are gratefully acknowledged.

Finally, I should like to acknowledge the efforts of Miss Elizabeth E. Force, of the Division of Medical Sciences, who served as manager for the Panel activities. Her rare combination of managerial ability, technical competence, and personal enthusiasm did much to promote the effectiveness of the entire operation.

CHARLES L. DUNHAM
Chairman, Division of Medical Sciences

Preface

The 1967 amendments to the Clean Air Act of 1963 required that the Secretary of Health, Education, and Welfare

> from time to time, but as soon as practicable, develop and issue to the States such criteria of air quality as in his judgment may be requisite for the protection of the public health and welfare. . . . Such criteria shall . . . reflect the latest scientific knowledge useful in indicating the kind and extent of all identifiable effects on health and welfare which may be expected from the presence of an air pollution agent. . . .

A critical step in implementing these requirements of Congress has been the issuance of Air Quality Criteria Documents by the National Air Pollution Control Administration (NAPCA), more recently designated the Air Pollution Control Office Technical Center of the Environmental Protection Agency (EPA). Air Quality Criteria Documents already published are on particulate matter, oxides of sulfur, hydrocarbons, carbon monoxide, photochemical oxidants, and oxides of nitrogen. Until recently, these documents generally were prepared by the combined efforts of inhouse NAPCA staff and consultants—both individuals and teams from private companies.

In the spring of 1970, the Division of Medical Sciences, National Academy of Sciences–National Research Council, entered into a contract with the EPA to produce background documents for pollutants, including particulate polycyclic organic matter (POM). To facilitate the development and production of these four documents and to ensure a critical "overview" in evaluating the content of the documents, the Academy set up the Committee on Biologic Effects of Atmospheric Pollutants (BEAP), under the chairmanship of Dr. Arthur B. DuBois. *Ad hoc* panels of experts were formed to evaluate each of the selected pollutants.

The comprehensive, evaluative, and multidisciplinary aspects of the problem faced by the Panel on Polycyclic Organic Matter are clearly illustrated in the following "Statement of Work," taken from the contract with the Academy:

Prepare an open-ended series of comprehensive state-of-the-art technical reports, which will reflect the latest scientific knowledge useful in indicating the kind and extent of all identifiable effects on human health and welfare which may be expected from the presence of a variety of pollutants in the ambient air. In developing these reports, consideration will be given to the sources, chemical and physical characteristics of the pollutants, the techniques available for their measurement in the ambient air, their prevalence in contaminated air and possible modifying conditions such as: reaction time, effects of other pollutants simultaneously present, and meteorological conditions. Documentation of the effects of these pollutants on human health and well-being, on animals, on vegetation, on materials and on man's environment in general is deemed to be of primary importance. These reports will contain detailed comment on dose/response relationship and margins of safety to be used in establishing air quality standards. They will indicate groups of persons in the general population known to be or likely to be particularly sensitive to exposure.

It is important to recognize that at no time have the scientists, physicians, economists, etc., who have dealt with previously prepared criteria documents felt that there was ample unequivocal evidence to support all their conclusions and recommendations. This was true even for perhaps the best understood of the common gaseous pollutants—carbon monoxide and oxides of sulfur. Despite the serious and widespread deficiencies in our knowledge of POM and its biologic effects on man, the Panel members and consultants attempted to make the most effective use of reliable knowledge available today, not only from the published literature, but also from the current unpublished experimental data from a number of laboratories. Furthermore, there was unanimous agreement that, although there are wide areas of total ignorance, misunderstanding, or disagreement in the literature on the carcinogenic activity of polycyclic organic compounds, recognizing and evaluating such gaps in vital knowledge serve the positive function of providing a basis for the preparation of specific recommendations for future research.

This document is not simply a review of the pertinent literature. Such reviews and surveys have already been produced (e.g., *Preliminary Air Pollution Survey of Organic Carcinogens. A Literature Review,* published in 1969 by the Department of Health, Education, and Welfare[768]). Although they are helpful in illuminating the scope of the literature, they are not intended to synthesize or evaluate information. This report attempts to interpret, evaluate, and reconcile the immense amount of information available, especially that concerning the carcinogenic effects of POM.

Contents

1

Introduction

Evidence of the induction of lung cancer via inhalation in man is extensive. The best-documented and most decisive evidence is related to occupational exposures, but there is also evidence that the generally low concentrations of some pollutants found in community air may be associated with an increased risk of lung cancer.

There are many examples—some strikingly unequivocal, others only suggestive—of human lung cancer caused by environmental factors. The major hazard is currently attributed to cigarette smoking, but other environmental sources are suspect. Indeed, the lung may lead the list of human organs for the variety and number of instances in which environmental agents are involved in cancer induction, with the skin second on the list. That the lung should share this role with skin reflects its direct contact with environmental agents.

Many etiologic agents in cigarette smoke have been proposed, e.g., polycyclic hydrocarbons, arsenic, nitrosamines, and polonium. Polycyclic aromatic hydrocarbons like benzo[a]pyrene are thus candidates for promoting lung cancer. This is not to say, however, that other initiating carcinogens do not contribute.

Laboratory studies can have a variety of purposes aimed at increased understanding of the human problem; among these are determining whether an environmental agent is related to lung cancer in man, pinpointing the active agent(s) among a number of suspected environmental agents, determining dose–response patterns or relations, and developing a better understanding of the biologic course of the disease.

Present knowledge indicates that fractions of particulate POM contain only two classes of compounds that are known animal carcinogens—the polycyclic aromatic hydrocarbons and their neutral nitrogen analogues, the aza-arenes (e.g., indoles and carbazoles). Numerous types of POM exist in urban air, such as pyrene, anthanthrene, benz[a]anthracene, benzofluoranthenes, dibenzanthracenes, chrysene, phenylenepyrene, benzoperylene, coronene, fluoranthene, and alkyl derivatives of these compounds, as well as benzopyrene. For this reason and because there are experimental data on the carcinogenic effects of benzo[a]pyrene and other polycyclic aromatic hydrocarbons, much attention has been directed in this report to the evaluation of these compounds in carcinogenesis.

Laboratory data indicate that cancer can be produced by the superimposition of ozonized gasoline on influenza in mice; and the combination of polycyclic aromatic hydrocarbons with sulfur dioxide produces cancer in rats. Thus, irritants may be very important co-reacting factors. Host-related and other co-acting factors are discussed and evaluated in this report. The problems inherent in the extrapolation of experimental data to humans are also considered.

The major thrust of this report has been the working hypothesis that there is an urban–rural difference in lung cancer rates that may stem from community air pollution and that may be attributable to polycyclic hydrocarbons known to be carcinogenic, such as benzo[a]-pyrene. Epidemiologic evidence linking community air pollution to lung cancer has been ambiguous to date, but a modest contribution of air pollution to lung cancer is possible. Much effort has been expended by some of the Panel members in the statistical re-evaluation of existing epidemiologic data in an attempt to delineate the association of cigarette smoking and urban factors with lung cancer.

For complete evaluation of the hazards imposed on man by polycyclic aromatic hydrocarbons in the air, it is necessary to identify the sources of pollution and the types of substances they emit, to evaluate the physical and chemical reactivity of these substances in the atmosphere, and to describe existing methods for collecting, separating, detecting, and quantitating polycyclic aromatic hydrocarbons. These areas of research are examined in detail in this report.

The report has been structured along the same functional lines as the Panel's activities. The environmental appraisal of POM (sources, characteristics, etc.) appears in Chapters 2–5. Studies of the effects of POM in animals, mammalian cells, and vegetation are discussed in the next nine chapters. The human effects of POM are described, on the basis of extensive epidemiologic studies, in Chapters 15, 16, and 17. The methodology of collecting and analyzing atmospheric POM is set forth in Appendixes A, B, and C.

2

Structure and Nomenclature of Polycyclic Aromatic Hydrocarbons and Aza-Arenes

The nomenclature used throughout this presentation is that adopted by the International Union of Pure and Applied Chemistry (IUPAC) and by Chemical Abstracts Service. The most important rules, all of which are described in detail in *The Ring Index,*[586] are the following:

1. The structural diagram is written to present the greatest possible number of rings in a horizontal row.
2. Horizontal and vertical axes are then drawn through the center of the horizontal row, and the molecule is oriented in such a way as to place the maximal number of rings in the upper right quadrant and the minimal number of rings in the lower left quadrant.
3. The carbon atoms are numbered in a clockwise direction starting with the carbon atom that is not part of another ring and is in the most counterclockwise position of the uppermost ring farthest to the right; carbon atoms common to two or more rings are not numbered.
4. The faces of the rings are now lettered in alphabetical order beginning with "a" for the side between carbon atoms 1 and 2 and continuing clockwise around the molecule; ring faces common to two rings are not lettered.

5. In naming a compound formed by the addition of a component, the numbers and letters are placed in square brackets and placed immediately after the name of the added component, showing where a substituent group is attached or where a ring is fused to the face of the molecule. If a ring is fused to more than one face of the molecule simultaneously, this is indicated by using the appropriate letters to denote the faces so involved.

6. The structural formulas used show aromatic rings as plain hexagons and a methylene group as CH_2.

Most of the polycyclic aromatic hydrocarbons and polycyclic azaheterocyclic compounds (aza-arenes) listed in Tables 2-1 and 2-2 are organic materials identified in the urban atmosphere, generally as suspended particles.[388]

The property of these compounds that is most relevant to the human health status of an exposed population and mainly discussed in this report is their potential carcinogenic activity. For quick reference, the carcinogenicity of a compound in this chapter is indicated by a simple code:

–	not carcinogenic
±	uncertain or weakly carcinogenic
+	carcinogenic
++, +++, ++++	strongly carcinogenic

The indications of carcinogenicity refer to the Public Health Service (PHS) survey of compounds tested for carcinogenicity.[346,699,700] Because many of these compounds are referred to in the PHS survey by the old Richter nomenclature, which differs from the modern nomenclature, the older names are given here in parentheses. Also, starred (*) compounds indicate disagreement with standard numbering.

TABLE 2-1 Polycyclic Aromatic Hydrocarbons

Compound	Structure	Carcinogenicity
Anthracene $C_{14}H_{10}$ m.p., 216 C; b.p., 340 C	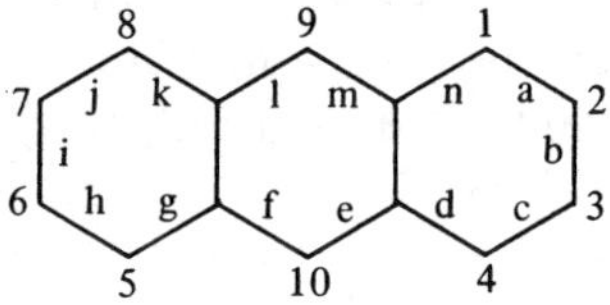	–
Benz[a]anthracene $C_{18}H_{12}$ m.p., 158 C; sublimes (1,2-benzanthracene)	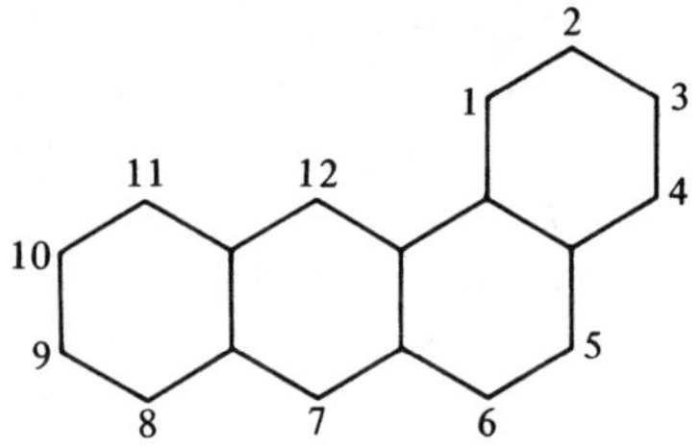	+
7,12-Dimethylbenz[a]anthracene (9,10-dimethyl-1,2-benzanthracene)	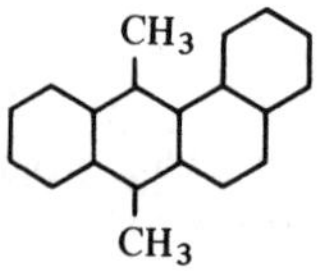	++++
Dibenz[a,j]anthracene (1,2-7,8-dibenzanthracene)	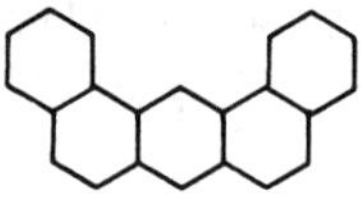	+
Dibenz[a,h]anthracene (1,2-5,6-dibenzanthracene)	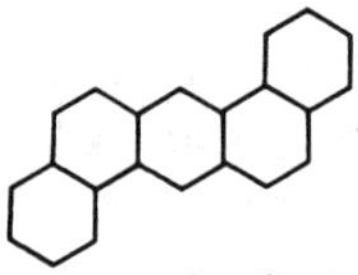	+++
Dibenz[a,c]anthracene (1,2-3,4-dibenzanthracene)	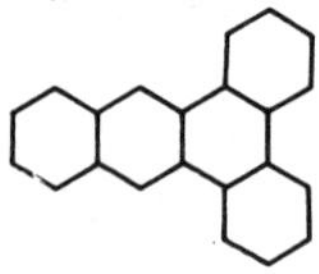	+
Phenanthrene $C_{14}H_{10}$ m.p., 101 C; b.p., 340 C	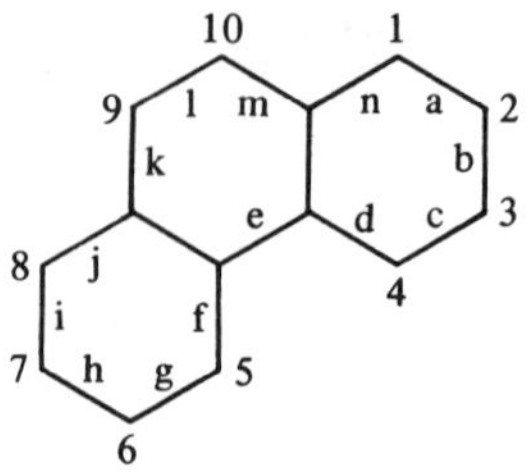	–

TABLE 2-1 Polycyclic Aromatic Hydrocarbons–Continued

Compound	Structure	Carcinogenicity
Benzo[c]phenanthrene (3,4-benzphenanthrene)	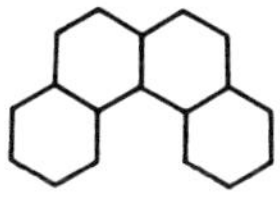	+++
Fluorene $C_{13}H_{10}$ m.p., 116 C; b.p., 293 C	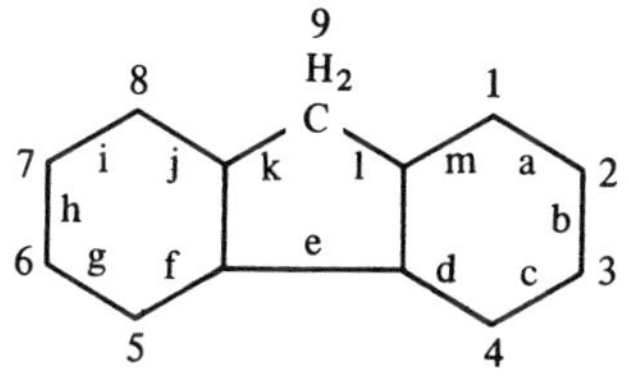 9 H_2 C; positions 1–8; bonds a–m	–
Benzo[a]fluorene (1,2-benzfluorene)	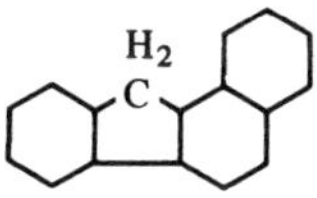 H_2 C	–
Benzo[b]fluorene (2,3-benzfluorene)	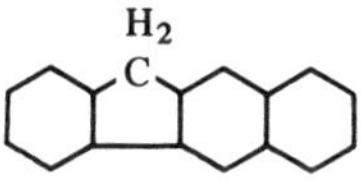 H_2 C	–
Dibenzo[a,h]fluorene] (1,2–6,7-dibenzfluorene)	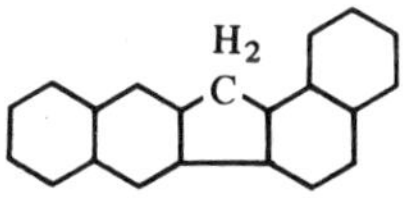 H_2 C	±
Dibenzo[a,g]fluorene (1,2–5,6-dibenzfluorene)	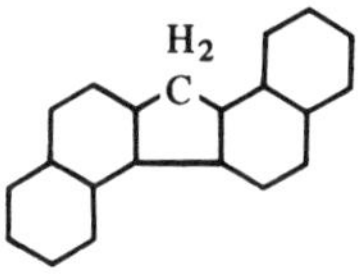 H_2 C	+
Benzo[c]fluorene (3,4-benzfluorene)	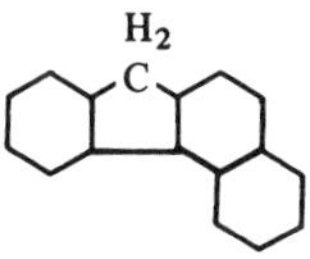 H_2 C	–
Dibenzo[a,c]fluorene (1,2–3,4-dibenzfluorene)	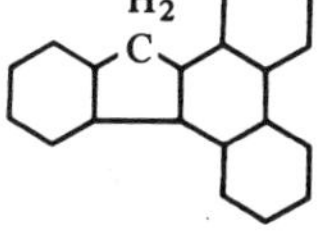 H_2 C	±

TABLE 2-1 Polycyclic Aromatic Hydrocarbons—Continued

Compound	Structure	Carcinogenicity
Fluoranthene $C_{16}H_{10}$ m.p., 110 C; b.p., 393 C	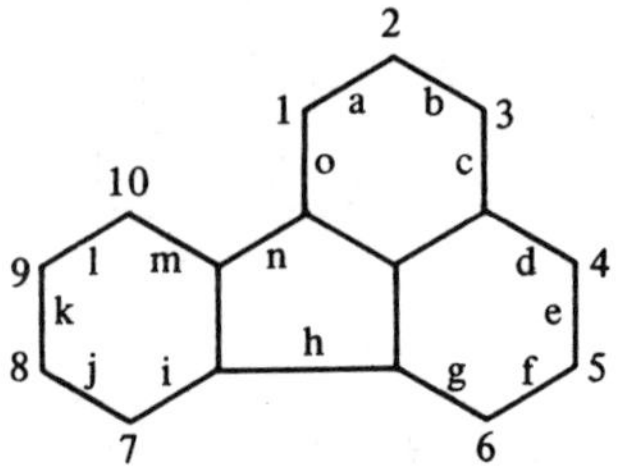 1 2 3 4 5 6 7 8 9 10; a b c d e f g h i j k l m n o	−
Benzo[b]fluoranthene (2,3-benzofluoranthene)	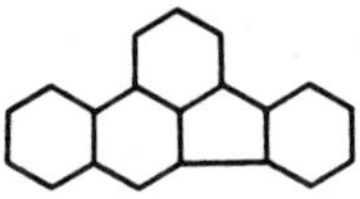	++
Benzo[j]fluoranthene (7,8-benzofluoranthene)	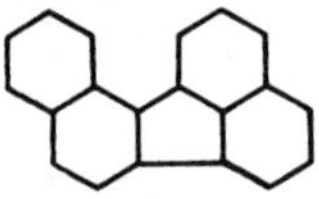	++
Benzo[k]fluoranthene (8,9-benzfluoranthene)	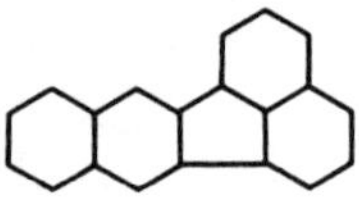	−
Benzo[ghi]fluoranthene	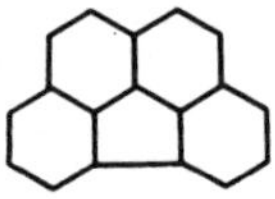	−
Aceanthrylene $C_{16}H_{12}$ m.p., 113 C	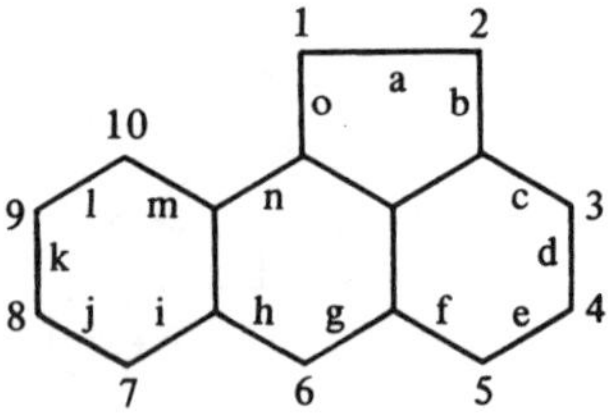 1 2 3 4 5 6 7 8 9 10; a b c d e f g h i j k l m n o	−
Benz[j]aceanthrylene = cholanthrene	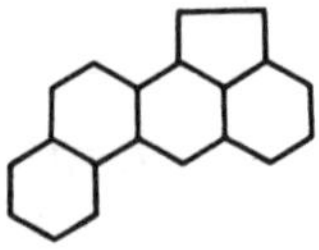	++

TABLE 2-1 Polycyclic Aromatic Hydrocarbons–Continued

Compound	Structure	Carcinogenicity
3-Methylcholanthrene	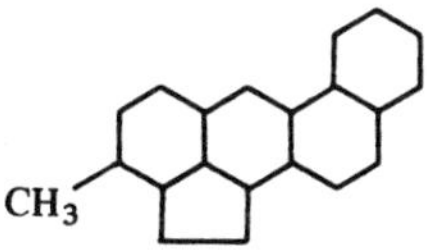 CH_3	++++
Naphthacene = benz[b]anthracene $C_{18}H_{12}$ m.p., 341 C; sublimes	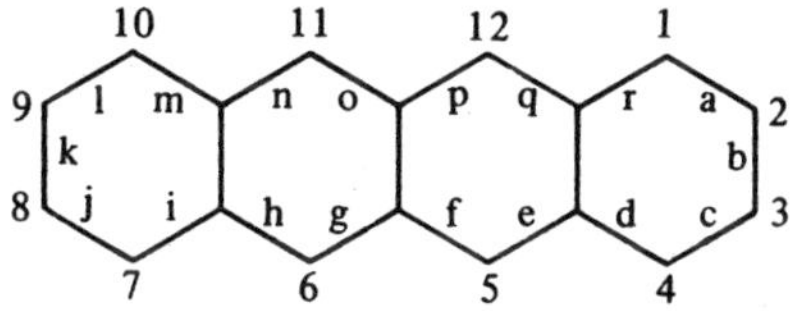	–
Naphtho[2,1,8-qra]naphthacene = naphtho[2,3-a]pyrene (2′,3′-naphtho-1,2-pyrene)	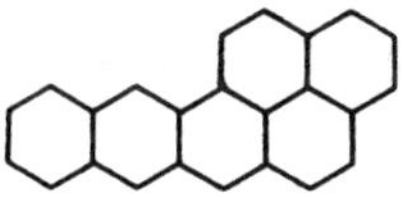	–
Pyrene $C_{16}H_{10}$ m.p., 150 C; b.p., >360 C	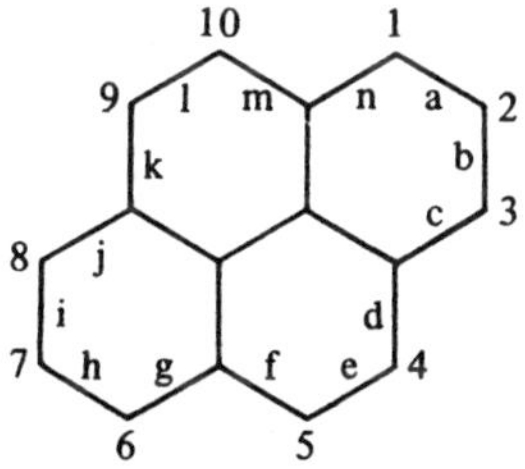	–
Benzo[a]pyrene (1,2-benzpyrene) (3,4-benzypyrene*)	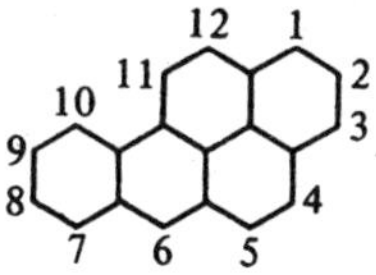	+++
Benzo[e]pyrene (4,5-benzpyrene) (1,2-benzpyrene*)	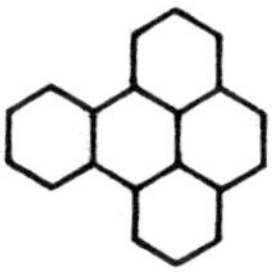	–
Dibenzo[a,l]pyrene (2,3–4,5-dibenzpyrene) (1,2–3,4-dibenzpyrene*)	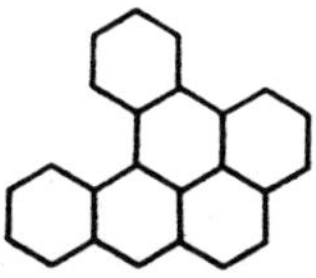	±

TABLE 2-1 Polycyclic Aromatic Hydrocarbons–Continued

Compound	Structure	Carcinogenicity
Dibenzo[a,h] pyrene (1,2–6,7-dibenzpyrene) (3,4–8,9-dibenzpyrene*)	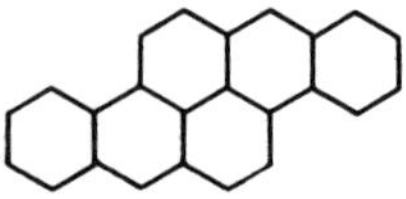	+++
Dibenzo[a,i] pyrene (2,3–6,7-dibenzpyrene) (4,5–8,9-dibenzypyrene*)	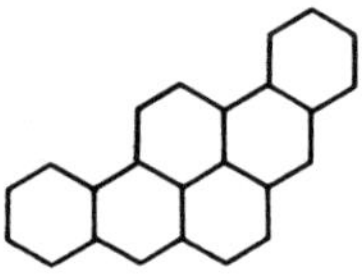	+++
Dibenzo[cd,jk] pyrene = anthanthrene		–
Indeno[1,2,3-cd] pyrene (*O*-phenylenepyrene)	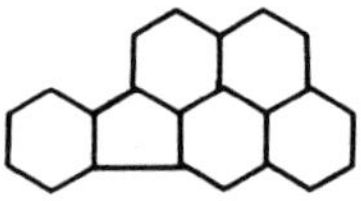	+
Chrysene = 1,2-benzophenanthrene $C_{18}H_{12}$ m.p., 254 C; b.p., 448 C	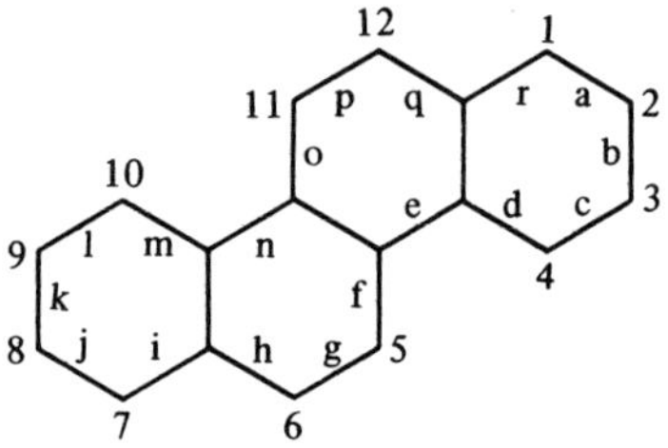 12 1 11 p q r a 2 o b 10 e d c 3 9 l m n 4 k f 8 j i h g 5 7 6	±
Dibenzo[b,def] chrysene = dibenzo[a,h] pyrene (3,4–8,9-dibenzpyrene*)		++
Dibenzo[def,p] chrysene = dibenzo[a,l] pyrene (1,2–3,4-dibenzpyrene*)	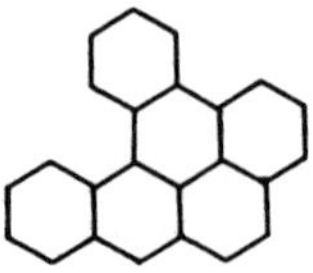	+

TABLE 2-1 Polycyclic Aromatic Hydrocarbons–Continued

Compound	Structure	Carcinogenicity
Dibenzo[def,mno] chrysene = anthanthrene = dibenzo[cd,jk] pyrene		–
Perylene $C_{20}H_{12}$ m.p., 273 C; b.p., ca. 500 C	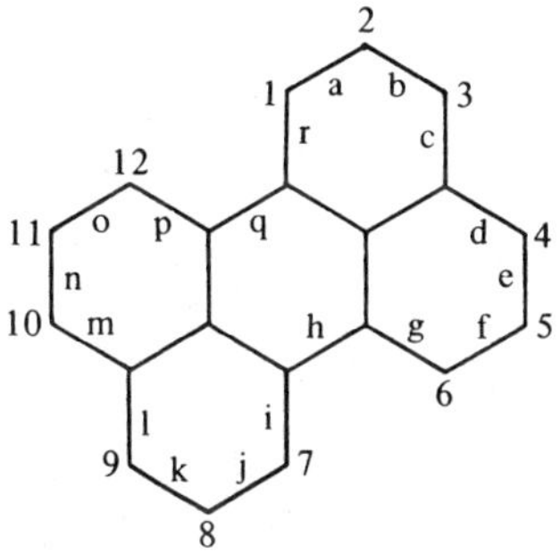	–
Benzo[ghi] perylene	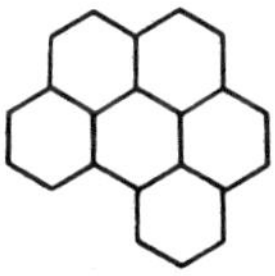	–
Coronene $C_{24}H_{12}$ m.p., 438 C; b.p., 525 C	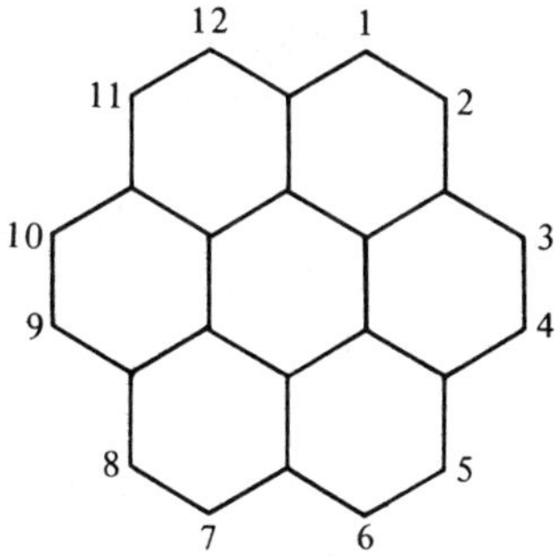	–

TABLE 2-2 Aza-Arenes

Compound	Structure	Carcinogenicity
Acridine $C_{13}H_9N$ m.p., 111 C; b.p., >360 C	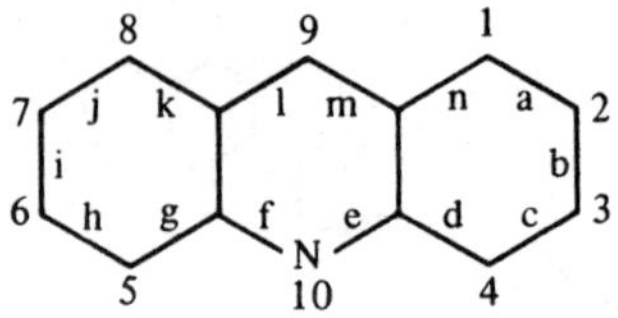	–
Dibenz[a,j]acridine (1,2-7,8-dibenzacridine)	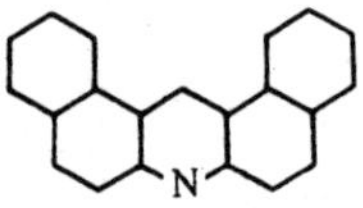	++
Dibenz[a,h]acridine (1,2-5,6-dibenzacridine)	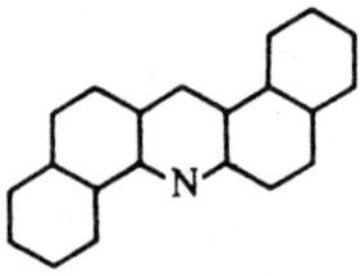	++
Dibenz[c,h]acridine (3,4-5,6-dibenzacridine)	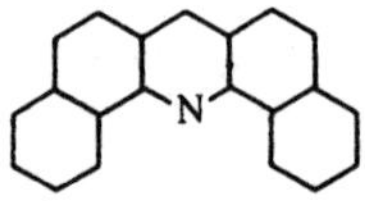	±
Carbazole $C_{12}H_9N$ m.p., 246 C; b.p., 355 C	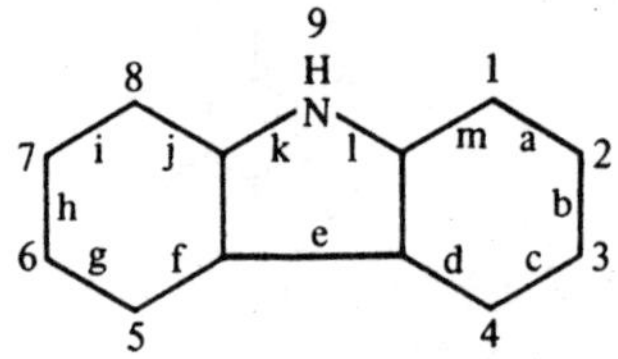	—
Benzo[a]carbazole (1,2-benzcarbazole)	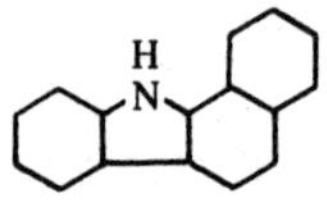	±
Dibenzo[a,g]carbazole (1,2-5,6-dibenzcarbazole)	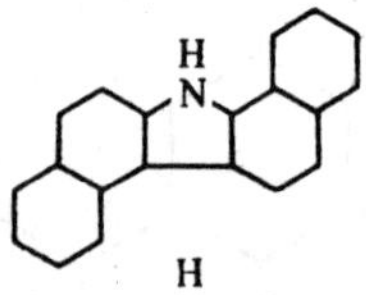	±
Dibenzo[c,g]carbazole (3,4-5,6-dibenzcarbazole)	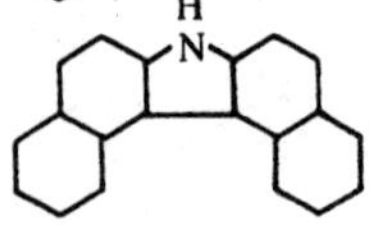	+++
Dibenzo[a,i]carbazole (1,2-7,8-dibenzcarbazole)	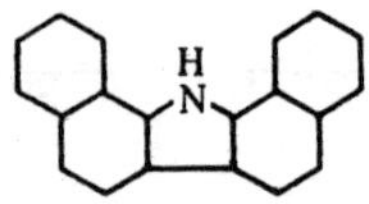	±

3

Sources of Polycyclic Organic Matter

Polycyclic organic matter (POM) can be formed in any combustion process involving fossil fuels or, more generally, compounds containing carbon and hydrogen. The amount of POM formed will vary widely; efficient, controlled combustion favors very low POM emissions, whereas inefficient burning favors high emissions. This chapter catalogs the more obvious sources of POM emissions to the atmosphere and points out the uncertainties in our information on various source contributions.

MECHANISM OF POM FORMATION

Although the mechanism of POM formation in combustion processes is complex and variable, a relatively clear picture of the overall reaction has emerged, owing primarily to Badger.[21] Chemical reactions in flames proceed by free-radical paths; in POM formation, a synthetic route is postulated, as shown in Figure 3-1. Radical species containing one, two, or many carbon atoms can combine in rapid fashion at the high temperatures (500-800 C) attained in the flame front. This pyrosynthesis of pyrolysis products is obviously a func-

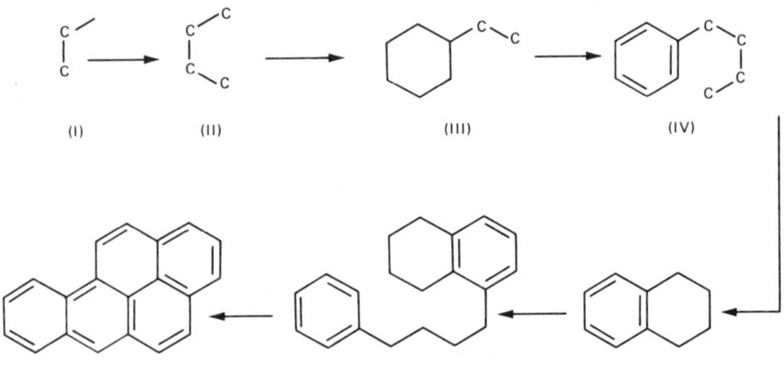

FIGURE 3-1. Mechanism of benzo[a]pyrene formation. (After Badger.[21])

tion of many variables, not the least of which is the presence of a chemically reducing atmosphere, common in the center of flames. In these conditions, radical chain propagation is enhanced, allowing the buildup of a complex POM molecule, such as benzo[a]pyrene. It is important to note that, although methane itself can lead to POM,[102] the formation of these large molecules is favored by the presence of higher-molecular-weight radicals and molecules in the fuel. Obviously, it is unnecessary to break the starting material down completely to a two-carbon radical in order to form benzo[a]pyrene. Any component of the combustion reaction that can contribute intermediate pyrolysis products of the structure required for benzo[a]pyrene synthesis would be expected to lead to increased yields of benzo[a]pyrene. Thus, Badger and Spotswood[22] have shown that in the pyrolysis of alkylbenzenes, including *n*-butylbenzene, enhanced benzo[a]pyrene formation is due primarily to increased concentrations of intermediate structures of types III, IV, and V (Figure 3-1).

Badger also showed conclusively that specific aromatic and diolefinic compounds serve as precursors for other polycyclic organic products. The mechanism in Figure 3-1 is a pathway to benzo[a]pyrene formation, but similar routes could be devised, with somewhat different intermediates, to lead to most of the known POM produced in combustion processes. Badger's work, with its reliance on calculated C–C and C–H bond energies to predict favored pathways and the experimental confirmation of these steps with radioisotopic labeling, provides a clear-cut mechanism for POM formation in the combustion process.

As pointed out by Hoffmann and Wynder,[388] the use of nitrogen atmospheres in Badger's experiments has been criticized for its lack of relevance to the actual combustion of organic molecules. Nonetheless, the conditions are similar to those of the oxygen-deficient environment in flames, and the data are in good qualitative agreement with observed POM combustion products.

NONTECHNOLOGIC SOURCES OF POM

Uncontrolled combustion, such as that in forest fires, would be expected to produce POM. Although the requirements for appreciable POM formation can be met in these fires, data on actual emission rates are lacking. The only other nontechnologic source of airborne POM is agricultural burning, but, because it is often planned by man, its contribution is covered below, under "Refuse Burning."

TECHNOLOGIC SOURCES OF POM

Large quantities of POM are generated in the vast number of technologic activities that prevail in our society. The contribution of any particular source depends on many factors, including geography, urbanization, and climate; thus, nationwide emission inventories can be misleading. Olsen and Haynes[572] have summarized the available data.

Man-made POM emission sources can be broadly separated into transportation or mobile sources and stationary sources. In the transportation category, a major emitter is the conventional gasoline-powered automobile, although all combustion engines contribute to the overall atmospheric POM burden. Because they are ubiquitous and are known to be contributors to POM concentrations in most urban areas, motor vehicles require close scrutiny. The category of stationary sources of POM encompasses a wide variety of processes that can be local contributors to POM concentrations. It has been customary to subdivide this category into heat and power generation, refuse burning, and industrial activities. Indoor POM emissions must also be considered in this assessment.

Transportation Sources

GASOLINE-POWERED VEHICLES

A significant mobile source of atmospheric POM is the conventional automobile, powered by a spark-ignited internal-combustion engine.

The technical literature on vehicular effects is sparse, mainly because of the difficulties associated with the type of experimentation required. In the last few years, results of investigation in this area have been published;[528] heightened interest in air pollution has resulted in the initiation of comprehensive programs, which are in various stages of completion. The available published literature and some of the preliminary results from current programs are reflected in this summary. Unpublished or incomplete data are included with the recognition that additional tests may vitiate some of the preliminary judgments.

An assessment of the current vehicular benzo[a]pyrene contribution, compiled from nationwide fuel-consumption data, is summarized in Table 3-1.

As will be apparent in the following discussion, most efforts have been directed at estimating the automobile contribution, with less emphasis on trucks and buses. One study[343] is available on benzo[a]-pyrene emission from gasoline-powered trucks; it shows a wide variation in emission factors, from 70 to 1,500 μg/gal.

The contributions of gasoline-powered vehicles can be separated into vehicular effects and fuel-composition effects. The first category includes the effects of air : fuel ratio or mixture stoichiometry, emission control devices, operating modes, deterioration, and combustion-chamber deposits. The second includes effects of such variables as aromaticity, fuel POM level, additives, and lubricants.

Effects of Vehicular Characteristics Efficient combustion is enhanced by the presence of excess air, i.e., air : fuel ratios greater than stoichiometric. Air : fuel ratios less than stoichiometric lead to

TABLE 3-1 Estimated Benzo[a]pyrene Emission in the United States

Vehicle Type	Fuel Consumed, gal/year	Benzo[a]pyrene Emission Factor, μg/gal	Benzo[a]pyrene Emission, tons/year
Gasoline-powered			
Automobiles	56.4×10^9	170[a]	10
Trucks	24.2×10^9	~500[a]	~12
Diesel-fuel-powered			
Trucks and buses	5.8×10^9	62[b]	0.4
Total			~22

[a] Data from Hangebrauck *et al.*[343]
[b] Data from Begeman and Colucci.[42]

the products of incomplete combustion, such as carbon monoxide and unburned and oxygenated hydrocarbons. Before the current concern for reducing vehicular emissions, most vehicles operated with fuel-rich carburetion to promote smooth performance and readily accessible power. Modifications in post-1967 vehicles have resulted in "leaner" fuel-air mixtures and, the data suggest, in significantly lower POM emissions. Table 3-2 shows a compilation of available data; vehicular variables will obviously influence the values shown but should not affect the trends indicated. Automobile exhaust POM is generally referred to in terms of the benzo[a]pyrene emission, primarily because of its cited carcinogenicity and the fact that more data are available on this material than on any other. Data are becoming available on prototype emission control devices, such as thermal reactors and catalytic converters, and a preliminary figure is included for comparison.

It is apparent from these data that the introduction of presently used emission control devices resulted in about an 85% reduction in benzo[a]pyrene emissions from the pre-1965 levels. The data in Table 3-2 have been selected from representative recent research on vehicles operating over cyclic test conditions that approximate driving patterns found in actual customer use. Prototype emission control devices, such as thermal reactors and catalyst systems, result in a continuing downward movement of vehicular POM emissions. Other variables discussed in this section are less important than the vehicle effects.

The effect of oxidizing and reducing atmospheres on incomplete combustion and POM formation is important in estimating vehicular emissions. Recent data[42] indicate that benzo[a]pyrene production at an air: fuel ratio of 10 : 1 is 30 times higher than at a ratio of 14 : 1.

TABLE 3-2 Automotive Benzo[a]pyrene Emission Factors

Source	Benzo[a]pyrene Emission Factors, μg/gal of Fuel Consumed
Uncontrolled car (1956–1964)	170[a]
1966 Uncontrolled car	45–70[b]
1968 Emission-controlled vehicle	20–30[c]
Advanced systems	≤10[d]

[a] Data from Hangebrauck *et al.*[343] and Begeman and Colucci.[42]
[b] Data from Gross.[316]
[c] Data from Begeman and Colucci[42] and Gross.[316]
[d] Estimated from Hoffman *et al.*[382] and Faust and Sterba.[264]

Hoffman *et al.*[382] suggest that benzo[a]pyrene emission is 10 times higher at "rich" carburetion (2.85% CO) than at "lean" carburetion (0.9-1.4% CO). This effect is the central reason for the lower POM emissions from current emission-controlled vehicles. The effect of engine operating temperature is closely related to this aspect; cold engines operate in a "choked" or "rich" condition, indicating that POM emissions would be maximized in cold starts and minimized in hot engine operation.

A clear trend toward higher POM emissions with increasing engine life has been documented by several workers. Hangebrauck *et al.*[343] observed a sharp increase in benzo[a]pyrene emission rates of automobiles as they approached the 50,000-mile age, the rates being about 5 times higher than those of lower-mileage (e.g., 5,000-mile) vehicles. Begeman and Colucci,[42] who studied oil consumption effects, report a tenfold increase in benzo[a]pyrene emission when oil consumption is increased from 1,600 miles/qt to 200 miles/qt. These workers also found that benzo[a]pyrene from the combustion chamber was preferentially concentrated in the crankcase; eight times more benzo[a]-pyrene entered the crankcase than left the exhaust system (at normal oil consumption rates). These data help to explain the higher POM emission rates of older vehicles. As cylinder wear increases, the lubricant concentration in the upper cylinder increases and the heavy lubricant molecules provide convenient intermediates for POM formation.

As vehicles accumulate mileage in normal consumer use, deposits form in the combustion chamber. The nature and composition of these deposits have been shown to influence total hydrocarbon exhaust emissions; as mileage is accumulated, total emissions increase until a stabilized condition is reached at several thousand miles. Gross[316] has shown that the condition of the deposit exerts a significant effect on POM emission levels; the POM emission levels are about twice as high in a vehicle with stabilized deposits from operation with leaded fuel as in the same vehicle with stabilized deposits from operation with unleaded fuel. Another study[382] sees no effect of combustion-chamber deposits on POM emissions.

Effects of Fuel Composition The presence in fuel of precursors of radical intermediates would be expected to facilitate POM formation; i.e., the pyrosynthetic path would be shortened. Conjugated dienes and aromatics in the fuel should provide the maximal enhancement of fuel-related POM formation. The literature does point to fuel composition as having an important effect, but simple judgments as to the advantages and disadvantages of compositional modifications are confounded by the number of variables in the vehicle-fuel-exhaust-system relation.

Early research on the effects of fuel components on POM emissions pointed clearly to increased aromatic content of fuel as a cause of higher POM exhaust emissions. For example, Boubel and Ripperton[79] showed that a benzene-fueled engine produced 10–30 times more POM than an engine using *n*-hexane, cyclohexane, or hexene-1. Hoffmann and Wynder[390] reported that higher emissions of benzo[a]-pyrene and benz[a]anthracene resulted from blends of 50% *o*-xylene and 50% benzene than from gasoline, pure paraffins, and pure olefins. Hoffman *et al.*[382] diluted unleaded, high-aromatic gasoline with pure isooctane and achieved dramatic reduction in benzo[a]pyrene emissions. Begeman[40] reported higher benzo[a]pyrene emissions with test fuels containing high POM and aromatic concentrations than with commercial gasolines. Most of these studies have been carried out with synthetic blends, as opposed to gasolines of conventional compositions. To assess accurately the role of fuel in the question of atmospheric POM, it is imperative to use realistic compositions in consumer driving conditions.

Gross[316] reports that, when full-boiling-range fuels were used in well-maintained vehicles operated under federal requirements for testing 1968–1971 vehicles, POM emissions increased by 36–74% in an uncontrolled vehicle and 8–34% in an emission-controlled vehicle as fuel aromaticity was increased from 12 to 46%.

A more realistic picture of the effect of gasoline composition can be obtained using Gross's data.[316] When engine tests using a leaded fuel of low to intermediate aromaticity, with stabilized combustion-chamber deposits, are compared with tests using an unleaded fuel of high aromaticity, also with stabilized deposits, no dramatic effect of fuel aromaticity is apparent. Thus, potential increases in POM emissions due to higher fuel aromaticity are offset by changes in the nature of the combustion-chamber deposits when unleaded fuel is used. These data must be regarded as tentative, because this program is in progress, and other research in progress does not support these conclusions.[382] However, the benefits to be gained with future control devices operating on unleaded fuel seem to outweigh greatly the effects of fuel composition, such as aromaticity. Research in this area is aimed at unraveling those effects, which now appear to be more complicated than previously assumed.

The POM content of gasoline has been shown to affect POM emissions from vehicles. Begeman and Colucci[41] estimate that as much as 36% of the benzo[a]pyrene in the exhaust gas can be attributed to the fuel benzo[a]pyrene content; Gross[316] estimates that 15–30% increases in POM emissions can be obtained when fuel POM is varied between the concentration extremes found in the field. The actual effect was smaller in the controlled vehicle than in the uncontrolled vehicle. These results are consistent with the Badger mechanism outlined previously: High-molecular-weight fuel components and lubricant losses to the combustion chamber will result in higher POM emissions.

DIESEL-FUEL-POWERED VEHICLES

Any critical discussion of the relative contribution of diesel engines to the atmospheric POM concentration must be qualified by the supposition that the vehicle is operated under rated load conditions, i.e., is not overloaded. In normal use, the most objectionable features of diesel operation (soot formation, odor, etc.) are apparent when the engine is overfueled. Begeman and Colucci[42] have determined that a diesel engine, operated on a bus-driving cycle, emitted 62 μg of benzo[a]pyrene per gallon of fuel. Reckner *et al.*[626] showed that POM emissions from a diesel test engine increased with load up to half-load, leveled off, and then dropped sharply at full load. Idle operation resulted in high POM emissions, presumably because of lower combustion-chamber temperatures.

Oil consumption in diesel engines can be somewhat higher than in spark-ignited engines. However, Begeman and Colucci[42] point out that, because diesel combustion chambers do not operate under vacuum, lubricating oil should not be drawn into the ignition area; thus, POM from this source is probably not significant.

The only published data[626] on the effects of fuel characteristics on diesel POM emissions indicate that fuel aromaticity is not related to exhaust POM levels. The test fuels ranged between 5 and 23% aromatic content, with the fuel POM consisting primarily of pyrene, anthracene, and fluoranthene. There was no detectable benzo[a]-pyrene.

Although the variables discussed above were evaluated in engine-laboratory conditions, it is apparent that actual on-the-road operation of diesel-powered vehicles can result in higher POM emissions, owing to overloading, poor maintenance, and so on. The objectionable features of diesel truck and bus operation, such as smoking and odor, might be associated with higher POM emissions, and additional research in this area should be fruitful.

MISCELLANEOUS TRANSPORTATION SOURCES

Quantitative data are generally lacking for diverse mobile sources of POM, such as aircraft engines and various nondiesel two-cycle engines, e.g., lawnmowers, outboard motors, and motorcycles. Aircraft and turbine engine operation has apparently never been surveyed for POM emissions.

A study of POM in the exhaust gas from two-cycle engines has been reported.[408] Two-cycle engines, which do not have crankcases, operate on a mixture of premixed oil and fuel, the oil being the sole source of lubrication in the system. The data suggest that these engines yield large amounts of benzo[a]pyrene, with an emission factor of 11,000 μg/gal for an oil : fuel ratio of 1 : 33, and that benzo[a]pyrene yields are a direct function of oil concentration in the fuel. These findings are consistent with the effect of oil consumption on exhaust-gas benzo[a]pyrene found in four-cycle engines.[42] In the most extreme conditions, the oil : fuel ratio reported by Begeman and Colucci[42] was 1 : 29. It seems obvious that the presence of higher-molecular-weight components than normally found in the gasoline boiling range has a positive effect on POM formation.

The major noncombustion transportation source of POM is probably the degradation of automobile tires in use. Carbon blacks, used in tire manufacturing, contain POM and other high-molecular-weight organic compounds (S. S. Epstein, personal communication). Marchesani *et al.*[519] estimate that 4.3 tons of rubber particles from tires are emitted per day per million people in the United States. The benzo[a]pyrene contribution from the degradation can be roughly estimated from the analytic data of Falk *et al.*,[260] an emission rate of 0.3 lb/day per million people is projected. Although tire degradation does not appear to be a significant source of benzo[a]pyrene, the ultimate burning of used tires and vehicles (which are categorized as refuse burning) may be of far greater importance.

EMISSION CONTROL PROCEDURES

The emission control devices on cars since the 1968 models have reduced benzo[a]pyrene emission factors by about 85%, compared with uncontrolled vehicles. This reduction is due to the more efficient combustion associated with "leaner" air–fuel mixtures. It is anticipated that these methods will continue in use until the mid-1970's, when more stringent controls on total emissions will be required. At that time, such devices as catalysts or thermal reactors will probably be required, and the mixture stoichiometry will depend on the particular systems chosen. These devices should result in additional reductions in POM emission, as shown in Table 3-2. Reductions in emission of POM to the atmosphere from automobiles can be projected through the 1980–1990 period, as the older cars are removed from service and a greater proportion of vehicles are equipped with advanced emission control systems. Although current work centers on the effects of control devices, operating conditions, and fuel composition, surveillance of vehicles in normal customer usage, including those in poor operating condition, might support the extrapolations made in this section. The emissions from heavier vehicles (such as gasoline- and diesel-powered trucks and buses) seem significant, but little work has been done on defining the emission factors of these vehicles. Knowledge of these factors is clearly needed. There should also be continued efforts to clarify the effects of fuel composition changes, such as trends toward lead removal from gasoline. Although these factors seem to be less important than the vehicle-emission-control system, close scrutiny of them is nonetheless desirable.

Stationary Sources

Polycyclic organic matter is emitted from a vast number of diverse stationary sources. Although the complexity and variety of POM preclude a rigorous assessment of their contribution, it can be seen from compilations of analytic data taken by the U.S. Public Health Service in most of the urban areas of the country that some urban areas close to significant POM sources are subjected to high atmospheric POM concentrations. A recent comprehensive review of POM sources by Hangebrauck *et al.*[343] summarizes the current knowledge of the relative contributions of the various stationary sources. Although emissions from stationary sources consist of a variety of chemical entities, the practice of using benzo[a]pyrene as an indicator of other POM is suggested, owing to the dual factors of the demonstrated carcinogenicity of benzo[a]pyrene and the relatively large amount of published data on it.

HEAT AND POWER GENERATION

Coal, oil, gas, and wood are burned in a variety of installations. Hangebrauck *et al.*[343] concluded that the most important source of benzo[a]pyrene of these four was the inefficient combustion of coal in hand-fired residential furnaces. Data on all four as producers of benzo[a]pyrene are shown in Table 3-3. That efficiency of combustion, and not the fuel used, is the controlling factor is emphasized by the low benzo[a]pyrene emission factor found in power plants burning crushed or pulverized coal. Oil- and gas-burning units used for institutional and home heating, as well as steam for process heating, were also shown to be sources of low POM emission. Although these data are consistent with our knowledge of POM formation processes—i.e., reducing conditions and insufficient oxygen—caution should be used in extrapolating data from some 75 individual sources to the nation as a whole. As will be pointed out, high ambient air

TABLE 3-3 Estimated Benzo[a]pyrene Emission from Heat and Power Generation Sources[a] in the United States

Type of Unit	Gross Heat, BTU/hr	Benzo[a]pyrene Emission Factor, μg/10^6 BTU	Benzo[a]-pyrene Emission, tons/year
Coal			
Hand-stoked residential furnaces	0.1×10^6	1,700,000–3,300,000	420
Intermediate units (chain-grate and spreader stokers)	$60–250 \times 10^6$	15–40	10
Coal-fired steam power plants	$1,000–2,000 \times 10^6$	20–400	1
Oil			
Low-pressure air-atomized	0.7×10^6	900	2
Other	$0.02–21 \times 10^6$	100	
Gas			
Premix burners	$0.01–9 \times 10^6$	20–200	2
Wood		50,000	40

[a] Data from Hamburg,[336] Hangebrauck *et al.*,[343] Muhich *et al.*,[550] U.S. Department of Agriculture,[764] U.S. Department of Health, Education, and Welfare,[767,774] U.S. Department of the Interior,[777] Wadleigh,[798] and L. McNab (personal communication).

POM concentrations in a particular region can be associated with local fuel practices. It is apparent that substitution of energy sources that are inherently more efficient than coal combustion in residential units may be a short-term solution in areas of high POM emission.

No firm data are available in the literature on the extent of POM emission from wood-burning combustion units. The growing popularity of home fireplaces, as well as rural heating demands, call for evaluation of this factor. An emission factor of about 50,000 μg of benzo[a]pyrene per million BTU is used to estimate the wood-burning contribution.[767]

The total benzo[a]pyrene emissions from heat and power generation sources shown in Table 3-3 must be regarded as speculative. This applies most directly to the coal- and wood-burning residential usage figures. The possible error in these approximations is such that the contribution of heat and power generation sources to the atmosphere cannot be quantified; an estimate of 500 tons of benzo[a]-pyrene emitted per year to the atmosphere appears justified as an upper limit.

REFUSE BURNING

The intentional combustion of solid wastes as a method of disposal, as well as accidental or naturally occurring uncontrollable combustion processes, can contribute significantly to overall POM emissions. Such sources of POM should come under increasing scrutiny in view of the increasing solid-waste disposal problem in the United States today. Unfortunately, the very diversity and nature of these sources has led to great uncertainties as to their actual contributions to atmospheric POM concentrations.

The review by Hangebrauck *et al.*[343] cites benzo[a]pyrene emission factors for municipal and commercial incineration of such wastes as those collected from households, business, and restaurants, as well as for burning of municipal and agricultural refuse and junked automobile parts. Benzo[a]pyrene emissions from these sources vary widely and reflect the importance of efficient combustion in reducing POM emissions. Large (50–250 tons/day) municipal incinerators had benzo[a]pyrene emission factors of 0.1–6 μg/lb of charged refuse, and commercial (3–5 tons/day) incinerators had factors of 50–260 μg/lb. Data show a benzo[a]pyrene emission factor of about 150 μg/lb of charged refuse for open burning of municipal wastes, as well as for grass clippings, leaves, etc. Significantly, the destruction of auto components in test "open-burning" facilities yielded a benzo[a]pyrene emission factor of 1.3×10^4 μg/lb of refuse.[343]

In their summation, Hangebrauck *et al.*[343] conclude that about 20 tons of benzo[a]pyrene are emitted from these sources per year. More recently,[336,550,764,774,777,798] significantly higher emissions from these sources have been suggested, reflecting higher estimates of total nationwide refuse burning, rather than appreciably different emission factors. These newer data are compiled in Table 3-4. The largest single identified contributor listed is coal refuse bank burning (L. McNab, personal communication), a commonplace occurrence in mining areas. These banks of coal-mining refuse (coal, shale, calcite) can be spontaneously ignited and will burn for long periods in sufficient combustion conditions.

In general, the tonnage figures ascribed to the various refuse burning classifications must be regarded as order-of-magnitude approximations. The highly speculative nature of the emission factors used in the publications cited does not inspire a high level of confidence in the derived estimates. An estimate of 600 tons of benzo[a]pyrene

TABLE 3-4 Estimated Benzo[a]pyrene Emission from Refuse-Burning in the United States

Source of Benzo[a]pyrene	Benzo[a]pyrene Emission, tons/year
Enclosed incineration	
Municipal	<1[a]
Commercial and industrial	23[a]
Institutional	2[b]
Apartment	8[a]
Open burning	
Municipal	4[c]
Commercial and industrial	10[b]
Domestic	10[b,d]
Forest and agricultural	140[d,e]
Vehicle disposal	50[f]
Coal refuse fires	340[g]

[a] Data from U.S. Department of Health, Education, and Welfare.[774]
[b] Data from Muhich *et al.*[550]
[c] Data from Olsen and Haynes.[572]
[d] Data from U.S. Department of Health, Education, and Welfare[774] and Muhich *et al.*[550]
[e] Data from U.S. Department of Health, Education, and Welfare,[774] Wadleigh,[798] and U.S. Department of Agriculture.[764]
[f] Data from U.S. Department of the Interior[777] and Hamburg.[336]
[g] Data from L. McNab (personal communication).

emitted per year appears to be the best available value on the basis of current knowledge.

INDUSTRIAL ACTIVITIES

The major direct petroleum-industry source of POM is the catalytic cracking process by which organic molecules in crude oil are broken down into the lighter components used in the manufacture of motor gasoline, heating oil, aviation fuel, etc. The cracking takes place in the presence of a catalyst, which can become deactivated through the deposition of carbon, or coke, on the active sites. It is in the regeneration of the catalyst, through the combustion of the coke on the catalyst surface, that benzo[a]pyrene and other POM are formed. These emissions are finally passed either to the atmosphere or to a carbon monoxide waste-heat boiler. The latter device, originally designed to make use of the waste heat from carbon monoxide gas, functions as a direct-flame afterburner and removes almost all the POM from the effluent being emitted to the atmosphere.

As can be seen from the data in Table 3-5, the contribution of catalytic cracking processes to the atmospheric POM concentrations is a function of the proportion of units equipped with carbon monoxide waste-heat boilers. The various catalytic cracking units listed in the table represent both the moving-bed catalytic systems [Thermofor (TCC) and Houdriflow (HCC)] and the fluidized-bed system [fluid catalytic cracking (FCC)]. A recent survey[627] suggests that a greater proportion of the units are equipped with carbon monoxide boilers than were so equipped in 1967.[343] The results of the later survey show an annual contribution of 6 tons of benzo[a]pyrene from refinery catalytic cracking operations. The total represents a reduction to about one third of previously published estimates.

Another petroleum-industry process of interest is the air-blowing of asphalt. This procedure is designed to yield materials of higher softening point for roofing applications. The effluent from air-blowing may contain many hydrocarbons, including POM. In the one test of an actual process,[343] however, very little benzo[a]pyrene was found in the benzene-soluble fraction of the particulate matter. The total contribution of asphalt air-blowing is estimated at less than 0.03 ton of benzo[a]pyrene per year.

Catalytic cracking of petroleum and air-blowing of asphalt are the most obvious sources of POM emission in the petroleum industry, but there may be miscellaneous other processes that have not been

TABLE 3-5 Estimated Benzo[a]pyrene Emission from Catalytic Cracking Sources[a] in the United States

Type of Cracking Unit[b]	Petroleum Consumption, million barrels/year	Benzo[a]pyrene Emission, tons/year
FCC		
no boiler	424	0.08
CO boiler	1,230	0.02
Subtotal	1,654	0.10
HCC		
no boiler	14	3.4
CO boiler	55	0.0
Subtotal	69	3.4
TCC (air-lift)		
no boiler	27	2.4
CO boiler	118	0.0
Subtotal	145	2.4
TCC (bucket-lift)		
no boiler	17	0.0
CO boiler	75	0.0
Subtotal	92	0.0
Total	1,960	5.9

[a] Data from Hangebrauck *et al.*[343] and *Oil and Gas Journal.*[627]
[b] FCC, fluid catalytic cracking; HCC, Houdriflow moving-bed system; TCC, Thermofar moving-bed system.

evaluated. For example, the common refinery practice of flaring waste gas might be a source of POM; modern combustion controls on flares would be expected to remove these sources from consideration.

The industrial emissions cited above were measured directly; that is, the effluent stream itself was analyzed for POM. Many other industrial sources are not amenable to such direct sampling, and indirect means—often imprecise—have been used to estimate their emissions.

Some of the processes considered include coke production in the iron and steel industry; carbon black, coal-tar pitch, and asphalt-hot-road mix processes; and general chemical processes. The analytic procedure used in evaluating all such sources has been to sample the atmosphere in the immediate vicinity of an expected emission source or complex of sources. This necessarily leads to less accurate estimates than direct sampling. Except for coke production, none of the industrial processes considered above contributes significant amounts of benzo[a]pyrene to the total atmospheric concentration. This conclusion is obscured somewhat by the presence of other local emission sources already discussed, such as residential coal-fired furnaces.

There is evidence[711] that high benzo[a]pyrene emissions are associated with the gaseous discharge of coke ovens. In the United States, recent activities of the National Air Sampling Network and the Pennsylvania State Department of Public Health support the belief that iron and steel works do contribute to higher atmospheric benzo[a]-pyrene concentrations in the areas surrounding them. Corresponding studies in areas outside the United States[459,740] lead to similar conclusions. An emission factor, admittedly crude, for benzo[a]pyrene emission from coke effluents has been calculated[711] at 1.8 g/ton of coke. Application of this factor to estimated nationwide coke discharges results in a predicted emission of 192 tons/year.

It is obvious that many more industrial processes may contribute to atmospheric benzo[a]pyrene. They will constitute localized sources and can be expected to lead to increased atmospheric concentrations at local sampling sites.

Industrial emission of benzo[a]pyrene is summarized in Table 3-6.

INDOOR POM EMISSION

Although the outdoor environment has received a fair amount of study in terms of POM, little is known of the sources and magnitude of the indoor burden. The possible sources in residential structures are improperly vented furnaces and incinerators, tobacco smoke, and leakage from the outdoors. In industrial plants, many of the processes referred to previously can, if not controlled properly, emit POM to the indoor environment.

TABLE 3-6 Summary of Estimated Industrial Benzo[a]pyrene Emission in the United States

Source of Benzo[a]pyrene	Benzo[a]pyrene Emission, tons/year
Petroleum	6
Asphalt air-blowing	<1
Coke production	200

In the only published work, Stocks and co-workers[726,733] studied polycyclic hydrocarbons and smoke in garages and offices and reported the concentrations of benzo[a]pyrene, benzo[ghi]perylene, pyrene, and fluoranthene. Their data suggest that office sites have 25–70% lower POM concentrations than found in the immediate outdoor environment; the POM concentrations in bus and car garages were at least as high as and usually somewhat higher than those of the ambient air.

One major source of nonindustrial indoor POM pollution is tobacco-smoking.[831] During smoking, the mainstream smoke is inhaled and, in the interval between puffs, the sidestream smoke escapes into the environment. The use of one unfiltered cigarette (85 mm) releases 30–50 mg of "tar," which contains 0.10–0.15 μg of benzo[a]pyrene, 0.20–0.30 μg of pyrene, and 0.25 μg of chrysene. In a medium-size room (40 m^3), three smokers can pollute the air with 2–4 μg of benzo[a]pyrene, 5–8 μg of pyrene, and 6 μg of chrysene per 1,000 m^3 of air. Depending on ventilation and smoking activities, the indoor pollution by cigarette sidestream smoke can be significantly higher; for example, Galůskinová[288] reported 28–144 μg of benzo[a]pyrene per 1,000 m^3 for a beer hall in Prague.

The most important source of personal air pollution is the mainstream smoke of tobacco products. For example, a smoker of 30 unfiltered cigarettes (popular 85-mm U.S. brand) inhales about 1.0 μg of benzo[a]pyrene daily. Stated differently, 1,000 m^3 of cigarette mainstream smoke contain about 100,000 μg of benzo[a]pyrene, compared with 0.01–74 μg/1,000 m^3 found in polluted air. Although the concentration of benzo[a]pyrene in a given volume of air and in the same volume of cigarette smoke are not directly comparable, they do provide some insight into the relative importance of tobacco-smoking in pollution with POM.

EMISSION CONTROL PROCEDURES

Efforts aimed at improving the combustion efficiency of most of the processes covered in this section would be obvious first steps toward control of POM emission. However, the main contributors to the POM emission of heat and power generation are small coal furnaces and wood burning. Neither is economically amenable to better controls, so alternate fuel sources would be the preferred solution.

With regard to refuse burning, more efficient incineration equipment in commercial, industrial, and apartment-building sources would be appropriate. Their relative contribution diminishes in comparison with open burning, especially coal refuse burning and, to a lesser extent, forest and agricultural burning. Coal refuse burning could simply be eliminated by proper attention to refuse accumulation practices, and intentional forest and agricultural burning could be discontinued.

Polycyclic organic matter emissions from catalytic cracking in the petroleum industry are well on their way to effective control through the increasing use of carbon monoxide waste boilers. The contribution from coking emissions in the iron and steel industry must be more accurately assessed.

The contribution of stationary sources to the total POM emission inventory, although poorly quantified, appears to be large. Latest estimates of gross tonnage are to be viewed as a first approximation and may be valid to within a factor of 10. Even with this qualification, the stationary-source contribution probably accounts for 90% of the total nationwide POM emission. It is important to note that emission by stationary sources is usually highly localized, in contrast with that by mobile sources, and results in high atmospheric POM concentrations in the vicinity of major emitters.

Owing primarily to these localized emissions, comprehensive epidemiologic studies should be initiated in geographic areas that are subject to high atmospheric concentrations of POM. A particularly fruitful study might be done in the Appalachian Mountain–Mississippi River area.

Stationary-source emission factors must be validated and the analyses extended to include as many additional kinds of POM (i.e., other than benzo[a]pyrene) as feasible. In particular, the importance of hand-fired furnaces burning coal or wood must be critically evaluated. Coal refuse burning is in the same category; the large, highly speculative value chosen for this contribution requires verification. Alternate disposal methods for coal refuse should be developed in the interim.

TABLE 3-7 Summary of Benzo[a]pyrene Emission by Stationary Sources in the United States

Source of Benzo[a]pyrene	Benzo[a]pyrene Emission, tons/year
Heat and power generation	~500
Refuse burning	~600
Coke production	200

Emission of POM from coke production also requires scrutiny; alternate manufacturing practices in the iron and steel industry should be developed in case such emission must be controlled.

The best available current data suggest the stationary-source benzo[a]pyrene emission shown in Table 3-7.

GENERAL NATURE OF POM EMISSIONS

Individual POM Emissions

The polycyclic organic molecule mentioned most prominently here has been benzo[a]pyrene. This material has been identified as a prominent constituent of most of the processes discussed and has also been shown to be a potent carcinogen. Although these facts confirm the importance of benzo[a]pyrene, many other materials emitted in the same processes have some carcinogenic activity.

It has been felt that benzo[a]pyrene could be used as an indicator molecule, implying the presence of a number of other components of similar structure. Several workers[153,667,670,673] have reported numerous types of POM in urban air, including pyrene, anthanthrene, benz[a]anthracene, benzofluoranthenes, dibenzanthracenes, chrysene, phenylenepyrene, benzoperylene, coronene, fluoranthene, and alkyl derivatives of these compounds, as well as benzopyrenes. (See Table 2-1 for some of these materials and their structures and properties.)

There have been attempts to develop relations between these individual compounds (such as the ratio of benzo[a]pyrene to pyrene and coronene to pyrene) as a function of their source. For example, ratios shown in Table 3-8 have been determined for vehicular emissions, industrial emissions, refuse burning, and heat generation.

It is obvious that these ratios can vary widely as a function of emission source. Before benzo[a]pyrene can be used as an accurate barometer of the entire class of POM, more information on these ratios, as well as on the carcinogenic significance of the other POM molecules, will be required.

Area-Concentration Relations

It is evident that three major stationary sources—coal-fired and wood-fired residential furnaces, coal refuse fires, and coke production—account for more than 90% of the annual nationwide benzo[a]pyrene emission. Of the remaining sources, the transportation contribution

TABLE 3-8 Ratios of Individual POM Molecules by Emission Source

Emission Source	Pyrene: Benzo[a]pyrene	Benzo[ghi]perylene: Benzo[a]pyrene	Benz[a]anthracene: Benzo[a]pyrene
Automobiles[a]	7:1–24:1	2:1–5:1	1:1–2:1
Trucks			
Gasoline-powered[b]	50:1–90:1	–	–
Diesel-fuel-powered[b]	<1:1–50:1	–	–
Catalytic cracking[c]	<1:1–23:1	0.3:1–3:1	–
Incinerators[c]	6:1–16:1	0.2:1–1:1	–
Heat generation[d]	1:1–1,000:1	–	–

[a] Data from Gross,[316] Hangebrauck *et al.*,[343] Hoffman *et al.*,[384] Kotin *et al.*,[455] and Sawicki *et al.*[674]
[b] Data from Hangebrauck *et al.*[342]
[c] Data from Hangebrauck *et al.*[343]
[d] Data from Falk *et al.*[260]

is probably the most significant, in that it pervades all segments of the nation. It is instructive to consider the predominant areas of the country with regard to these stationary sources, as in Table 3-9.

When the areas of major emissions are grouped, it is obvious that POM emissions are very high through the southeastern region along the Appalachian Mountains, as well as in the area to the immediate west as far as the Mississippi River and north to the Great Lakes. Although more quantitative extrapolations are not warranted, we can consider the aerometric data now available through the National Environmental Research Center of the Environmental Protection Agency as indicative of the major urban areas in which POM concentrations may constitute health problems. A survey of the data for the winter of 1969, in which benzo[a]pyrene concentrations are reported,* is enlightening. Of the 40 U.S. cities in which the winter benzo[a]pyrene concentration exceeded 5 $\mu g/1{,}000\ m^3$, 34 are in the region just defined, as are 44 of the 53 cities with concentrations in excess of 4 $\mu g/1{,}000\ m^3$. The densely populated Northeast and the Far West are conspicuous by their relatively low atmospheric benzo[a]pyrene concentrations; the Los Angeles Basin, with its high vehicle and human population densities, has concentrations of 1.5–3 $\mu g/1{,}000\ m^3$.

In the only other set of determinations of relative source contributions to the atmospheric POM burdens, Colucci and Begeman[153] have calculated that, in Detroit, motor vehicles contribute 5% of the

*Data from National Aerometric Data Bank, P.O. Box 12055, Research Triangle Park, North Carolina 27709.

TABLE 3-9 Contributions to National Totals of Benzo[a]pyrene by Source and State[a]

Benzo[a]pyrene Emission Source	State	Fraction of U.S. Total, % State	Fraction of U.S. Total, % Group of States
Coal-fired furnaces	Illinois	22	58
	Ohio	12	
	Wisconsin	10	
	Michigan	8	
	Indiana	6	
Coal refuse burning	W. Virginia	45	90
	Pennsylvania	25	
	Kentucky, Colorado, Virginia	20	
Coke production	Pennsylvania	29	79
	Ohio	16	
	Indiana	14	
	Alabama, Maryland, W. Virginia	20	

[a] Based on data from Hamburg,[336] L. McNab (personal communication), Muhich *et al.*,[550] U.S. Department of Agriculture,[764] U.S. Department of Health, Education, and Welfare,[767,774] and U.S. Department of the Interior.[777]

benzo[a]pyrene in the downtown area, 18% of that in the freeway area, and 42% of that in the atmosphere in the suburbs. Similar studies were made in New York City[151] and Los Angeles,[152] but the same types of calculations were not possible. In both cases, the data would permit only correlation techniques, which indicate positive statistical relations of benzo[a]pyrene with both automotive and stationary sources.

These data indicate that, in the absence of other major sources, as in some suburban locations, the vehicular contribution may be as high as 50%. Aerometric data indicate that, when this relative vehicular contribution is high, the local atmospheric POM concentrations are low.

The implications of these trends are evident: Epidemiologic data should be obtained in areas of high and low POM concentration to establish the effect of atmospheric contamination by POM. Until such studies are made, the nature and degree of source controls required will be unknown.

SUMMARY AND RECOMMENDATIONS

Polycyclic organic matter can be formed in any combustion process involving hydrocarbons. Naturally occurring POM emission to the atmosphere does not appear to be significant. The major technologic emissions include those from transportation sources and such stationary sources as heat and power generation, refuse burning, and industrial processes.

The internal-combustion engine is a ubiquitous source of POM. Current efforts to reduce total vehicular emissions have reduced POM emissions, and projections of future control levels point toward a continuing and marked decline. However, such projections presuppose properly maintained and operated vehicles; close scrutiny should be directed at the effects of deterioration of automobile emission control devices and the use of diesel-fueled vehicles in overloaded conditions. Research efforts to determine the effects of fuel composition and of advanced emission control devices should be continued. Polycyclic organic matter emissions from aircraft should be assessed, as well as those from local mobile sources, such as two-cycle engines.

POM emissions from major stationary sources are poorly quantified. Available data suggest that coal-fired residential furnaces, coal refuse bank burning, and coke production from the iron and steel industry are responsible for the bulk of the nationwide POM emission. However, serious reservations may be expressed as to the validity and magnitude of these data. It is noted that atmospheric concentrations of POM are high in areas in which the cited sources are concentrated. In addition, effective control procedures for these processes are lacking. Substitution of alternate fuels or more efficient combustion processes and discontinuance of coal refuse storage practices seem to be the only appropriate methods for the restriction of coal-related emission; the emission associated with coke production requires additional research on control procedures and source analysis.

Current data suggest the following relative contributions of major source categories to the total POM emission inventory (expressed in terms of annual estimated benzo[a]pyrene emissions): heat and power generation, 500 tons/year; refuse burning, 600 tons/year; coke production, 200 tons/year; and motor vehicles, 20 tons/year.

These data represent nationwide estimates based on extrapolations from individual source emissions. In specific areas, the relative contribution of any given source may differ significantly from that implied by the nationwide figures. For example, the vehicular source may be the major contributor in suburban areas where other major sources are absent. Epidemiologic studies using source inventory data and ambient atmospheric concentrations are required to assess the importance of control measures in both high and low atmospheric POM areas.

4

Atmospheric Physics of Particulate Polycyclic Organic Matter

The presence of aerosols in the atmosphere, even in locations very remote from population centers, is well known. Although many sources of natural aerosol particles have been identified,[420] their magnitude remains uncertain. Some typical natural sources of particles are listed in Table 4-1, which classifies material as "primary" in origin (emitted as particles directly from a source) or as "secondary" in origin (formed by condensation or chemical reaction in the atmosphere itself). From a preliminary assessment indicated in Table 4-1, it appears that, globally, three major sources are dust rise caused by wind, sea spray, and vegetation. The hydrocarbon aerosols from vegetation are believed to be primarily terpenes.[874]

Superimposed on the natural background material is a significant amount of aerosol produced by man. Some major identified sources of anthropogenic aerosols are listed in Table 4-2. (The table excludes cigarette smoke, which, although probably small in tonnage, represents a significant hazard.) This list suggests the importance of combustion as a contributor to the particle population, especially in urban areas. Although polycylic organic matter represents only a very small fraction of the total amount of particulate matter in the atmosphere or associated with combustion sources, it constitutes a very important

TABLE 4-1 Some Sources of Natural Aerosols in the Atmosphere[a]

Source	Estimated Aerosol Production Rate, tons/day
Primary	
Dust rise by wind	20,000–1,000,000
Sea spray	3,000,000
Forest fires (intermittent)	400,000
Volcanic dust (intermittent)	10,000
Extraterrestrial (meteoritic dust)	50–550
Secondary	
Vegetation: hydrocarbons	500,000–3,000,000
Sulfur cycle: $SO_4^=$	100,000–1,000,000
Nitrogen cycle: NO_3^-	1,000,000
NH_4^+	700,000
Volcanoes: volatile SO_2 and H_2S (intermittent)	1,000
Maximal total	~10,000,000

[a] Modified from Hidy and Brock.[372]

minor fraction, because of its potential hazard to human and animal life. In urban areas, it is expected that localized anthropogenic sources will far exceed natural sources, in contrast with the proportions of the

TABLE 4-2 Some Important Sources of Anthropogenic Aerosols in the Atmosphere[a]

Source	Estimated Aerosol Production Rate, tons/day
Primary	
Combustion and industry (potentially containing traces of POM)[b]	100,000–300,000
Dust rise by cultivation (intermittent)[c]	100–1,000
Secondary	
Hydrocarbon vapors (incomplete combustion, etc.; may involve traces of POM)	7,000
Sulfates (SO_2, $H_2S \rightarrow SO_4^=$)	300,000
Nitrates ($NO_x \rightarrow NO_3^-$)	60,000
Ammonia	3,000
Maximal total	~700,000

[a] Modified from Hidy and Brock.[372]
[b] Assumes 90% emission control of emissions from coal-burning installations.
[c] United States only.

atmosphere as a whole, as indicated in Tables 4-1 and 4-2. Unfortunately, there is little quantitative information on the relative contributions of sources in polluted areas.

Because of the high melting and boiling points of materials classified as POM (see Chapter 2), the bulk of POM is believed to be linked with aerosols. It is not known in what form such material generally is present in aerosols. Available evidence, however, indicates that benzo-[a]pyrene is identified primarily with soot particles.[154,751] POM may exist as particles of relatively pure material, or it may be adsorbed in small amounts on other particles. Adsorption is especially important because of the possibility that POM is more easily assimilated biologically if it is associated with soot, dust, or other particles (see, for example, Kotin and Falk[449]).

Because much anthropogenic POM is identified with combustion (see Chapter 3), it is likely that such material is emitted as vapor from the zone of burning. Either in a stack or in an exhaust pipe, it will cool and condense on existing particles or form very small particles itself. If it enters the atmosphere as vapor, it will undoubtedly be adsorbed on existing particles while undergoing condensation. In fact, small particles in the atmosphere may act as nucleation centers for forming POM aerosols when such vapors are supersaturated.

As POM is mixed with aerosols in the atmosphere, it is spread among particles of widely varied sizes by collision processes. POM-containing particles are dispersed in air by turbulence and may be transported great distances from their origin by winds. They are eventually removed from the atmosphere by sedimentation or deposition, such as on rocks, buildings, and vegetation. Removal is enhanced by washout from under rain clouds and by rainout from within clouds.

All factors involved in the aging of atmospheric aerosols may be important in the consideration of POM as a health hazard. Collision processes change the original size of POM-containing particles, whereas transport tends to disperse such material over broad areas. Dispersion tends to dilute POM, making exposure to it less likely, while at once increasing the likelihood of exposure to it in areas remote from the original source. Removal processes are significant, in that they rid the air of POM, taking it out of circulation and making it less likely to be inhaled; however, such removal may contaminate edible vegetation, and POM may thus be ingested by man and animals.

The U.S. Public Health Service has recently published a general review of the properties of atmospheric aerosols, their measurement, and their health hazard;[775] the bulk of that material will not be repeated here.

Keeping in view the possible significance to health of POM-containing aerosols, however, the physical properties of POM and related materials are reviewed here, and then the fate of POM in the atmosphere in relation to physical changes, transport, and removal of aerosols.

PHYSICAL PROPERTIES

Concentration and Particle Size Distribution

The concentration of aerosols with respect to a range of particle size is crucial in the evaluation of the penetration of material into the respiratory system.

Evaluation of particle size distribution is somewhat arbitrary, in that particle dimensions are difficult to define uniquely. Aerosols in the atmosphere contain particles of many different shapes, ranging from spheres to long fibers. Particle size may be defined by two general methods. The first refers to the dispersed material in terms of an equivalent geometric, projected area. The second defines size in terms of some property of the particles, such as settling rate, optical scattering cross section, or ratio of electric charge to mass. Reported data on size distributions differ somewhat even for the same sample, depending on the detector used. In this respect, comparisons of data taken at different times and places may be ambiguous. Much of the size distribution information is reported in terms of a radius or diameter of an "equivalent sphere," without careful definition of this dimension in relation to the sampling or detection method. Despite its ambiguities, "equivalent" size is adopted here for simplicity in discussing recent results of size distribution observations.

Because of the great variety of urban sources for aerosols, it would be difficult, if not impossible, to assess and classify individual emissions. Some very limited measurements have been made from major sources like automobile exhausts. However, these have been rather crude and have remained largely unreported by investigators.

In the case of automobile exhaust, Mueller *et al.*,[548] using different kinds of samplers, indicated that about 68% by weight of aerosol emitted was smaller than 0.3 μm. Recent preliminary data from the University of Minnesota aerosol laboratories suggest that automobile exhaust contains more than 10^5 particles/cm^3, with a mean radius as small as 100 Å. The automobile, then, contributes significant amounts of aerosols that will remain in the air for extended periods, although the total mass emitted may be small.

Most of the data in the literature[775] on aerosol distributions from stationary sources are confined to particles larger than the submicrometer range. However, it is known that combustion processes produce copious quantities of nuclei that measure a few tens of Ångströms. Almost nothing is known about either the size distribution or the chemical composition of particles from stationary sources that are smaller than 0.1 μm.

Only recently have measurements become available that afford a reliable idea of the range of particle size in urban atmospheres. It has been found, for example, that aerosols may be found in city air ranging in size from a few tens of Ångströms to hundreds of micrometers. It is not yet known whether there is significant fractionation with respect to size of hydrocarbon material.

To measure aerosol particles in the atmosphere over the wide range of particle size known to exist, more than one instrument is required. The large-particle fraction—greater than 0.5 μm in diameter—may be observed with a variety of techniques, involving either indirect sampling (optical counters) or direct collection of particles (impactors, filters, or centrifuges). Such equipment has been described in detail elsewhere.[449] For particles less than 0.5 μm in diameter, there are only a few methods for observation. Among them are the Aitken nuclei counters,[409] small adiabatic expansion chambers allowing nuclei 50 Å or more in diameter to grow sufficiently in supersaturated water vapor to be counted optically, and the electrostatic analyzer,[409] a device relying on the specificity of the relation between an equilibrium electric charge and particle size (mass) for very small particles. The latter device appears to operate most satisfactorily at an equivalent radius of 0.02–0.1 μm.

Using a hybrid analyzer system, Husar and colleagues[409] have developed a method for measuring the aerosol size distribution from about 100 Å to over 10 μm in diameter. Some results of this system are shown in Figure 4-1 for the urban atmosphere in Minneapolis and in Figure 4-2 for Pasadena. Some typical nonurban data from other methods[374] have been added in Figure 4-2 for comparison. Pasadena and Minneapolis have their own characteristic patterns, the main difference between them being in the extent and intensity of photochemical smog production. The Los Angeles area (which includes Pasadena) has a much more photochemically reactive atmosphere than Minneapolis. In both cases, the curves illustrate that urban atmospheres contain enormous numbers of very small particles, compared with nonurban atmospheres. During the day, the Los Angeles

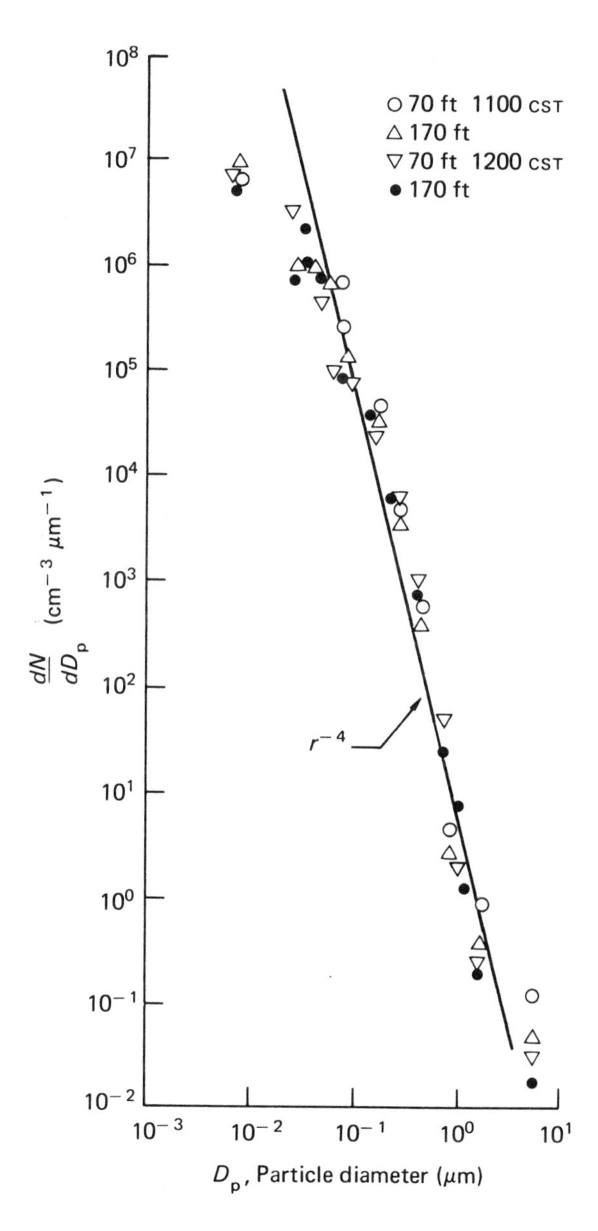

FIGURE 4-1 Particle size distribution of aerosols in Minneapolis, Minnesota, taken February 22, 1967. Data are shown for two different heights above the ground at 11 a.m. and 12 noon. (Derived from Peterson *et al.*[591])

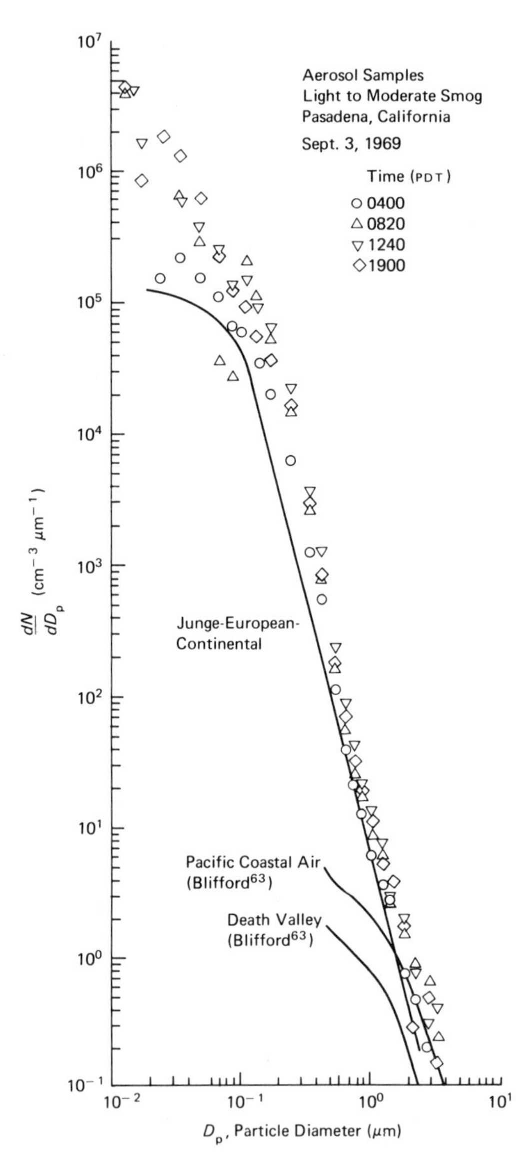

FIGURE 4-2 Particle size distribution in Pasadena, California, on the basis of the Minnesota Aerosol Analyzing System. Data of K. T. Whitby. (Reprinted with permission from Hidy and Friedlander.[374]) Results of Blifford[63] for nonurban air as measured with an impactor are shown for comparison.

atmosphere appears to be enriched in small particles, compared with the night. However, the small-particle concentration in Los Angeles is generally below that in Minneapolis. The diurnal difference in Los Angeles is speculatively associated with increased man-made production during the day. However, the difference between Los Angeles and Minneapolis appears to be linked with the high photochemical reactivity of the Los Angeles atmosphere. Evidently, the condensation of vapors on particles larger than 0.1 μm in radius during the daylight evolution of photochemical smog overshadows the production of Aitken nuclei in more reactive air (see, for example, Whitby *et al.*[815]).

There is considerable natural variation from place to place and from time to time, at least over portions of the observed aerosol size distribution. However, it has been observed that the upper end of the spectrum often, but not always, has a remarkable regularity, particularly in data taken for many cases averaged together. The large-particle portion of the spectrum (greater than 0.1 μm) often tends, on the average, to follow a power law form:

$$n(D_p) = \frac{dN}{dD_p} = \text{const}\ \phi\ D_p^{-4}, \qquad (1)$$

in which D_p = particle diameter; $n(D_p)dD_p$ = the number of particles in the diameter range, D_p to $(D_p + dD_p)$; N = total number of particles; and ϕ = volume fraction. On the basis of many observations[141,581] of urban aerosols, the constant in this equation is approximately 0.40, but it may vary from 0.24 to 0.56. The power of the diameter ranges between -3 and -5.[421] There is no theoretical explanation, as yet, for the apparent regularity in the upper portion of the "average" aerosol particle size spectrum. However, some speculation has been reported.[371,421]

Because $n(D_p)$ has some regularity, particularly above 0.05 μm in diameter, it is possible to estimate crudely the dosage to various parts of the respiratory system expected in urban areas with limited knowledge of the size distribution. Serious questions remain, however, about the relation between an equivalent spherical diameter, as measured by the optical and electric devices of the Whitby analyzer, and a particle size and shape more relevant to deposition on the surface of the lung.

Particle Shape and Density

The characterization of a particle with respect to shape permits transformation of projected area to, for instance, total area. With information for particle density, one can calculate particle weight. Because of the applicability of different particle dimensions to atmospheric conditions or to the effects produced in biologic receptors by particulate pollution, there is great interest in characterizing particle shapes to transform from one particle size dimension to another. The significance of the different classifications of particle size, shape, and weight in the evocation of biologic responses is discussed by Davies.[183] Hodkinson[381] reviewed the effects of particle shape on measurement of size and concentration. Cartwright[119] has illustrated the utility of a specific particle shape—the ellipsoid of revolution—in characterizing quartz dust. To measure the particle radius directly associated with deposition by aerodynamic forces, such devices as the aerosol centrifuge have been built. Perhaps the best known is that designed by Goetz *et al.*[300] Other promising devices have also been developed.[380,725]

The most accurately reported characterization of particle shapes is probably that of Kotrappa,[457] for which a spinning spiral duct[725] was used to classify particles of coal, uranium dioxide, and thorium dioxide aerodynamically. Kotrappa's work illustrates the use of shape factors to describe irregularly shaped particles. The dynamic shape factor, κ, of an irregularly shaped particle moving at its terminal settling velocity is the ratio of the drag force action on it to the drag experienced by a spherical particle of the same mass and density moving at the same velocity. The Stokes diameter, D_{st}, of a particle is defined as the diameter of a spherical particle of the same density and terminal settling velocity as the irregular particle. The aerodynamic diameter, D_{ae}, of an irregularly shaped particle is defined as the diameter of a sphere of unit density with the same terminal settling velocity as the particle. The volume shape factor, α_v, is defined by:

$$\text{volume of particle} = \alpha_v D_p{}^3 \ldots, \qquad (2)$$

in which D_p is the diameter of a circle of the same area as the projected area of the particle.* The surface shape factor, α_s, is given by:

$$\text{surface of particle} = \alpha_s D_p{}^2 \cdot \qquad (3)$$

In Kotrappa's studies, the volume shape factor, α_v, remained relatively constant at 0.38 for the coal sample whose projected area diameter was 0.5–4.5 μm. The dynamic shape factor, κ, was relatively constant at 2.0 for particles with diameter less than 2.5 μm but in-

*This is identical with D_p used in Figures 4-1 and 4-2 for spheres only.

creased slowly for larger particles. The ratio of the projected area diameter to Stokes diameter (D_p / D_{st}) was constant at 1.45 for D_p less than 2.5 μm but decreased slowly to 1.35 for D_p of 4.0 μm. For coal, α_s was 31.5 and α_v was approximately 0.35.

The only work reported for shape factors[722] of particles found in an urban aerosol indicated large variations in individual particle shapes and in the average shape factors for the distribution of sizes greater than $D_p = 1.0$ μm.

As important as particle shape in the rate of aerosol deposition is particle density. Although this property is essential to evaluation of particle settling in the respiratory system, little information on it can be found in the literature. Because particles may agglomerate loosely, forming very porous aggregates, their apparent or effective density may be less than their actual density. Some early work suggests that the apparent density of aggregates of smoke particles and crystalline material could range from 0.1 to 2, depending on the packing.

A study by Lane and Stone[471] has indicated that the ratio of the effective density to the density of a unit of an aggregate of polystyrene spheres approaches an asymptotic value of about 0.65. The effective density is calculated from the settling rate and the particle mass. The aggregate diameter was estimated from the Stokes terminal settling velocity. Similar experiments on fly ash from a power station indicated an effective aggregate density of 1.58 g/cm^3, where the average ash density was measured independently to be 2.3 g/cm^3.

Some data,[166] recently reported on the effective density of urban aerosol collected in Pittsburgh, Pa., indicate that the density of such material ranges from 1.8 to 2.1 g/cm^3.

Moments of Size Distribution

In addition to the ambiguities in its interpretation, measurement of aerosol size distribution is too time-consuming and complicated for routine air monitoring. Therefore, most local monitoring is based on one of the integral moments of the size distribution—for example, the zero moment, that is, the total number (concentration) of particles (N); or the third moment of the radius distribution, that is, the volume fraction (ϕ). If particles exist as equivalent spheres of uniform density, the volume fraction should be proportional to the mass concentration, m, often given in terms of micrograms per cubic meter. Because of the significance of adsorption of POM on suspended particles, the second radius moment—the total surface area per unit volume—is also important; but it has not been measured directly, because satisfactory techniques are largely unavailable. Once the size distribution is known, however, the superficial surface area can be estimated.

Because of the shape of the aerosol size distribution curve, it has been found that the total number concentration is strongly associated with particles less than 0.1 μm in radius, whereas the volume (or mass fraction) is associated mainly with particles larger than 0.1 μm. The large-particle fraction provides the largest dosage in mass, so it has generally been accepted as the convenient observable quantity for monitoring. It is usually measured by means of filters or impactors.[775]

Some typical values for particle number (N) and mass concentration (m) are listed in Table 4-3. The organic fraction of the mass concentration as measured by the benzene-soluble component is also listed, with the benzo[a]pyrene fraction for comparison.* Of the organic fraction, a variety of organic compounds have been identified, including some materials classified as POM.[164] However, the identified fraction represents only 10% of the organic components of urban aerosol. Although the total number concentration is often very large in cities, the mass concentration varies less and rarely exceeds about 200 $\mu g/m^3$ in the United States. The benzene-soluble fraction of this is only a few percent of the total mass, and the concentration of benzo[a]pyrene is far lower. Even in remote areas, there is a contribution of organic material, as expected from Table 4-1. However, it is uncertain whether the benzo[a]pyrene in these areas comes from natural or anthropogenic sources.

Data from the National Air Surveillance Network (NASN) from 1967 to 1969 indicate that the concentrations of benzo[a]pyrene are highest during the first and last quarters of the year (Table 4-4), which is consistent with increased fuel consumption in stationary sources during these periods. In contrast, maxima in the nonurban stations occur more or less randomly during the year.

The frequency distribution of benzo[a]pyrene concentrations found for different NASN sites in 1969 is shown in Figure 4-3. Comparison of the annual average distribution with the distribution for the first quarter shows the relation of increased benzo[a]pyrene concentration with the increase in fuel burning that is associated with the winter months. The curves in Figure 4-3 indicate that the average urban con-

*The benzene-soluble extract is not necessarily equivalent to the total amount of organic material in the sample, but it is taken to be representative of such a fraction.

TABLE 4-3 Typical Values of Aerosol Concentration for Different Geographic Areas (Annual Averages)[a]

Location	N, particles/cm^3 [b]	Mass Concentration (m), $\mu g/m^3$ [c]	Benzene-Soluble Fraction of m, $\mu g/m^3$	Benzo[a]pyrene Fraction of m, ng/m^3
Nonurban				
Continental				
General	10^3–10^4	20–80	1.1–2.2	–
California	10^3–10^4	39	2.8	0.48
Oregon	–	47	0.9	0.09
Colorado	10^2–10^4	14	1.1	0.11
Indiana	–	39	2.1	0.25
Maine	–	18	1.2	0.12
New York	–	29	1.8	0.25
So. Carolina	–	40	2.7	0.43
Maritime				
General	10^2–10^4	–	–	–
Pacific offshore	10^2–10^4	19–146	1.5–6.1[d]	–
Oahu, Hawaii	10^2–10^4	10–49[d]	0.7–6.3[d,e]	–
Urban				
Continental				
General	10^3–10^6	>100	7	–
Los Angeles	10^3–10^6	93	12.5[f]	1.87
Portland	–	72	6.6	2.60
Denver	10^3–10^5	110	9.0	2.52
Minneapolis	10^3–10^5	70	6.1	1.18
Chattanooga	–	105	6.9	4.18
New York	–	105	8.9	3.63
Greenville, S.C.	–	76	7.4	7.49
Maritime				
Honolulu, Hawaii	10^2–10^4	40	2.3	0.59
San Juan, Puerto Rico	–	77	6.9	1.42

[a] Data based on 1969 National Air Surveillance Network observations, except for maritime data, which are based on Junge,[420] Holzworth,[391] Barger and Garrett,[32] and G. M. Hidy (unpublished data).
[b] Aitken nuclei.
[c] Geometric means.
[d] Short-term data.
[e] Chloroform-extractable.
[f] Recent measurements suggest that this is only about half the noncarbonate carbon fraction in the Los Angeles aerosol.[549]

centration of benzo[a]pyrene in the first quarter is about 2.5 ng/m^3 and is an order of magnitude larger than the nonurban concentration.

The average benzo[a]pyrene concentration in U.S. cities has decreased since 1950 by a factor of nearly 3. Over the same period, the

TABLE 4-4 Seasonal Observations of Benzo[a]pyrene and Benzanthrone in Some U.S. Cities[a]

	Mass Concentration, ng/m^3							
	Benzo[a]pyrene				Benzanthrone			
	Quarter				Quarter			
Site	1	2	3	4	1	2	3	4
Los Angeles	2.98	0.79	0.64	3.05	4.48	1.87	1.74	6.10
Medford, Oregon	2.60	2.18	1.45	9.97	8.14	1.69	2.92	9.69
Albuquerque, N.M.	1.02	0.23	0.29	2.95	1.47	0.57	0.67	3.34
Ashland, Ky.	21.17	6.38	6.21	9.80	12.17	3.69	5.47	6.64
Chicago	7.20	3.21	1.60	3.52	4.86	3.38	2.31	3.78
Nashville, Tenn.	5.73	1.76	0.77	2.93	4.76	2.04	1.62	5.68
Philadelphia	6.33	1.69	1.41	6.68	11.02	1.64	1.60	3.65
Pittsburgh, Pa.	21.32	18.27	6.04	9.37	9.28	4.75	3.10	3.91
Greenville, S.C.	19.60	2.84	0.66	4.91	15.52	2.70	1.56	8.46
Missouri (nonurban)	0.24	0.16	0.17	0.08	0.47	0.23	0	0.47
Pennsylvania (nonurban)	2.52	0.83	1.04	0.54	0.83	0.57	1.01	1.00

[a] Preliminary data from the National Air Surveillance Network, 1969 (J. B. Clements, Environmental Protection Administration, personal communication).

average concentration in nonurban sites has also decreased somewhat. The overall reduction in average benzo[a]pyrene concentration is probably associated with nationwide changes in fuel usage and improved furnace design in stationary sources, with more burning of natural gas and fuel oil instead of coal.

The Environmental Protection Agency has found that benzanthrone (7H-benz[d,e]anthracen-7-one) may be an additional indicator of POM from combustion processes. During 1969 and 1970, samples from some NASN stations have been analyzed for benzanthrone. For comparison, some results for seasonal variation over 1969 are tabulated in Table 4-4. In many cases, the benzanthrone concentrations are larger than the benzo[a]pyrene concentrations. The highest concentrations of benzanthrone occur during the winter and fall quarters. To date, there are no data that describe benzanthrone as a carcinogenic agent.

A recent study of Colucci and Begeman[152] is an example of a more detailed short-term urban survey of POM than is available from NASN. From 1964 to 1965, these investigators found that the concentrations of benzo[a]pyrene and benz[a]anthracene were 4½ times greater in central Los Angeles than at two suburban sites. However, the subur-

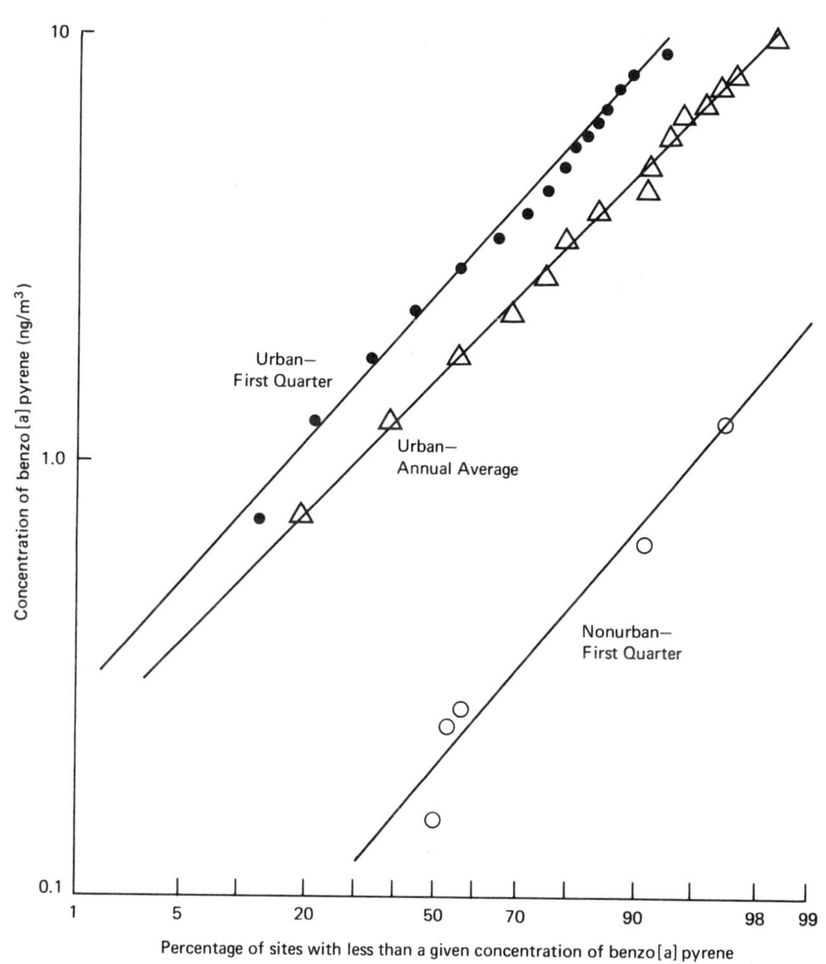

FIGURE 4-3 Frequency distribution of benzo[a] pyrene concentration in air for all U.S. urban and nonurban sites in first quarter with annual average urban distribution (1969 NASN data).

ban site downwind of the downtown area (on the average) appeared to have systematically higher benzo[a] pyrene concentrations than the upwind site. Daily concentrations reported in the Los Angeles area ranged from 0.1 ng/m^3 to over 10 ng/m^3, depending on the season. Benz[a] anthracene concentrations were 1½ times larger than the benzo[a] pyrene concentrations. Annual average benzo[a] pyrene concentrations were similar to the NASN data for downtown Los Angeles. The POM concentrations increased substantially in winter. Benzo[a] pyrene concentrations were higher at night, in contrast with those of other pollutants. All pollutants were higher on weekdays than on weekends. Benzo[a]pyrene concentration was found to be correlated with carbon monoxide and lead concentrations, with coefficients ranging from 0.6 to 0.9. Benzo[a]pyrene concentration was also related significantly to those of hydrocarbon vapors, oxides of nitrogen, and vanadium (a nonautomotive pollutant). Despite the strong relation to lead, the statistics in the study failed to reveal a clear identification of benzo[a]pyrene emissions with automotive or stationary combustion sources.

Because of the regularities in the size distribution function, it appears possible to use correlations of the type suggested by Pasceri and Friedlander[581] and Clark and Whitby[141] to estimate properties of the distribution crudely. The aerosol distribution may be shown in dimensionless variables. In Figure 4-4, all the observed data for the distribution in terms of concentration $n(D_p)dD_p$ in the range D_p to $(D_p + dD_p)$ are scaled to N and ϕ, as defined by the spectrum itself. Thus, in principle, the entire distribution function may be derived from observations of the total number concentration of particles and the total volume fraction (or mass concentration). Because this idea has been tested with only very limited data, it cannot yet be accepted as a "universal" law. It is of interest in this connection that the direct correlation between light scattering from aerosols and the mass concentration[129] depends on the existence of the Junge subrange, or a distribution approximating this power law over the range from 0.1 μm to about 10 μm in radius. Tests of this correlation in a number of cities suggest that the integrating nephelometer, for example, may be a promising instrument for observing mass concentration because of the relation between light scattering and the size spectrum.

An important feature of the Junge subrange, given by Eq. (1), is that the mass concentration should be constant where this form holds. A test of the sensitivity of size distribution observed as measured by such devices as those of Husar *et al.*[409] is to measure the mass identified with each size range directly. Corn[165] has reported such data; he found that, at least in Pittsburgh, the Junge subrange with a power law exponent of −4 was not verified. Instead, the distribution appeared to fall off with an exponent less than −4 from 0.5 to 3 μm in radius.

There is very little information about the actual surface area of aerosols in urban air. One of the few studies that has been made has been carried out by Corn *et al.*,[166] who found that the specific sur-

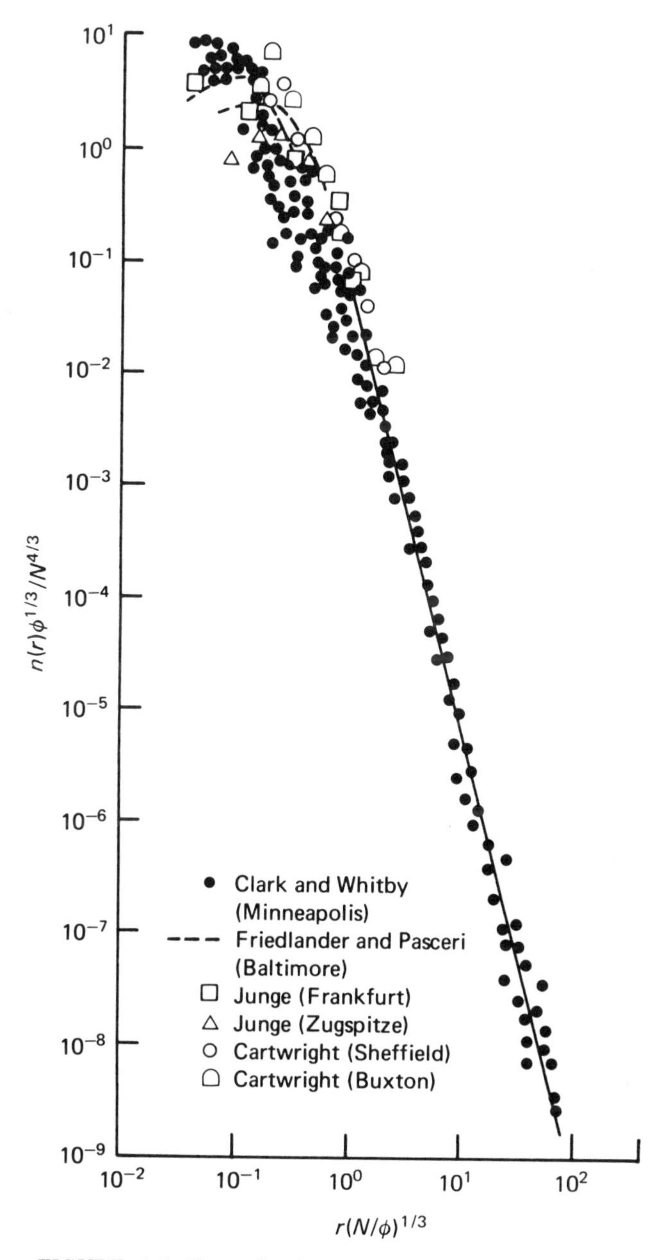

FIGURE 4-4 Normalized correlations of aerosol size spectra observed in polluted or partially polluted air at the ground. (Modified from Clark and Whitby.[141])

face area of aerosols in Pittsburgh varied seasonally from 1.90 m^2/g in the spring to 3.05 m^2/g in the winter.

POM and Particle Size Distribution

There is practically no information on the particle size or mass distribution of POM in an aerosol. In one study, DeMaio and Corn[193] found that more than 75% of the weight of selected polycyclic hydrocarbons was associated with aerosol particles less than 2.5 μm in radius. However, Thomas *et al.*[751] determined that the amount of benzo[a]pyrene per unit weight of soot was constant in the sources they tested.

A potentially important factor in limiting the size of the smallest POM-containing particles is the so-called Kelvin effect—the equilibrium vapor pressure varies with the radius of curvature of the particle and with the surface energy of the volatile material. Unfortunately, there is no information available on the surface energy at the POM–air or POM–solid interfaces. Therefore, it is not possible to calculate the Kelvin effect.

A problem in determining the size fractionation in POM-containing aerosols may arise from current sampling methods. It has been found, for example, that polycyclic hydrocarbons have high volatilities in the presence of air flow.[639] It is therefore possible that some of the POM may be lost by vaporization from smaller particles, owing to the Kelvin effect during the sampling process.

Adsorption and Elution of POM

Mechanisms for the adsorption of POM on particle surfaces have not been studied extensively. However, one investigator[814] has speculated that the benzo[a]pyrene and other arenes appear to be adsorbed primarily on the surface of soot by hydrogen bonding.

Perhaps as important as the ability of POM to be adsorbed on aerosols is its ability to be desorbed or eluted from particles in biologic fluids. Studies of Falk *et al.*[263] have indicated that POM cannot be readily eluted by plasma from soots less than 0.04 μm in diameter. Larger soot will release adsorbed polycyclic aromatic hydrocarbons in the presence of plasma and cytoplasmic proteins. As particle size increases, elution becomes more rapid and more extensive.

More recent studies of Kutscher *et al.*[469] have dealt with the ability of bovine serum to separate benzopyrene from three kinds of soots:

fine-grained Corax L (mean size, 0.028 μm), inactive MT soot (0.4 μm), and flame soot 101 (0.115 μm). The results indicate that a maximum of 10% benzopyrene was eluted from Corax L, whereas soots MT and 101 yielded 20% and 13%, respectively. The first traces from Corax L were detected after 7 hr; it took only 15 min for soots MT and 101 to appear in the serum. Incubations of single components of albumin, alpha, beta, or gamma globulin, and Corax L that had adsorbed benzopyrene showed that only albumin would elute the adsorbed material.

An investigation by Pylev[613] has demonstrated the significance of particle surface area in lung retentivity. This investigator adsorbed different amounts of benzo[a]pyrene on soots ranging in size from 128 to 3,047 Å, with specific surface areas of 10–250 m^2/g. He exposed rats to the soots and evaluated retentivity after dissection. The amount of benzo[a]pyrene left in the rats' lungs differed markedly, depending on the particle surface area. The higher the specific surface area of particles, the more difficult the POM was to elute. Pylev concluded that it is necessary to know, in addition to particle size, more about the composition of the material on which POM may be adsorbed.

DYNAMICS OF AEROSOLS NEAR THE GROUND

Once in the atmosphere, suspended particles will evolve by the mechanisms listed in Table 4-5. Each of the processes listed can influence

TABLE 4-5 Processes Affecting Evolution of Aerosols in a Unit Volume of Lower Atmosphere[a]

1. Growth or change in particles by homogeneous or heterogeneous chemical reactions of gases on the surface of particles
2. Change in particles by attachment and adsorption of trace gases and vapors to aerosol particles
3. Net change by collision between particles undergoing Brownian motion or differential gravitational settling
4. Net change by collision between particles in the presence of turbulence in the suspending gas
5. Gain or loss in concentration by diffusion or convection from neighboring air volumes
6. Loss by gravitational settling
7. Removal at the earth's surface on obstacles by impaction, interception, Brownian motion, and turbulent diffusion
8. Loss or modification by rainout in clouds
9. Loss by washout under clouds

[a] After Hidy.[371]

the observed size distribution of aerosols, as well as their chemical composition as a function of size.

Collision Processes

The coagulation of aerosol particles tends to create large particles while continually depleting the smaller particles. Coagulation requires that two or more particles collide and stick together. Relatively little is known about adhesion during collision, so coagulation of aerosols is generally assumed to occur after collision, with a probability of near unity. There is no sound basis for such an assumption, but the hypothesis is indirectly supported by the concept of Brownian motion in many (not all) experiments.

Collisions between particles may result from their Brownian motion if they are small enough. Otherwise, differences in their velocities in air are required. Other than collision by thermal agitation, some mechanisms include turbulent air motion, differential settling, phoretic forces, and electric forces. Collisions by turbulence involve the fact that particles, because of their large inertia, cannot follow exactly the local eddying motion and may experience a local shear as a result of interaction between eddies of different size. During the fallout of large particles they may sweep out smaller particles in their path; this constitutes the common "scavenging mechanism" in the atmosphere. Relative motion between particles resulting from thermal or concentration gradients during evaporation or condensation of water vapor is possible, but is believed to be a second-order effect. Electric forces between particles also may play an important role in coagulation, particularly for particles less than 10 μm in radius. Although considerable effort has been devoted to development of an understanding of electrification, no extensive calculations have quantitatively evaluated the significance of electric forces, compared with other collision processes.

Because the atmosphere is frequently in a turbulent state, particles tend to be transported rapidly in and out of fixed regions of air space. Thus, particles are transferred in and out of an aerosol cloud by the winds and by diffusion processes associated with turbulence and Brownian motion. In the atmosphere, diffusion by turbulence far exceeds that by thermal agitation. In addition to these diffusion processes, particles will disappear from a volume of air by sedimentation. Particles are much more massive than the suspending gas, so there is always a tendency for aerosol clouds to lag behind air motion and to settle out to the earth's surface in the absence of upward air motion.

Removal from the Atmosphere

Aerosol may be removed from the atmosphere by several mechanisms. Deposition of large particles by gravitational settling is important. Over the radius range of 0.5–20 μm, inertial forces acting on aerosols cause departures in particle trajectories from the air flow around obstacles, thereby allowing deposition by impaction. Obstacles may include rocks, buildings, fences, and vegetation. Because of the finite size of aerosols, they may be intercepted by obstacles even if their trajectories do not depart significantly from the motion of the suspending air. If particles are less than 0.2 μm in radius, their thermal agitation may be sufficient to allow them to diffuse around an obstacle if they pass very near it, so turbulent diffusion around obstacles is important in particle removal.

It is possible to estimate roughly the deposition rate of particles on an obstacle if the air-flow field near it is known. Methods have been discussed in recent monographs.[282,373] Little information is available, however, from which one could deduce the removal rate by buildings or other solid structures in actual atmospheric conditions. Some limited data have been reported for deposition rates on vegetation. Chamberlain,[128] for example, has measured the deposition of radioactive particles on flat grass surfaces, and Neuberger *et al.*[561] and Rosinski and Nagamoto [642] have investigated deposition on trees. The study of Neuberger *et al.* suggests that conifers are better natural filters than deciduous trees.

Role of Rain Clouds

The development of rain clouds, superimposed on the removal mechanisms just discussed in an otherwise dry atmosphere, sometimes influences aerosols containing POM significantly. In contrast with other removal mechanisms, rain clouds are intermittent and their geographic frequency varies widely. Nevertheless, their effects have to be accounted for in any evaluation of the fate of POM in the atmosphere.

Aerosols provide centers for nucleation of water droplets in the atmosphere after the air becomes supersaturated with water vapor. Aerosols inside clouds are captured in droplets by a variety of mechanisms, ranging from scavenging to electric and phoretic interactions. As a result of these processes, called "rainout," the size distribution and the chemical composition as a function of size may be modified. Because of the great difficulties in interpreting and observing natural aerosols inside and outside clouds, practically nothing is known about the nature of such in-cloud modifications.

When precipitation begins to fall from clouds, the hydrometeors will sweep out smaller particles during their fall toward the ground, as in the case of the dry scavenging mechanism. This washout is believed to be significant in removing many pollutants, including POM, from the atmosphere.

There remains some controversy about the importance of washout, compared with other processes, based on current theoretical and experimental work. Theory indicates that the collection efficiency of spheres falling through other spheres is a direct function of particle size. In particular, for Stokes's flow, only very few particles smaller than about 20 μm in radius are collected on falling spheres of much greater size. Some laboratory data support this conclusion, but observation in the atmosphere suggests a stronger action of precipitation in removal of aerosols. Of course, the theory for interacting spheres can be only qualitatively extrapolated to nonspherical material. For example, the collection of smoke particles on snowflakes is undoubtedly different from rain falling through liquid spheres. There is little quantitative information on the effectiveness of washout with interacting nonspherical particles.

Relative Significance of Aging and Removal Processes

Until recently, efforts have been devoted mainly to identifying and characterizing mechanisms involved in the evolution of aerosols in the atmosphere. There remains considerable uncertainty regarding some of these mechanisms, particularly those related to rain clouds and those requiring chemical transformations. Nevertheless, it is possible to evaluate the relative significance of many mechanisms.

A recent study,[371] using semiquantitative theoretical arguments, has indicated the importance of some factors in aerosol behavior near the ground. The data in Table 4-6, from a typical urban atmosphere, suggest that coagulation by Brownian motion, turbulence, and diffusion are most important factors in shaping the size spectrum for particles less than 0.1 μm in radius. For the large-particle fraction, the collision mechanisms are weak, and turbulent diffusion with sedimen-

TABLE 4-6 Processes Contributing to the Aging of Urban Aerosols at the Ground Level (particles lost/cm³ per sec)[a]

Process	Order of Magnitude of Contribution		
	0.05 μm in radius	0.5 μm in radius	5 μm in radius
Convective diffusion[b]	10^{-1}–1	10^{-3}	10^{-6}
Thermal coagulation[b]	1	10^{-3}	10^{-6}
Scavenging by particles (10 μm $N_p = 10^{-1}/cm^3$)[c]	10^{-3}	–	–
Turbulent coagulation ($\epsilon = 1{,}000\ cm^2/sec^3$)[d]	10^{-3}	10^{-3}	10^{-4}
Sedimentation removal[b]	10^{-6}	10^{-6}	10^{-7}
Washout (100 μm $N_p = 10^{-3}/cm^3$)	10^{-3}	10^{-6}	–

[a] Derived from Hidy.[371]
[b] 0.05 μm $N_p = 10^5/cm^3$; 0.5 μm $N_p = 10^2/cm^3$; 5 μm $N_p = 10^{-1}\ cm^3$.
[c] N_p = concentration of scavenging particle.
[d] ϵ = dissipation rate of turbulent kinetic energy.

tation should dominate aerosol behavior. In the middle range of size, collisions by Brownian motion and other mechanisms are weak, but turbulent diffusion remains important. Although weak, turbulent coagulation may be important in allowing particles to be transmitted up the size spectrum from the range dominated by Brownian motion to a range effectively terminated by removal through sedimentation.

The effect of washout by precipitation is evaluated tentatively as less important than dry mechanisms near the ground.[371] However, above an altitude of about 300 m through typical cloud-top levels, washout is likely to become more and more important in removing atmospheric aerosols. These conclusions remain speculative, but further studies should improve the accuracy of evaluation.

ATMOSPHERIC DISPERSION OF POM

Microscale Diffusion

It is important in evaluating the hazards of POM to have some knowledge of the extent of dispersion of this material far from its sources. Classic calculations[802] using the Gaussian plume diffusion model without chemical reaction of the contaminant have been applied extensively in evaluating dispersion a few miles downstream from single sources. These computations can be made for steady wind patterns over smooth terrain for various classes of atmospheric density stratification, provided empirical diffusion coefficients can be estimated. This model always predicts a considerable reduction in ground-level concentration of a pollutant coming from a ground-level or elevated source.

One of the great difficulties with the single-plume models has been their inability to account for topography quantitatively. Some recent work of Hinds,[377] however, has suggested that the diffusion coefficient in such calculations can be modified empirically (on the basis of observations) to evaluate better the influences of rough topography.

Recently, attempts have been made to extend micrometeorology-scale diffusion calculations to include chemically reactive pollutants. An illustrative model was posed by Friedlander and Seinfeld,[281] who investigated the behavior of bimolecular reactions of pollutants undergoing diffusion modeled by the Lagrangian similarity hypothesis. In this calculation, secondary pollutant concentrations at ground level vary according to both the reaction rate and the diffusion rate. In particular, it is sometimes possible to quench some reactions by dilution.

An alternative approach to the Friedlander and Seinfeld analysis has been proposed for photochemical smog by Eschenroeder and Martinez,[251] who follow the chemical reactions along streamlines of the wind field.

Neither of these models is realistic in the sense of having great practical utility for estimating reaction rates and dispersion of materials like POM, but they do represent a first step in recognizing the importance of linking chemistry and air motion in evaluating air pollution.

Larger-Scale Diffusion

On meteorology scales encompassing major urban areas, the Gaussian plume models cannot supply an adequate quantitative assessment of atmospheric dispersion. Account must be taken of the local wind field and the surface distribution of sources. An example of a first attempt to simulate the dispersion over a large urban area, the Los Angeles Basin, has been reported recently by Lamb and Neiburger.[470] Their experimental results suggest that inert contaminant levels can be verified qualitatively. However, the data suffer from simplifications of the local meteorology that will undoubtedly have to be considered in more realistic studies. In particular, some effects, like surface heating in cities, tend to change the mixing and wind field in air over urban areas. The disturbance in the local meteorology caused by a city has been studied in a preliminary way,[78,270,598] but further work must be done to appreciate this factor fully in the dispersion of pollutants.

On scales of tens of kilometers, attempts have been made to evaluate transport and pollutant accumulations over air sheds. One example of Reiquam[630] uses a calculation that breaks up the air-shed volume into well-mixed boxes with local sources. The analysis, again, is highly simplified but leads to qualitative verification with sample air sheds.

There is concern currently for the mechanisms of dispersion over synoptic meteorology scales (1,000 km horizontally). Some efforts to evaluate such problems are beginning to appear in the literature. Kao and Henderson,[425] for example, have applied turbulence theory to investigate classes of synoptic-scale diffusion in the atmosphere. Holzworth's study[392] offers some estimates of large-scale weather influences on community air pollution in the United States.

Qualitatively, one would expect an increasing penetration of pollution into remote areas as time goes on. Indeed, this is suggested in some respects for aerosol, in the trends of increasing mass concentrations measured by the National Air Surveillance Network over the last 10 years.[503] In contrast with the total particle loading increase in remote areas, NASN data for the last 3 years suggest a decrease in benzo[a]pyrene in many of the remote sampling areas. This may be due to many factors, with meteorology being only one. Perhaps equally important are improvements in combustion processes and the use of different fuels.

In any case, it is clear that significantly increased concentrations of POM, as measured by benzo[a]pyrene, remain largely localized in urban areas. Even in the cities, however, POM will be considerably diluted in the ambient atmosphere from its source concentrations in most circumstances.

LIFETIME OF POM IN THE ATMOSPHERE

On the basis of very limited evidence, aerosols are expected to remain in the lower atmosphere for 5–30 days (see, for example, Junge[420]). More recent evidence from Esmen and Corn[252] suggests that, for urban aerosols in Pittsburgh air, the residence time without precipitation is some 4–40 days for particles less than 1 μm in diameter and 0.4–4 days for particles 1–10 μm in diameter. With removal associated with rainfall, these times are believed to be somewhat shorter.

Chemical reactivity of benzo[a]pyrene in the atmosphere, associated with degradation taking place on soot in sunlight, yields a chemical half-life of less than a day, as indicated by data of Thomas *et al.*[751] However, half-lives of several days have been reported, apparently without solar radiation, in earlier work of Falk *et al.*[258] These results for the chemically determined lifetime of benzo[a]pyrene in the atmosphere are ambiguous. However, one may guess that exposure to sunlight of POM on the external surfaces of soot or dust could lead to rapid degradation, whereas POM adsorbed in pores of particles may persist for considerably longer periods.

Exposures to relatively large doses of POM occur in urban areas, where combustion sources are most highly concentrated. Air masses probably remain over a city for less than a day in most circumstances. A residence time of 1 day for air in the Los Angeles Basin can be roughly estimated from emissions of carbon monoxide and carbon monoxide concentrations given by Lemke *et al.*,[489] assuming an inversion height of 500 m and assuming that carbon monoxide is conserved in such an air mass. Thus, it appears that the time for chemical transition of POM will be roughly the same as or longer than the meteorologic residence time in urban centers. One can then expect some contamination of suburban and sparsely populated areas as a result of dispersion and transport of POM-containing particles by the winds before chemical transition can take place fully. This conclusion appears to be consistent with at least some of the NASN data in Table 4-3.

IMPORTANT AREAS OF UNCERTAINTY

From the standpoint of evaluating POM as a health hazard, the principal missing information lies in the chemical nature of POM-containing aerosols. There is a need to know how this material is distributed, with respect to particle size. In addition, more information is required on the physical and chemical evolution of POM as it interacts with other matter in the atmosphere. Of particular interest is the nature of POM adsorbed on different classes of atmospheric aerosols. As can be seen from this brief survey, very little is known about the actual form of POM in the atmosphere. The connection with any measure of adverse health effects is extremely difficult to evaluate in the absence of specific information on POM after it has aged in the air. If there is indeed a link between aerosols of soot or similar materials and hematite in carcinogenesis, it is essential to identify such synergistic materials for control purposes.

SUMMARY AND RECOMMENDATIONS

Form of POM in Air

Polycyclic organic matter detected in the atmosphere is associated exclusively with particulate matter, especially soot. It is uncertain whether POM condenses out as discrete particles after cooling or condenses on surfaces of existing particles after formation during combustion. In any case, knowledge of the behavior of aerosols is essential to understanding the fate of POM in the atmosphere.

Sources and Properties of Atmospheric Aerosols

As a first approximation, natural sources of aerosols contribute approximately 10 times as much as anthropogenic sources to the global burden of suspended particulate matter. However, localized emission inventories of these pollutants, particularly in urban areas, indicate that anthropogenic sources predominate. It is necessary to refine the estimates for specific categories of sources, particularly for the global inventory of aerosols.

Particle size is the physical property with the greatest influence on the behavior of POM-containing aerosols. Generally, the particle size spectrum of atmospheric aerosols extends from less than 0.01 μm to greater than 10 μm. There is considerable variation in the size–concentration distribution with location in space and time, but there is some regularity, on the average, in the particle equivalent diameter range $0.1\ \mu\text{m} \leqslant D_p \leqslant 10\ \mu\text{m}$. In this range,

$$n(D_p) = \frac{dN}{dD_p} = \text{constant}\ \phi\ (D_p)^{-4}, \qquad (4)$$

where ϕ is the volume fraction of particles. The "constant" has been reported to vary from 0.24 to 0.56 but can be approximated as 0.40. This formula is useful for interpreting broad features of aerosol size distributions, but it cannot be considered adequate for detailed description of geographic and temporal variations.

Polycyclic organic matter appears to be associated largely with particles less than 5 μm in diameter. Although large local variations are tabulated by the National Air Surveillance Network, concentrations of suspended particulate matter in U.S. urban areas range from

100 to 200 $\mu g/m^3$, as measured by high-volume sampling. The benzene-soluble portion of this material is approximately 10% by weight, but can vary from 8 to 14% in urban areas. The POM component is much smaller than the benzene-soluble fraction. The density and specific surface of the urban aerosol, on the basis of very limited data from one city, are 1.8–2.1 g/cm^3 and 1.90–3.05 m^2/g. Specific surface appears to vary with season; density does not.

Additional data relative to the physical properties of the atmospheric aerosol are needed. Simple and inexpensive instrumentation is required to obtain, in particular, size–weight concentration data during short intervals (minutes).

Lifetime of POM in Air

Because POM is carried by suspended particulate matter, its longevity in air depends on the lifetime of the carrier aerosol in air and on chemical alteration of POM itself. Initial estimates of atmospheric residence times of particles less than 5 μm in diameter exceed 100 hr in dry atmospheric conditions. Chemical reactivity in the presence of sunlight may lead to transition of POM adsorbed on soot to other material in several hours. Without sunlight, its lifetime may be much longer. Meteorologic factors suggest that air will remain over a city for less than a day. Therefore, it appears that the time for clinical transition will be about the same as or greater than meteorologically controlled residence times in a particular urban atmosphere.

5

Chemical Reactivity of Polycyclic Aromatic Hydrocarbons and Aza-Arenes

Chemical reactions of polycyclic organic matter in the atmosphere are important because such reactions appear to represent a major mode of removal of polynuclear compounds and because the products of the reactions may in some instances be health hazards themselves. The low vapor pressure of most polycyclic compounds has restricted laboratory studies on them to solutions, with few exceptions. In the atmosphere, however, most polynuclear compounds are adsorbed on particulate matter. Reactions of these compounds in the adsorbed state appear to occur particularly readily, and the general nature of the reactions is predictable on the basis of the few studies that have been conducted. There is less information on reactions in the vapor phase or in aerosol solution, but there is no reason to expect them to differ from those in solution.

The following review is not comprehensive; rather, it is limited to types of reactions likely to be of atmospheric importance. Each type is illustrated by only one or two examples, which were chosen because they illustrate important features of the reactions or because they are significant from a health standpoint. Tipson has reviewed the oxidations of polycyclic aromatic hydrocarbons,[753] and the general reactions of polycyclic hydrocarbons have been surveyed by Clar.[139]

GENERAL REACTIVITY CONSIDERATIONS

A great deal of theoretical and experimental work on the reactivity of polycyclic aromatic hydrocarbons has been carried out because of suggested correlations between reactivity and carcinogenicity.[94,181,196,611] Only a few relevant conclusions will be cited here. Reactions may be classified as substitution (in which a hydrogen atom is replaced by another atom) or addition (in which unsaturation is destroyed). Multiple substitution may occur, and addition is often followed by elimination (giving net substitution). Many of the primary products of these reactions undergo further reactions, resulting in more complex changes, such as quinone formation or bond cleavage. The various types of reaction are summarized below.

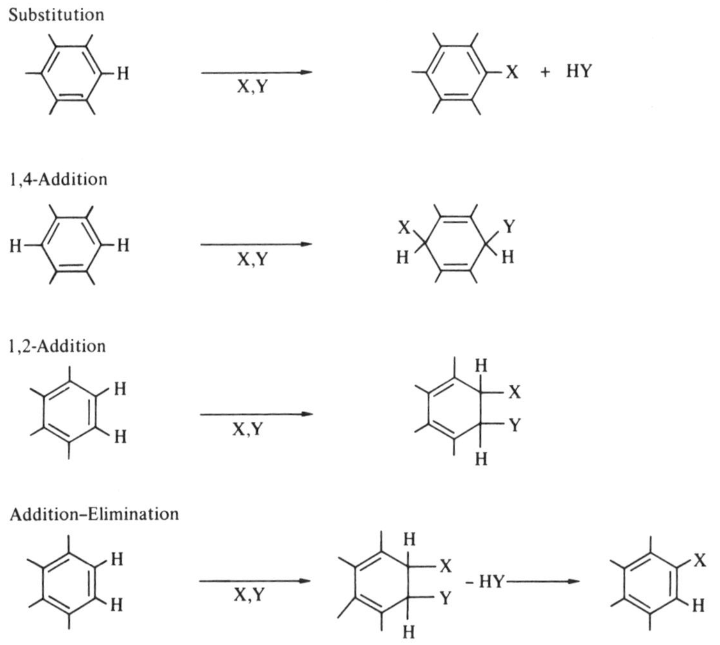

Substitution

-H —X,Y→ -X + HY

1,4-Addition

H- -H —X,Y→ X, H ... Y, H

1,2-Addition

H, H —X,Y→ H, X, Y, H

Addition–Elimination

H, H —X,Y→ H, X, Y, H - HY→ X, H

The type of reaction that a given compound will undergo depends both on the reagent and on the compound.[94,139,181,196,611,753] Reactions of linear hydrocarbons tend to take place at the anthracene 9,10-like position (1,4-addition or substitution) or the 1,2 position (1,2-addition). Reactions of angular hydrocarbons tend to take place at the anthracene 9,10-like position (1,4-addition or substitution) or the phenanthrene 9,10-like double bonds (1,2-addition). Reactions of more condensed ring systems tend to take place at positions adjacent to ring fusions. These tendencies are illustrated below.

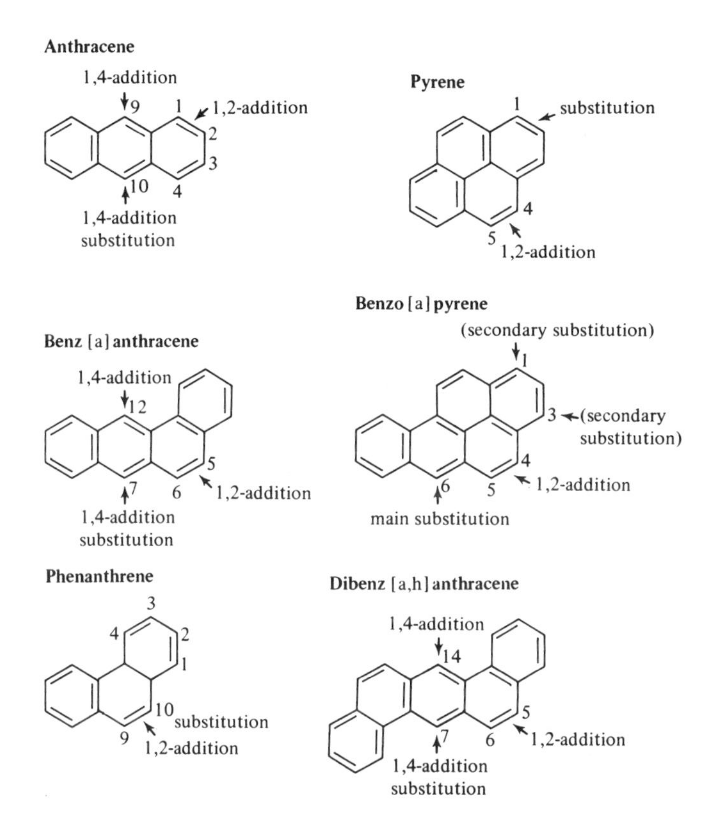

An alkyl or alkoxyl substituent strongly increases the reactivity in most of the reactions described for these compounds, although it will often prevent substitution at the position at which it is attached. Electron-withdrawing substituents generally make reactions more difficult.

PHOTOOXIDATION

Gollnick and Schenck have reviewed photooxidation of polycyclic aromatic hydrocarbons and dienes.[304] Tricyclic or larger hydrocarbons have strong absorption in ultraviolet radiation at wavelengths longer than 300 nm (present in solar radiation at ground level), and most are very readily photooxidized. Photooxidation is probably one of the most important processes in the removal of polycyclic hydrocarbons from the atmosphere, and it is the only one on which substantial data have been obtained under simulated atmospheric conditions, using adsorbed hydrocarbons.

Photooxidation in Solution

The most common photooxidation reaction of polycyclic aromatic and aza-aromatic hydrocarbons in solution is formation of endoperoxides. For example, 9,10-dimethylanthracene (1) yields the 9,10-endoperoxide (2),[716] 7,12-dimethylbenz[a]anthracene (3) yields the 7,12-endoperoxide (4),[158] and dibenz[b,h]acridine (5) yields the 1,4-endoperoxide (6).[253]

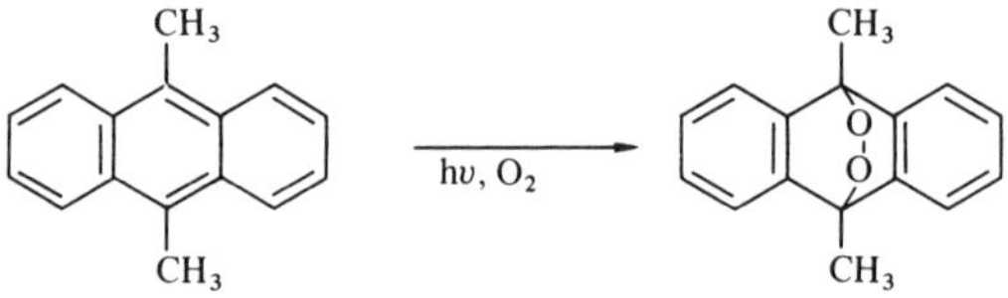

(1) $\xrightarrow{h\upsilon,\ O_2}$ (2)

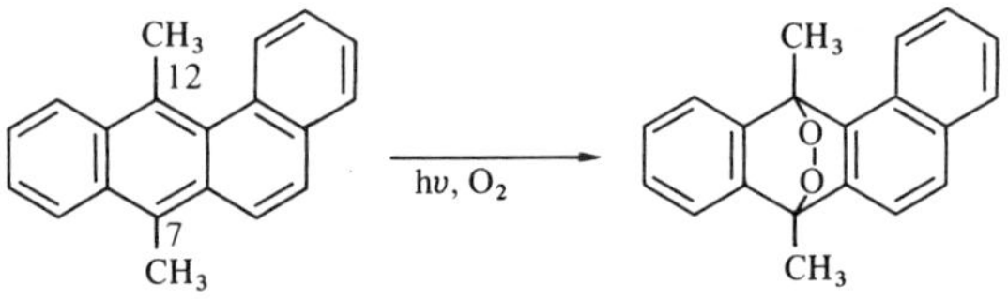

(3) $\xrightarrow{h\upsilon,\ O_2}$ (4)

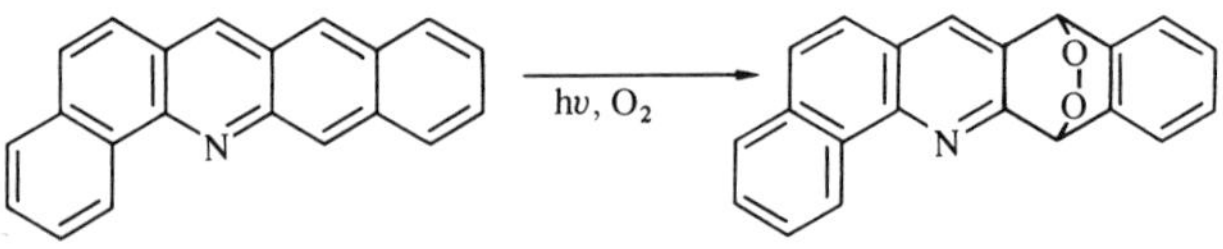

(5) $\xrightarrow{h\upsilon,\ O_2}$ (6)

Photolysis or pyrolysis of peroxides of this sort produces a variety of results, including dealkylation and ring cleavage.[716] This process proceeds by a free-radical mechanism (cleavage of the O–O bond) and initiates autoxidation.[139] This is shown by the possible products of pyrolysis of 9,10-endoperoxide (2) below.

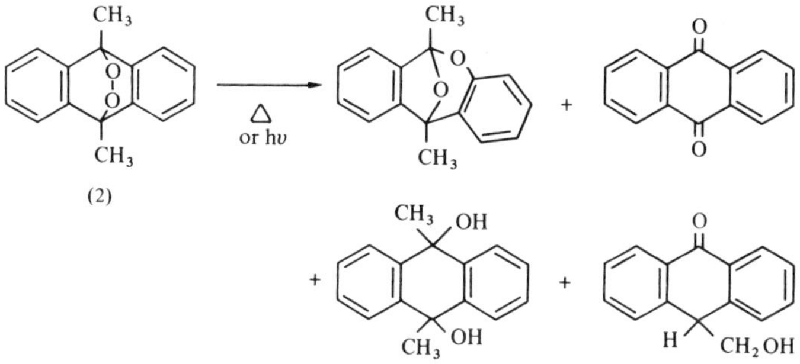

(2)

Alternatively, some endoperoxides on pyrolysis regenerate oxygen in the singlet state, essentially in a reversal of the formation reaction.[805] Some compounds undergo this reversal reaction rapidly at room temperature.[304] When the bridgehead carbon atoms bear hydrogen atoms, the initial endoperoxides are readily converted to quinones. In many cases, only the quinones can be isolated. For instance, compound (6) rearranges readily to the quinone (7), and only the quinone (9) was isolated from photooxidation of benz[b]acridine (8).[253]

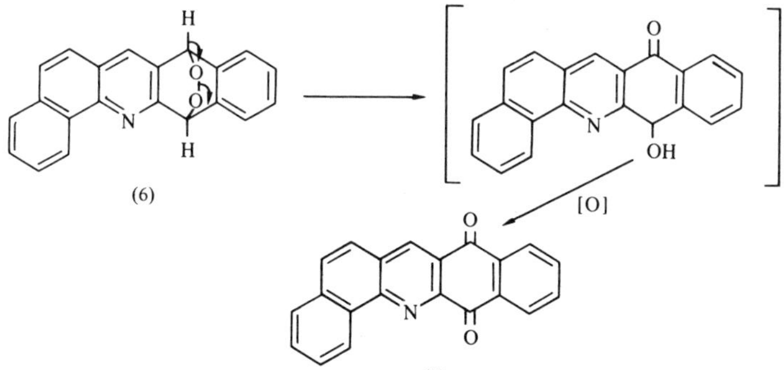

(6)

(7)

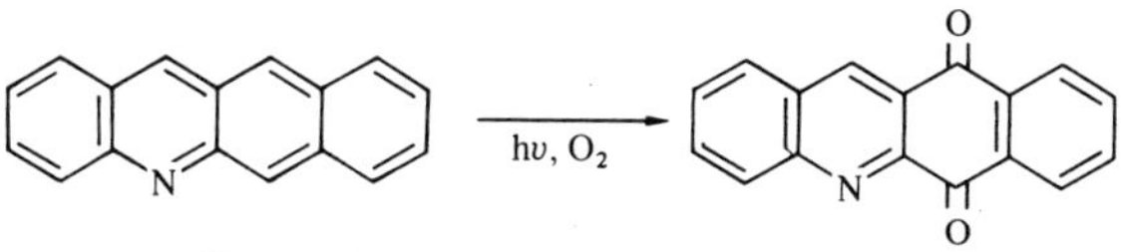

(8)

(9)

It should be pointed out that diones can also be produced when, for steric reasons, no endoperoxide can be formed. For example, benzo[a]pyrene (10) yields a mixture of the 6,12-dione (11), the 1,6-dione (12), and the 3,6-dione (13).[522] The mechanism of this oxidation may well differ from that of the usual one and may involve electron transfer. Endoperoxides are rarely formed from hydrocarbons like benzo[a]pyrene; normally, two anthracene 9,10-like positions, not at ring junctures, are required.

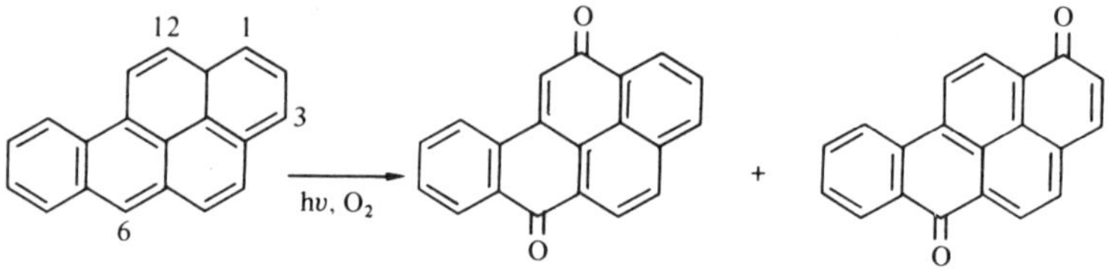

(10) (11) (12)

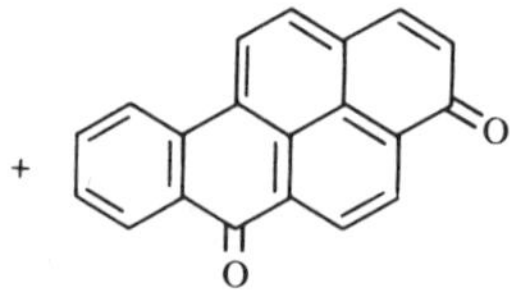

(13)

Photooxidations involve energy transfer from the triplet state of the aromatic compound (A), yielding singlet oxygen ($^1O_2$), which reacts with compound A, yielding peroxide (AO_2)[276] [see Eqs. (1) and (2)]. Singlet oxygen generated by various chemical sources also reacts with anthracenes to yield endoperoxides.[276] Singlet oxygen, directly detected in gas-phase systems, has been formed by energy transfer from benzene,[271,721] naphthalenes,[431,803] and benzaldehyde.[159,462] Although no photooxidizable aromatic has been reacted in the vapor phase, singlet oxygen reacts with cyclohexadiene[552] and with dimethylfuran[299] to yield the normal endoperoxide adducts in the vapor; therefore, aromatic compounds would be expected to undergo reactions in the gas phase identical with those undergone in solution. Further research is needed on this point.

$$A \xrightarrow[h\nu]{} {}^1A \longrightarrow {}^3A \xrightarrow[{}^3O_2]{} {}^1O_2 + A \qquad (1)$$

$$\begin{array}{l} {}^1A \\ \downarrow A \\ A_2 \end{array}$$

$$^1O_2 + A \longrightarrow AO_2 \qquad (2)$$

If the 9,10 positions of an anthracene (14) are no more than monosubstituted (14), a photodimer (A_2) may be formed by reaction of the excited singlet of A with ground state A (15). The extent of dimerization depends on the solvent; dimerization normally does not take place in conditions in which anthracene does not fluoresce.[304]

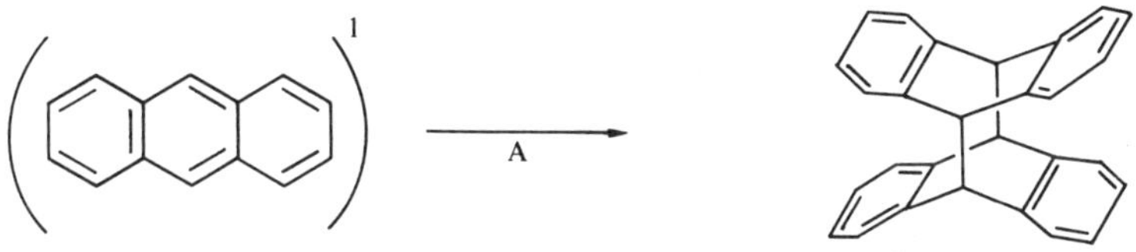

Monosubstituted anthracene (14) (15)

Photooxidation of Adsorbed Aromatic Hydrocarbons

Only a few reports of photooxidation of adsorbed aromatic hydrocarbons have appeared, but they are sufficient to demonstrate that the reactivity of adsorbed hydrocarbons is considerably greater than that of hydrocarbons in solution. Anthracene (16) adsorbed on silica gel or alumina is very rapidly oxidized to anthraquinone (17); the endoperoxide does not seem to be intermediate in this oxidation.[446] Further oxidation gives 1,4-dihydroxy-9,10-anthraquinone (18). Even naphthalene (normally inert) can be oxidized under these conditions. The degree of oxidation of anthracene depends somewhat on the adsorbent.[797]

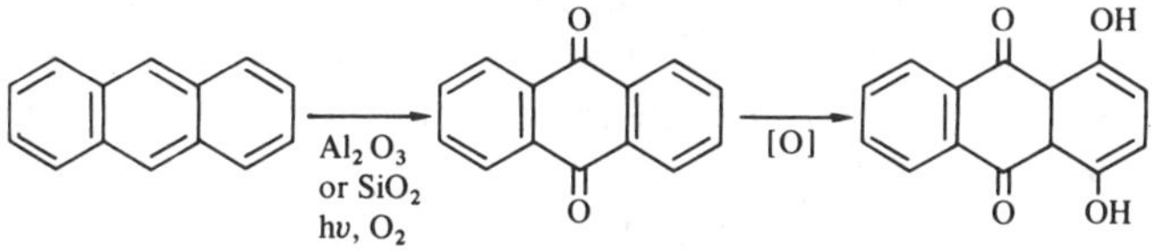

Al_2O_3 or SiO_2 $h\nu$, O_2 [O]

(16) (17) (18)

On thin-layer chromatography (TLC) plates (either alumina or silica gel as adsorbents), in room or ultraviolet light, anthracene, naphthacene, benz[a]anthracene, dibenz[a,c]anthracene, dibenz[a,h]-anthracene, pyrene, benzo[a]pyrene, benzo[e]pyrene, benzo[ghi]-perylene, and coronene are oxidized; phenanthrene, chrysene, and triphenylene are inert. Pyrene (19) is oxidized to the 1,6-dione (20) and the 1,8-dione (21) under these conditions.[413] These studies provide potential models for oxidation of hydrocarbons adsorbed on particulate matter in the atmosphere and indicate the difficulties faced by the analytical chemist.

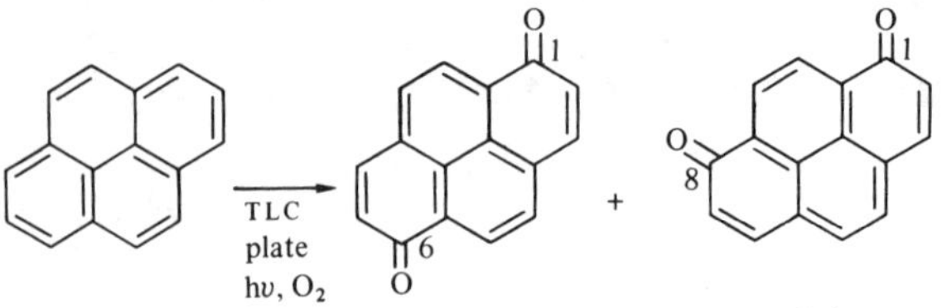

TLC plate $h\nu$, O_2

(19) (20) (21)

The results gain significance from the reports of formation of radical cations from hydrocarbons adsorbed on various substances; many of these oxidation reactions may occur via the radical cations.

There have been several reports of destruction of polycyclic aromatic hydrocarbons adsorbed on soot or smoke.[260,262,745,751] Most of the destruction requires light, although it is accelerated by synthetic photochemical smog.[260,262] It is difficult to calculate specific rates from the data presented, but exposure for roughly 40 min to light of one fourth the intensity of noon sunlight caused 35–65% loss of benzo[a]pyrene in airborne smoke samples in an irradiated flow reactor.[745,751] Exposure for 6 hr to sunlight caused 15–50% loss of benzo[a]pyrene in smoke on filters.[745] In a different study, 10% of benzo[a]pyrene adsorbed on soot or a filter was destroyed in 48 hr of exposure to light of unstated intensity, but 50% was destroyed in 1 hr of exposure to light and synthetic smog (containing the unnaturally high oxidant concentration of 30 ppm).[262] These studies imply that the chemical half-life of benzo[a]pyrene and of other polycyclic aromatics[260,262] may be limited to only hours or days in the atmosphere. Further experimentation in simulated atmospheric conditions with careful attention to wavelength distribution of the light used and chemical characterization of the products is highly desirable.

ACTION OF OZONE

Polycyclic Aromatic Hydrocarbons

Ozone reacts readily with polycyclic aromatic hydrocarbons.[24,541] Several modes of reaction have been identified. One important reaction is cleavage of phenanthrene-like double bonds, which in oxidative conditions eventually results in the formation of diacids; another involves oxidation at anthracene 9,10-like positions to yield quinones; a third is a more complex nuclear oxidation; and a fourth involves side-chain oxidation. Of the large body of literature, only a few particularly pertinent examples will be given here. Laboratory studies are normally carried out at low temperature in an inert solvent and are followed by either reductive or oxidative workup, but atmospheric reactions probably take a very similar course and would be expected to yield products similar to those of oxidative workup. Most of the primary products are also subject to further oxidation. Material balances are often rather poor.

Ozonolysis of benz[a]anthracene (22) provides a typical example; the products are compounds (23)–(25) and oxygen.[25]

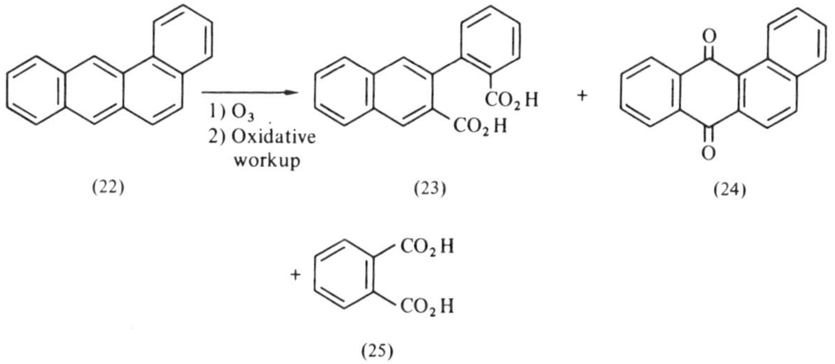

1) O_3
2) Oxidative workup

CO_2H CO_2H +

(22) (23) (24)

+ CO_2H CO_2H

(25)

Product (23) is produced by cleavage of the reactive phenanthrene-like double bond. The fate of the initial unstable molozonide (26) depends somewhat on the conditions; under atmospheric conditions, it would almost certainly be converted to compound (23). Aldehydes and various polymeric peroxides may be intermediate.

Benz[a]anthracene

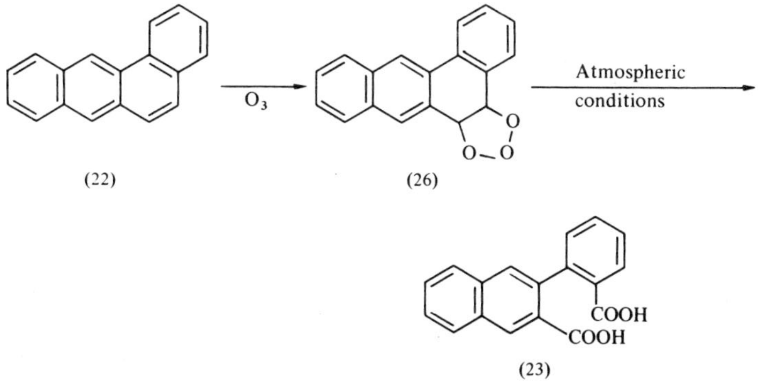

O_3 Atmospheric conditions

(22) (26)

COOH
COOH

(23)

Quinone (24) and phthalic acid (25) are believed to be, in part, products of decomposition of an initial trioxide (27) formed by addition of oxygen at the meso positions of benz[a]anthracene (22). A trioxide of some stability has indeed been isolated from the ozonolysis of 9,10-dimethylanthracene.[250]

Benz[a]anthracene (22) $\xrightarrow{O_3}$ (27) $\longrightarrow$ Quinone and Phthalic acid

However, a second route to quinone is also postulated, involving electrophilic attack of ozone on aromatic hydrocarbons, giving hydroxylated compounds and oxygen; the hydroxylated compounds undergo rapid further oxidation, leading ultimately to quinone (24). It is likely that the oxygen formed is in the singlet excited state.

Benz[a]anthracene (22) $\xrightarrow{O_3}$ [H, O, O, O $\longrightarrow$ OH + O_2] $\xrightarrow[\text{or } O_2]{O_3}$ Quinone (24)

This mode of attack may be the most important one in the oxidation of benzo[a]pyrene (10) to yield a mixture of the 3,6-dione (13) and the 1,6-dione (12) in a 3 : 1 ratio, accompanied by a trace of the 4,5-dione (28);[540] presumably, the 4,5-dione comes from the unstable molozonide, analogous to the formation of compound (26). A series of one-electron oxidations should not be ruled out as the mechanism of the ketone-forming processes.

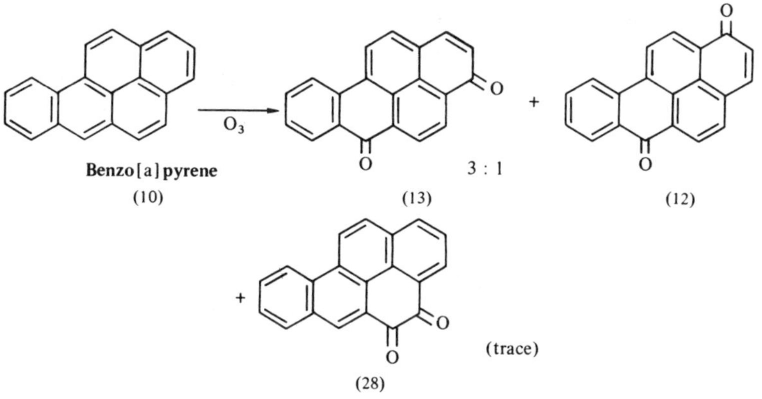

O_3

Benzo[a]pyrene (10)

(13) 3 : 1 (12)

(28) (trace)

The ozonolysis (O_3) of 3-methylcholanthrene (29) illustrates the side-chain oxidation that may occur with alkylated aromatics; in addition to minor products of more extensive oxidation [O], the major products are compounds (30) and (31).[544] Similar dealkylation occurs with 7,12-dimethylbenz[a]anthracene.[543]

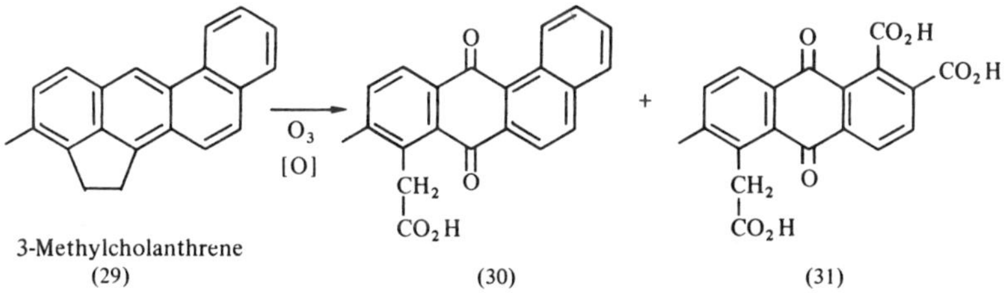

O_3 [O]

3-Methylcholanthrene (29)

(30)

(31)

Another type of oxidation involves attack at saturated C–H bonds; for instance, fluorene is oxidized to fluorenone; anthrone, to anthraquinone.[125]

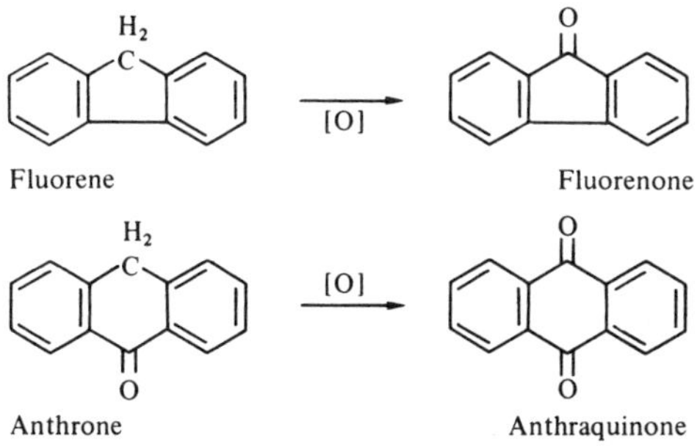

[O]

Fluorene → Fluorenone

Anthrone → Anthraquinone

Autoxidation of alkanes is initiated by ozone at ambient temperatures.[679-681] This reaction probably involves abstraction of hydrogen, yielding an initial hydrotrioxide, which can break down to yield radicals that can serve as initiators;[816] this may be the mechanism of side-chain oxidation, as given below.

$$RH \xrightarrow[O_3]{} R\cdot + \cdot O_3H \rightarrow RO_3H \rightarrow RO\cdot + HO_2\cdot \rightarrow \text{Chains} + O_2$$

Some of the oxygen produced from the reaction of isopropyl alcohol or diisopropyl ether with ozone is in the singlet state.[553] This oxidative reaction also occurs in the gas phase (J. N. Pitts, Jr., personal communication). At least one other reaction of ozone, that with tertiary phosphites, produces oxygen in its singlet state.[804]

Polycyclic Aza-Arenes

Aza-arenes and aromatic hydrocarbons react similarly, but aza-arenes are generally less reactive. Acridine yields compounds (32) and (33); phenanthridine yields compounds (34) and (35).[542] Trialkylamines, which have strongly basic nitrogen, are oxidized to N-oxides and undergo various oxidations of alkyl groups attached to the nitrogen.[26]

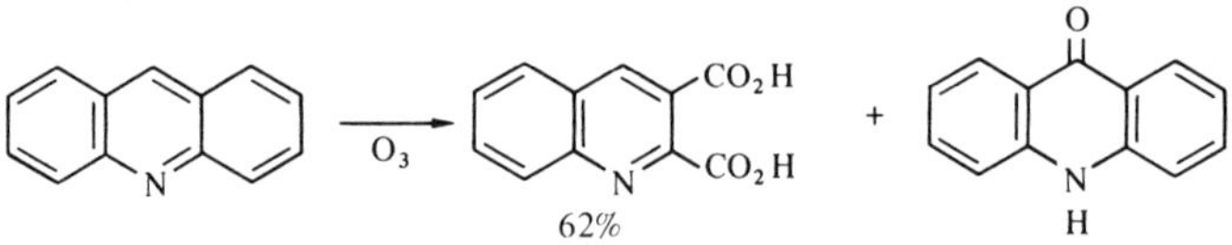

Acridine $\xrightarrow[O_3]{}$ (32) 62% + (33)

Phenanthridine $\xrightarrow[O_3]{}$ (34) + (35)

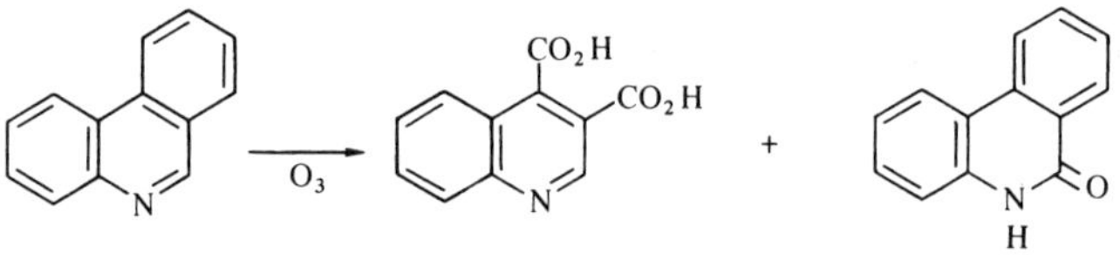

ACTION OF MISCELLANEOUS OXIDANTS

One-Electron Oxidation

One-electron oxidation of many polycyclic aromatic hydrocarbons occurs readily;[11,619] the primary products in most cases are radical cations.[5,6,811] These radical cations are unstable and react rapidly with water or other nucleophiles (including unoxidized hydrocarbon) or with oxygen. Acridines react similarly.[6,811] The overall reaction leads to formation of quinones (with water and oxygen), nucleophile adducts, or dimeric hydrocarbons.[5,6,811] Benzo[a]-pyrene (illustrated below) is oxidized anodically to the radical cation, which reacts with water to yield easily oxidized intermediates that are further oxidized to the 1,6-dione and two other quinones; some dimer is also formed, mostly at the electrode surface. The similarity of the overall process to photooxidation and to part of the ozonolysis process should be noted.

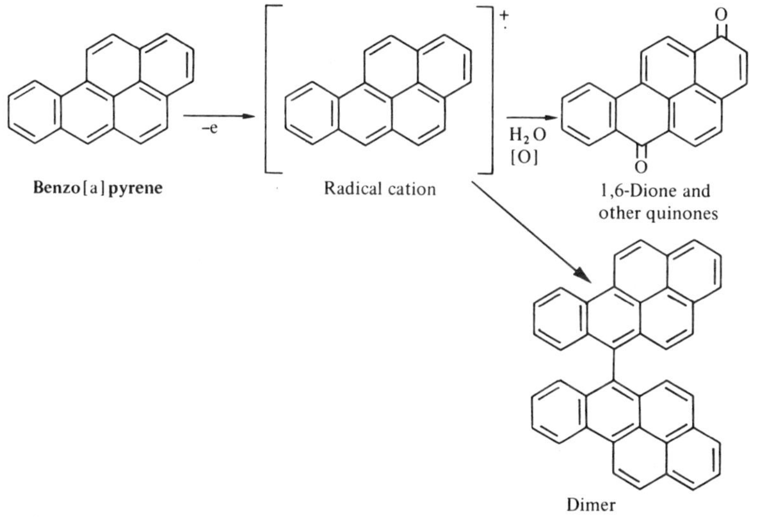

Benzo[a]pyrene —(−e)→ [Radical cation]$^{+\cdot}$ —(H_2O, [O])→ 1,6-Dione and other quinones

Radical cation → Dimer

Side-chain oxidation of alkyl-substituted aromatics can occur.[5,6,811] As illustrated below, durene is anodically oxidized ($-e^-$) in acetonitrile (CH_3CN) to side-chain substitution product (36), with acetonitrile acting as a nucleophile.[226] If water is present, the corresponding benzyl alcohol (37) is formed.

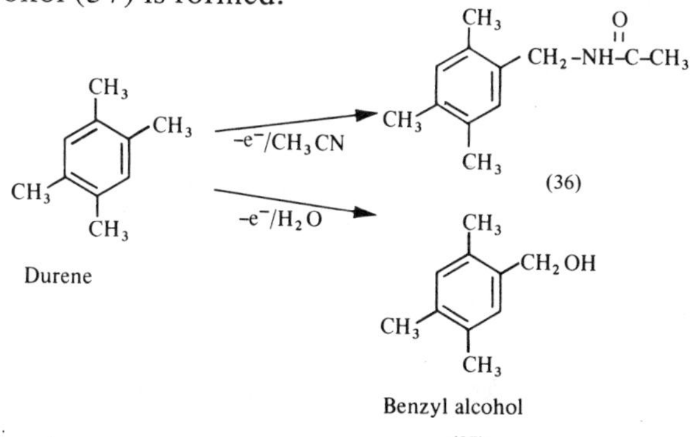

Durene —($-e^-/CH_3CN$)→ (36) [CH_2-NH-C(=O)-CH_3 substituted]

Durene —($-e^-/H_2O$)→ Benzyl alcohol (37) [CH_2OH substituted]

Although these reactions involve complex sequences of oxidations, nucleophilic attack, and proton loss, the primary product in most cases is the radical cation.[5] Similar reactions occur with a wide variety of one-electron oxidants. For example, 7,12-dimethylbenz[a]anthracene [compound (3)] with any one of a variety of one-electron metal salt oxidants yields mixtures in various proportions of the 7,12-endoperoxide [compound (4)] and compounds (38)–(43).[279]

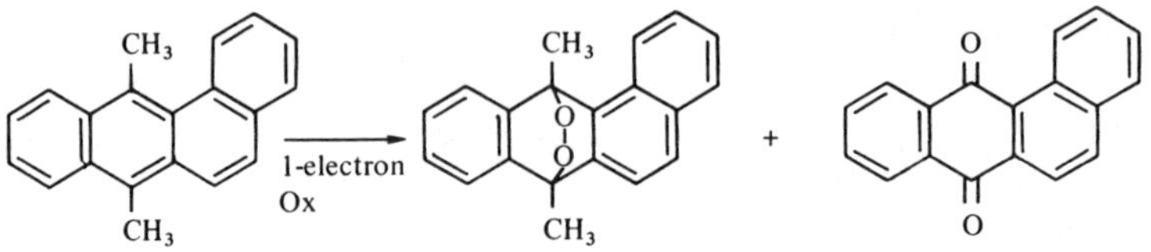

(3) —1-electron Ox→ (4) + (38) 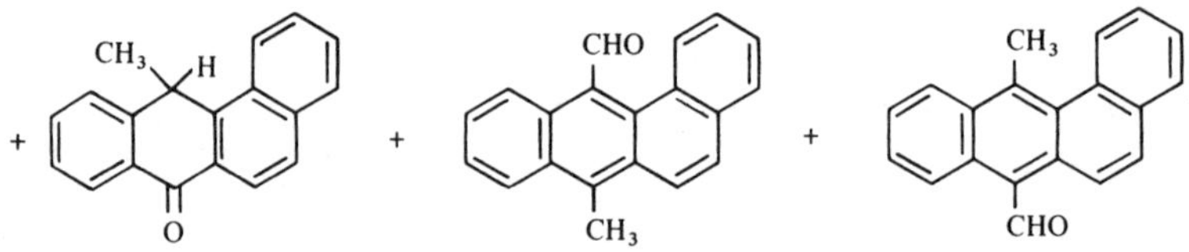 + (39) + (40) + (41) 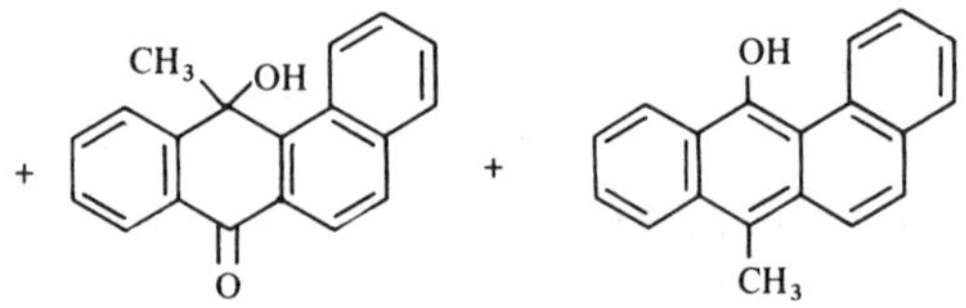 + (42) + (43)

Radical cations are formed from many polycyclic aromatic hydrocarbons simply by treatment with strong Lewis acids, particularly in the presence of oxygen.[2] Irradiation greatly increases the degree of electron transfer, so light can directly cause the formation of radical cations.[1] Adsorption on an alumina or silica surface can produce radical cations directly, particularly in the presence of oxygen or iodine.[335,641,817,818] The adsorbed radical cations react to yield dimers or react with nucleophiles, as described above; for example, benzo[a]pyrene (10) yields quinones and dimer. In the presence of nuclear bases or pyridine, substitution products are formed, such as compound (44).[817,818] This reaction with nuclear bases may be important in the binding of carcinogenic hydrocarbons to biologic materials.[817,818]

(10) Benzo[a]pyrene → Silica gel, I_2 or O_2 → [Radical cation]$^+$ → H_2O/O_2 → Quinones and dimer

Radical cation + Pyridine → (44)

Many reactions of polycyclic aromatic hydrocarbons probably proceed through radical cations. As a note of possible biologic importance, reaction of radical cations with electron donors regenerates the hydrocarbon, partly in an excited state;[176,361] this process represents a mechanism for producing excited hydrocarbon and (by reaction of the excited hydrocarbon with oxygen) singlet oxygen in the absence of light [see Eq. (3)]. This conceivably accounts for the formation of endoperoxide (4) in the one-electron oxidation of 7,12-dimethylbenz[a]anthracene (3). Furthermore, singlet oxygen is known to oxidize nucleic acid derivatives; such reactions with bound carcinogenic hydrocarbons would produce singlet oxygen near genetic material and thus might cause genetic damage.[275] This possibility requires further research.

$$R^{\overset{+}{\cdot}} \xrightarrow[e^-]{} R^* \xrightarrow[O_2]{} {}^1O_2 \qquad (3)$$

Peroxides, Radicals, and Other Oxidants

Polycyclic aromatic hydrocarbons react readily with peroxides; the products are those of substitution or of further oxidation to yield products similar to those described in previous sections. Benzoylperoxide [$(C_6H_5CO_2)_2$] reacts with benzo[a]pyrene, which is extremely reactive, to yield the 6-benzoyloxy derivative (45).[637] In the presence of oxygen, the reaction might well lead to further oxidation.

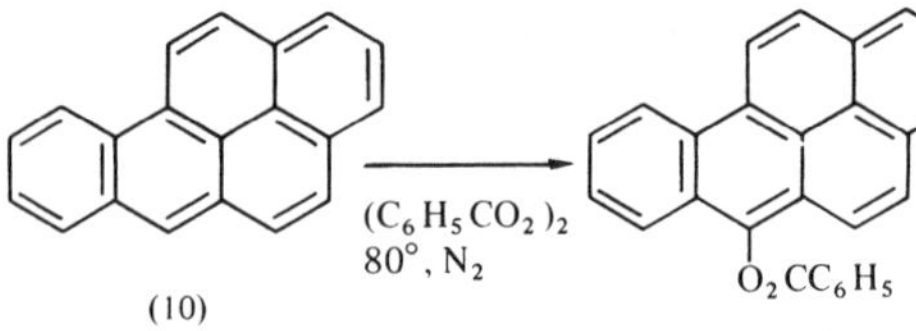

$(C_6H_5CO_2)_2$
80°, N_2

(10)
Benzo[a]pyrene

$O_2CC_6H_5$
(45)

Perbenzoic acid ($C_6H_5CO_3H$) oxidation of dibenz[a,h]anthracene (46) produces the 7-ketone (47) and the 5,6-epoxide (48), which reacts further. Oxidation with the more reactive peracetic acid (CH_3CO_3H) yields the 7,14-quinone (49), a diacid (50), and the 5,6-quinone (51).[785]

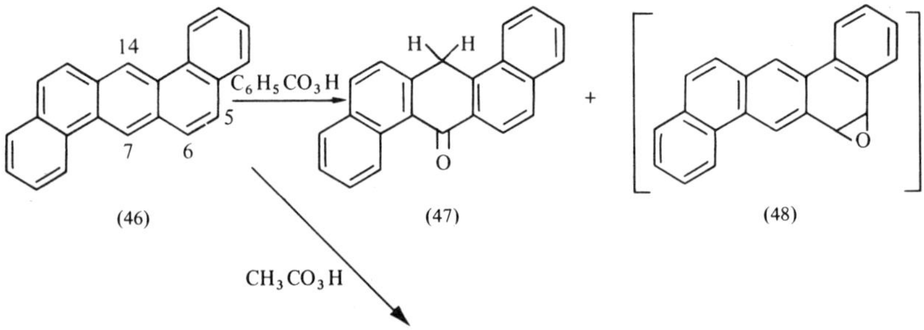

$C_6H_5CO_3H$

CH_3CO_3H

(46) (47) (48)

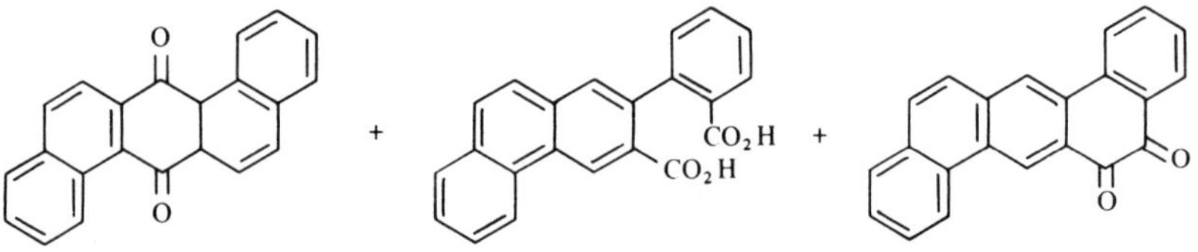

(49) (50) (51)

The products of attack at the 5,6 double bond are a consequence of the tendency of peracids to induce the formation of epoxides. The attacks at the 7 and 14 positions of dibenz[a,h]anthracene with peracids and at the 6 position of benzo[a]pyrene with benzoylperoxide may represent either electrophilic or free-radical substitution reactions. Radicals react rapidly with polycyclic aromatic hydrocarbons.[222,445] Anthracene reacts with benzoylperoxy radicals, formed in benzaldehyde autoxidation, to terminate the chain and produce anthraquinone.[222]

Anthracene (16) $\xrightarrow{C_6H_5CO_2^{\cdot}}$ Anthraquinone (17) (final product)

Oxygen atoms react rapidly with aromatic compounds and behave like electrophilic reagents.[319] Although polycyclic aromatics have not been studied in this regard, alkylbenzenes yield phenols. Toluene is typical, yielding a mixture of cresols and phenol (dealkylation). Ethylbenzene yields a small amount of acetophenone, in addition to similar products. Polycyclic aromatic hydrocarbons will certainly react similarly.

Considerable evidence indicates that destruction of aromatics in soot is accelerated by photochemical smog, although which constituents are responsible is not known.[260,262] Soil bacteria oxidize benzo[a]pyrene; in some conditions, 80% can be destroyed in 8 days.[597]

ACTION OF NITROGEN OXIDES

Polycyclic aromatic hydrocarbons, especially the larger ones, are extremely sensitive to electrophilic substitution and to oxidation. Nitrogen oxides or dilute HNO_3 can either add to, substitute in, or oxidize polycyclic aromatic hydrocarbons. Anthracene (16) is oxidized by dilute aqueous HNO_3 or nitrogen oxides to anthraquinone;[793] NO_2 adds to and substitutes in anthracene, yielding compound (53) [by loss of HNO_2 from compound (52)].[33] Benzo[a]pyrene is nitrated in minutes at room temperature with nitric acid diluted by acetic acid and benzene to give the mononitro compound.[267]

Anthracene (16) $\xrightarrow{\text{dil. aq. } HNO_3 \text{ or } NO_x}$ Anthraquinone (17)

Anthracene $\xrightarrow{NO_2}$ (52) $\xrightarrow{-HNO_2}$ (53)

H, NO_2; H, NO_2

NO_2

ACTION OF SULFUR OXIDES

Reactions of polycyclic hydrocarbons with atmospheric SO_2, SO_3, or H_2SO_4 should be very facile, particularly in aerosols or when adsorbed. Sulfonation of pyrene proceeds readily at room temperature with concentrated H_2SO_4 to produce a mixture of disulfonic acids;[795] the concentration of H_2SO_4 in aerosol droplets may be high enough to sulfonate the more reactive aromatics. SO_2 reacts with aromatic compounds in smoke, with or without light.[745] The products of reaction with SO_2, SO_3, and H_2SO_4 are sulfinic and sulfonic acids; these compounds are water-soluble and hence will no longer appear in the benzene-soluble fraction.

PHOTODYNAMIC ACTIVITY AND SINGLET OXYGEN

Photodynamic toxicity (ability to kill or damage organisms in the presence of light and oxygen) is associated with many carcinogenic and some noncarcinogenic hydrocarbons.[233] The photodynamic action of many compounds has been reviewed.[717] Some part of the photodynamic toxicity is probably caused by singlet oxygen, generated by energy transfer from the sensitizer.[275,277] Singlet oxygen is also formed by several reactions of ozone and possibly by reactions of radical cations. In addition, some photoperoxides dissociate thermally to yield singlet oxygen,[805] and peroxyacetylnitrate hydrolysis also produces singlet oxygen.[720] All these reactions provide the possibility of causing nonphotochemical biologic damage of the sort associated with photodynamic action.

It has been concluded that singlet oxygen is probably not an important atmospheric oxidant of olefins,[367] but that conclusion does not take into account any of the methods of formation of singlet oxygen mentioned here. Furthermore, the likelihood that some of these mechanisms may produce singlet oxygen near or within the human organism suggests that it cannot be ignored as a matter of environmental concern.

SUMMARY AND CONCLUSIONS

Polycyclic aromatic compounds are highly reactive. There is evidence that they are degraded in the atmosphere by photooxidation, by reaction with atmospheric oxidants, and by reaction with sulfur oxides. Comparative data on reactions in solution, vapor, and adsorbed phases are very limited, and most of the work has been done in solution. In the few cases for which there are comparative data, the reactions are similar in the different phases.

Reactions may be particularly facile when the compounds are adsorbed on particulate material, such as soot. Chemical half-lives may be only hours or days under intense sunlight in polluted atmospheres. Further research is needed on the products of chemical reaction of POM in typical atmospheric conditions and on the possible biologic activity of these products. Most of the likely atmospheric reactions of hydrocarbons produce oxygenated compounds. Such oxygenated compounds as 7H-benz[d,e]anthracen-7-one (54) are found in urban air,[662] and oxygenated fractions of air extracts seem to be carcinogenic.[233] More definite information on chemical half-lives in various conditions is essential.

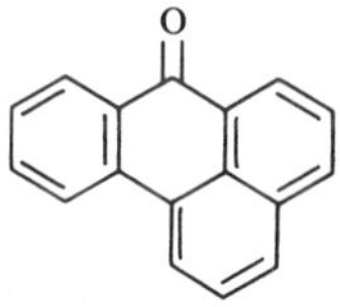

(54)

Several mechanisms involving polycyclic aromatic hydrocarbons and other pollutants may cause reactive oxidizing species to be delivered to genetic and other biologic material. These mechanisms deserve further study.

6

Historical and Theoretical Aspects of Chemical Carcinogenesis

Chemical carcinogenesis was first discovered in 1776, when the British physician Percival Pott called attention to the high incidence of cancer of the scrotum in the chimney sweeps of London and correctly attributed the disease to their continual contact with soot. In the almost 200 years since then, pure chemicals have been shown to induce cancer in a wide variety of animals and sites. As our civilization has become more industrialized, it has become increasingly contaminated with a number of cancer-producing chemicals in particulate, as well as nonparticulate, form. Polycyclic aromatic hydrocarbons are found in particulate air pollutants and in condensates of the cigarette smoke used as a self-pollutant. Among these compounds, benzo[a]pyrene is a powerful skin carcinogen and has been detected and analyzed in samples of polluted air. This compound is by no means the only polycyclic aromatic carcinogen found in polluted air. If these and other types of carcinogenic compounds were removed from the environment, many human cancers would be prevented.

Not all the chemical carcinogens present in polluted air have been identified. And a dominant role of the polycyclic aromatic hydrocarbons in the causation of human cancer has not yet been established.

A great deal has been learned about the mechanisms of chemical carcinogenesis, but the final answers still elude us. Although cancer is a disease of organized tissues, it appears to be more productive at this point to concentrate on the possible cellular and molecular mechanisms of chemical carcinogenesis, with particular reference to polycyclic aromatic hydrocarbons.

Chemical carcinogens can be considered to act by two cellular mechanisms: transforming or converting normal cells into cancer cells[349,350] and selecting for pre-existing cancer cells, as proposed by Prehn.[602] Although it may not seem too difficult to determine which of these mechanisms is correct, the matter could not be settled by studies in whole animals and had to await the development of reliable and quantitative systems for producing chemical carcinogenesis in cells in culture. Mondal and Heidelberger[537] have succeeded in transforming individual single normal cells to malignant cells with 3-methylcholanthrene with very high efficiency, thus ruling out the selection hypothesis in this system.

The possibility that the carcinogen activates or "switches on" a latent oncogenic virus has recently been proposed in its most comprehensive form by Huebner and Todaro.[399] They postulate that all chemical carcinogens act through intermediary or oncogenic viruses or their informational precursors. There is evidence that in some cases chemical carcinogens do lead to the appearance of detectable oncogenic viruses. These cases appear to involve primarily leukemogenesis and some sarcoma formation. In many other situations, there is no evidence of the participation of an oncogenic virus in chemical carcinogenesis. However, it may be impossible to disprove such participation. It should be pointed out that only the mouse mammary tumor virus is now known to induce carcinomas. If viruses are "switched on" by chemical carcinogens and give rise to the many carcinomas that chemicals are known to induce, then they must be viruses that are as yet unknown. The possibility that oncogenic viruses are involved in chemical carcinogenesis must always be kept in mind; but, even if the participation of a virus turns out to be ubiquitous in chemical carcinogenesis, it is still the chemical that triggers the process of cancer cell formation. Therefore, for human health protection, environmental chemicals should be reduced or eliminated to prevent potential carcinogenesis.

If it is assumed that chemical carcinogens themselves produce cancer without the intervention of an oncogenic virus, then two major molecular mechanisms are possible—mutational and nonmutational.

The somatic mutation theory of cancer was first proposed by

Boveri[81] in the 1920's. In modern terms, the mutational mechanism of chemical carcinogenesis requires that the chemical interact with the genetic material of the cell (DNA) in such a way as to alter its primary sequence to lead to a nonlethal perpetuated change that would be inherited by all progeny cells. This has been discussed by many, including Potter.[601] It has been shown with polycyclic aromatic compounds that there is a weak physical binding to DNA, which can be converted into covalent binding by ultraviolet irradiation and various chemical treatments, as studied by Ts'o and his colleagues.[758] Probably more significant with respect to the process of carcinogenesis is the fact that every chemical carcinogen of every type that has been properly studied has been found to bind covalently with the DNA, RNA, and protein of target tissues; this subject has been thoroughly reviewed by Miller.[532] Because the hydrocarbons themselves cannot undergo such covalent binding, it is clear that they must be metabolically converted by target cells into chemically reactive molecules, which may be considered as the "ultimate" carcinogens. In the case of polycyclic hydrocarbons, their binding to the DNA of mouse skin after topical application has been measured by Brookes and Lawley[93] and Goshman and Heidelberger.[307] They found that there was a rough quantitative correlation between the carcinogenic activities of various hydrocarbons and the amount bound to mouse skin DNA. The binding of carcinogenic hydrocarbons to DNA does not, however, prove that they act through a mutagenic mechanism. The mutagenic activities of many carcinogenic and noncarcinogenic compounds have been determined in many systems, including recent studies in bacteriophage T4 by Corbett *et al.*[163] In none of these studies has the correlation between mutagenesis and carcinogenesis been good, but that may be a result of deficiencies in the various test systems used, rather than of an intrinsic lack of correlation between the two biologic processes.

A nonmutational molecular mechanism for chemical carcinogenesis would require a perpetuated epigenetic change, which would result in altered gene expression. This would most likely be a derepression of genetic information already present in the genome of the cell. A theoretical molecular model for such a derepression was proposed in 1963 by Pitot and Heidelberger,[596] on the basis of the observation that carcinogenic hydrocarbons are covalently bound to a specific protein fraction in mouse skin.[744] The derepression of an oncogene has also been proposed by Huebner and Todaro[399] in their comprehensive theory of carcinogenesis. Braun[87] also supports the view that carcinogenesis results from a nonmutational mechanism. Immunologic evidence for derepressions accompanying carcinogenesis has been obtained by Abelev[3] in mouse hepatomas and by Gold *et al.*[301] in human gastrointestinal carcinomas. These workers demonstrated that the tumors contained fetal antigens that were repressed in the adult animal but reappeared in the tumors. If carcinogenesis is the consequence of a nonmutational mechanism, then there is a possibility of reversion of cancer cells to normal. Indeed, there is some evidence for this in a few cases, including neuroblastomas and the teratocarcinomas studied by Pierce.[594]

Another intriguing phenomenon accompanying carcinogenesis, for which there is at present no satisfactory theoretical explanation, is the acquisition of new transplantation antigens on the surfaces of chemically induced tumors. This has been extensively studied with hydrocarbon-induced sarcomas in mice by Prehn,[604] Klein,[438] and Old *et al.*;[571] with hydrocarbon-induced mouse skin carcinomas by Pasternak *et al.*;[584] with azo dye-induced rat hepatomas by Baldwin and Barker;[28] and with *in vitro* hydrocarbon carcinogenesis by Mondal *et al.*[538] These tumor-specific antigens of chemically induced tumors are individual and non-cross-reactive. Although the total number of such antigens is not known, the number exceeds 25 in the case of hydrocarbon-induced sarcomas. The significance of these antigens to the process of carcinogenesis remains unexplained.

CONCLUSIONS

1. Chemical carcinogens appear to transform normal cells into cancer cells directly.
2. Chemical carcinogens may or may not "switch on" a latent oncogenic virus that is responsible for the cancer induction.
3. If chemical carcinogens transform normal to cancer cells without the intervention of an oncogenic virus, they can do so either by a mutational or by a nonmutational mechanism. There is some evidence to support each possibility.
4. Hydrocarbon-induced tumors have individual transplantation antigens.
5. The response of a host to carcinogenic stimuli is determined by its immunologic and hormonal status, its exposure to some drugs, and its nutritional state. A variety of unknown factors in its environment may also alter its response to carcinogenic stimuli.

RECOMMENDATIONS

Further and more definitive work must be done on the interrelations between chemical carcinogens and oncogenic viruses. The question of whether carcinogenesis involves a mutational or a nonmutational mechanism can be settled only by genetic experiments. Hence, much development must be carried out aimed at understanding the genetics of mammalian cells. Further biologic, biochemical, and molecular biologic research must be carried out in whole animals and cell cultures undergoing chemical carcinogenesis and in various cell-free systems. An understanding of the fundamental mechanisms of chemical carcinogenesis should provide new means for the eradication of cancer through prophylaxis, induction of loss of malignancy, and perhaps even reversion of malignant cells to normal cells.

7

Experimental Design in Carcinogenesis Tests

STATISTICAL DESIGN AND ANALYSIS

Laboratory testing of carcinogens has been rather straightforward because of its emphasis on identifying strong carcinogens; experiments could be short and could use small numbers of test animals. However, the role of weak environmental carcinogens in imposing subtle but potentially important threats to human populations has required changes in experimental procedures.

Mantel and Bryan[517] describe a method of extrapolating to a conservative safe dosage of an agent, whether or not carcinogenic activity was found. The dosages and number of test animals used may influence the result of the extrapolation. In the extrapolation procedures, a worst true risk at the test level used consistent with the observed risk is determined; e.g., no tumors observed among 100 animals would be consistent at the 99% level of probability with a true risk of 4.5%. This worst true risk is extrapolated backwards by some arbitrarily shallow rule (e.g., 1 probit per tenfold change in dose) to an arbitrary level of "virtual" safety (e.g., a risk of one in 100 million). The "safe" dosage, where the observed risk is none in 100, would be equal to 1/8,300 times the test dosage. An alternative, but parallel, approach to setting "safe" dosage, which can be yet more conservative, is described in Gross *et al.*[317]

A more general purpose of an experimental design is to identify the carcinogenic properties of the agent tested. For strong carcinogens, it should reveal that tumors are elicited early and at a relatively low dosage; for weak carcinogens, that few tumors are ever obtained, even at a high dosage. Data resulting from such a general-purpose design could be variously used, e.g., for identifying weak or strong carcinogens and environmental or nonenvironmental carcinogens and for correlating carcinogenic activity with other physicochemical properties of agents. Such a general-purpose design seems appropriate if one wishes to identify carcinogenic activity in atmospheric pollutants, using either crude particulate matter or specific extracts.

A general-purpose design used by Hadidian *et al.*[327] and outlined in Mantel[516] permits testing over a wide dosage range to identify both weak and strong carcinogens, with test animals (rats) followed for 1½ years, so that both early- and late-appearing tumors would be found. In this design, testing was begun by using an agent at five different, but increasing, log dilutions. On the basis of the observed 30-day toxicity, six additional half-log dosages were inserted, depending on the toxic level, above, between, or below the original test dilutions. For both initial and additional test groups, treatment was maintained for a full year. The study was intended to use three male rats and three females per dosage. The small groups were expected to be satisfactory, because the information in the experiment would be based on the total number of rats in all the groups, not on those at a single dosage. Because of a particular interest in weak carcinogens, the test design was modified so that 15 rats of each sex were used at the next-highest dilution of the retained nontoxic dosages. From a review of the resulting data, however, it was clear that the unmodified and modified experiments led to essentially the same conclusions.

The complex results, which involved a wide variety of tumors, days of tumor appearance, days of death or sacrifice, and so on, were analyzed by a simplifying approach. A reference tumor rate was established for controls that survived at least 1½ years, so that the expected number of tumors, overall or for a particular site, among 30 male or 30 female rats could be determined. If the 30 rats of a sex treated with a particular agent (for all dosages combined) showed clearly above-expected numbers of tumors, any adjustment for early mortality among the rats could only increase, rather than diminish, the difference. The fact that tumors are competitive with each other, so that death from a tumor at one site can preclude seeing a tumor at another site, was handled by considering the total of tumors or tumorous rats. Data on actual tumor sites were, however, available for interpretation. It became necessary to eliminate such spontaneous tumors from the totals in order to bring out neoplastic effects for the agents tested; thus, a comparison of 21 tumors observed versus 18.5 expected could become 16 observed versus 1.6 expected when interstitial cell tumors were excluded. This modification served as a crude form of age adjustment.

The foregoing illustrates the importance of and need for flexibility in the analysis of data from a long-term carcinogenesis experiment. The resulting data will ordinarily not be neatly packaged, and there will probably be no completely valid way of dealing with all the complications. In the present instance, the method of analysis cut across the complications so as to fasten on a simple indicator of carcinogenic effect. Flexibility was illustrated in the elimination of spontaneous interstitial cell tumors, whose high incidence in a long-term experiment had not been anticipated. If an agent showed active carcinogenicity, eliciting early tumors, the rats would be prevented from surviving to the more advanced ages at which spontaneous tumors arose; overall, then, the total number of tumors might be increased only moderately. The remedial device of eliminating the spontaneous interstitial cell tumors brought out the carcinogenic effect clearly.

Another large-scale, long-term test of carcinogens, in mice, was reported by Innes *et al.*[412] The interest focused on pesticides and some other industrial chemicals. Agents were first tested for toxicity to determine a maximal tolerated dosage for each. With the emphasis on bringing out the existence of carcinogenic effects, only this maximal dose was used in testing, which was done with 72 mice divided among two strains and two sexes. Summary totals of mice with tumors were used in making comparisons with controls, but detailed information of kinds of tumors observed was also kept. The statistical analysis reported included both a significance test—i.e., whether the observed increase in tumor incidence was real—and a relative-risk measure of that increase. The relative-risk measure varied with the dosage administered for testing and, because of dosage differences, is not strictly comparable between agents.

In the two studies just described, the agents tested were administered orally to weanling rats or week-old mice. For other studies, a wider variety of routes could be of interest. The use of more sensitive neonatal test animals or of less sensitive young adult animals may be of value. It may be of interest to test some agents in conjunction with known carcinogens. With limited resources, just which agents to test constitutes a problem of priorities. In a general chemical-testing program, does one give highest priorities to chemicals already in use—with priority related to the extent of such use—or does one give highest priorities to new chemicals, which can bring about an abrupt change in the environment? The answer may be to have a program large enough to test all agents, in time, with more detailed testing of particular agents. In the case of atmospheric pollutants, the priority problem applies to the program as a whole, rather than to individual pollutants.

A characteristic of both studies described is that they emphasized testing at relatively high dosages, although the first involved some low dosages as well. For various reasons, low-dosage studies will yield unsatisfactory and perhaps ambiguous results. For example, suppose one tests at dosages comparable with actual human exposure; one would likely get negative results in a reasonably large experiment, even if the risk for humans were intolerably high—say, 0.1 or 0.01%. (Because of the large number of agents in the environment, the risk for each, when not otherwise justified, must be kept extremely small.) A low-dosage experiment cannot resolve the question of the existence of thresholds. At a very low dosage, the absence of induced tumors would not demonstrate a threshold, because it could be a chance occurrence where the tumor risk is low; and the occurrence of tumors would not belie the existence of a threshold, because they could be spontaneous, rather than induced, tumors. If testing is done at several dosages and the induced tumors are rather few, compared with spontaneous tumors, proponents of a threshold theory could find support in the apparent constancy of the tumor rate. An alternative and equally mistaken interpretation could be that the dose–response curve is so shallow that extreme extrapolation to much lower dosages would be necessary to attain a safe level. Experimentation to prove or disprove the existence of thresholds seems pointless.

In summary, a carcinogenicity-testing program is shaped by a variety of factors, including the purpose of the program. The purpose may be to determine safe dosages, to identify agents with interesting properties for future investigation, to identify specifically potent carcinogens, or even to identify specifically weak carcinogens. Screening experiments with many dosage levels can be informative, even if the numbers of subjects at individual dosages are small. Data from a carcinogenicity experiment can be complex, but there may be ways of simplifying the data for analysis. Flexibility of analysis may be required to overcome unanticipated problems, such as a high rate of spontaneous tumors among controls. The existence or nonexistence of thresholds in carcinogenesis cannot be established solely by testing programs, and experimentation solely to such ends would be wasteful.

IS THERE A THRESHOLD DOSE IN EXPERIMENTAL CHEMICAL CARCINOGENESIS?

Although dose-response studies have been carried out with many chemicals and the problem of quantitative carcinogenesis has been extensively studied for a few selected carcinogens,[308,440] it has not been possible to reach agreement as to whether there is a threshold dosage above which carcinogenesis is produced. The dilemma is related to the shape of the dose–response curve. Curve A in Figure 7-1 represents an idealized linear relation, and curve B an idealized threshold relation. Because it is impossible, at very low doses, to obtain reliable data without enormous numbers of animals, a curve like C (which is a hybrid between A and B) might be obtained with insufficient numbers of animals. Therefore, the concept of a threshold dose is probably meaningless, and it would be prudent, because of these uncertainties of measurement, to extrapolate dose–response curves to zero in a linear fashion.

To gain insight into this situation, it should be remembered that there are probably at least two events involved in chemical carcino-

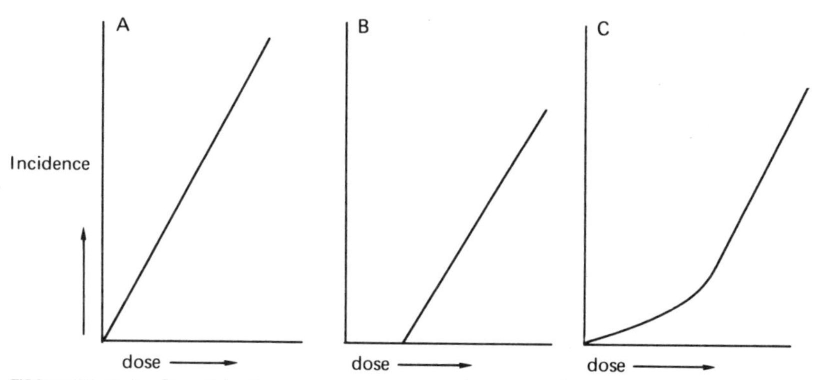

FIGURE 7-1 Possible dose–response curves for agent dose and carcinogenesis. A, idealized linear relation; B, idealized threshold relation; C, hybrid curve between A and B.

genesis. The first is the primary insult induced intracellularly by the carcinogenic chemical. This molecular process is (according to the available evidence) rapid and irreversible. This phase is followed by the biologic process or processes involved in the conversion of the premalignant cell(s) to malignant cell(s) or clone(s) of cells, which in turn results in a tumor. With some carcinogens, the amount of material required to bring about the primary molecular event is so small that it is experimentally difficult to determine an apparent no-effect dosage.

Some investigators maintain that for the so-called second phase of carcinogenesis–i.e., the actual induction of a tumor–there is a threshold dosage. Druckrey[216] studied the response to dimethylaminoazobenzene given orally to rats and found that the lowest dosages, 0.1 and 0.3 mg/day, in 150 rats did not induce any tumors; at 1 mg/day and at higher dosages, tumors were induced. Similar studies were also carried out with aromatic hydrocarbons.[97] However, several factors have to be considered in experiments aimed at determining no-response dosage. Because it is the usual pattern that the lower the dosage the longer the induction time, it follows that the age of the animal at the time of carcinogen treatment is important. A low dosage or a weak carcinogen may require a period of treatment beyond the life-span of the animal; hence, the duration of treatment, especially with low dosages of weak carcinogens, is important.

The group size is extremely important; as far as human exposure is concerned, an agent cannot be regarded as harmless on the grounds that it did not induce tumors in a group of 50 humans (this figure is used because in many carcinogenicity assays only 50 animals are used per group). If one considers a weak carcinogen as one that causes tumors in 0.1% or 0.01% of the test animals, 1,000 and 10,000 animals, respectively, will have to be used to detect these agents, and that assumes that a finding of only one animal with a tumor will be regarded as significant. It is impractical at present to test substances on groups of this size, so compounds are usually tested at much higher dosages than normally encountered in the environment and much smaller test groups are used; this is necessary in an attempt to reduce the gross insensitivity of animal test systems as a function of restricted sample size. Positive results thus obtained are justifiably extrapolated to human populations exposed to marginal doses.

The problem of threshold values is further complicated by the considerable evidence that the effect of a single exposure is irreversible and that the effect of several exposures is cumulative. In addition, in environmental exposure the human population comes into contact with carcinogens, cocarcinogens, and less well-defined cofactors. Cigarette-smoke condensate constitutes a good example of an environmental cocarcinogen or tumor-promoting agent.[792]

Neither epidemiologic nor experimental data are adequate to fix a safe dosage of any chemical carcinogen below which there will definitely be no tumorigenic response in humans.[95,370,518] For these reasons, synthetic chemicals, such as food additives and pesticides that are known to be carcinogenic, must not be deliberately added to the environment. With regard to air pollutants, which contain a variety of defined and undefined carcinogens, the lowest possible exposure must always be insisted on.

ATTEMPTS TO EXTRAPOLATE ANIMAL DATA TO MAN

Attempts to extrapolate from experimental data to human exposure are fraught with problems. One is that animal test groups are necessarily limited in numbers to, say, 50 per dosage group. As noted before, restricted test groups necessarily are grossly insensitive. For example, if an environmental agent produces cancer in one of 10,000 men, and if it is assumed that sensitivity to the carcinogen in question is similar in man and rodents, then test groups of 10,000 rats or mice would be required to obtain one cancer. For statistical significance, 50,000 rodents would be required. Furthermore, in any particular instance, humans may be more or less sensitive to the carcinogen in question than are rodents.

Apart from the gross insensitivity of animal test systems as a function of restricted sample size, a wide range of possible interactions and synergisms would obtain in natural human exposure, in contrast with artificial laboratory systems. Thus, it is impossible to determine safe levels of human exposure to any known or unknown carcinogens on the basis of supposed no-effect levels in practical numbers of animals. Human experience has provided valuable *post hoc* information from epidemiologic studies.

For these reasons, animal test systems using the latest sensitive procedures are the only methods available for defining and anticipating human hazards from POM. A number of factors that influence such tests are described in Chapters 8 and 9.

Classic test methods have used adult rodents, but the use of newborn animals appears to offer greatly increased sensitivity, which will result in the conservation of valuable and scarce test materials. Routes of administration, such as inhalation and intratracheal instillation of materials adsorbed to particles, may serve as simulant models for human exposure to air pollutants. Preliminary information suggests that carcinogenesis experiments in primates are practical and might yield results that can be extrapolated to man. Coordinated studies in experimental animals of toxicity, teratogenicity, mutagenicity, and carcinogenicity offer an optimal evaluation of the hazards of air pollutants.

The use of *in vitro* cell or organ cultures has the potential of providing additional, and possibly more sensitive, methods to detect toxic, carcinogenic, and mutagenic activities of air pollutants. Determination of the levels of production of hydrocarbon-metabolizing enzyme in various selected human cells and tissues may also assist in assessing the sensitivity or resistance of humans to air pollutants.

In all test systems, it is essential not only to test for the activity of benzo[a]pyrene and other polycyclic organic air pollutants, but also to test crude materials and their fractions and to isolate and characterize their carcinogenic components.

8

In vivo Tests for Carcinogenesis and Cocarcinogenesis

PRACTICAL ASPECTS OF TESTING FOR HYDROCARBON CARCINOGENESIS IN MICE AND RATS

A major purpose of experimental air pollution carcinogenesis studies has been to determine whether urban air pollutants might prove to be carcinogenic to different animal species and tissues and, if so, which of the compounds is primarily responsible for the activity. Once agents that are carcinogenic to experimental animals have been identified, attempts can be made to reduce and, if possible, to eliminate their emission into the environment.

Established Polycyclic Carcinogens

Bioassays on mouse skin,[110,147,388,404,450,453,582,632] subcutaneous mouse tissue,[404,487,488] and mouse cervix[72] and in newborn mice[241] have shown that the particulate matter of city air can be carcinogenic to experimental animals. Fractionation studies on urban pollutants suggest that polycyclic aromatic hydrocarbons may play an important role in the overall carcinogenicity and tumor-initiating activity of urban pollutants in experimental animals.[388,404,633,829] Carcinogens are also present in nonpolycyclic aromatic compound fractions, such as oxyneutrals or aliphatics. At present, experimental evidence that neutral and basic N-heterocyclic hydrocarbons contribute to the overall carcinogenic activity of urban air is lacking. Nevertheless, the well-established carcinogenicity of some N-heterocyclic hydrocarbons[346,699,700] and their identification in urban air require that they be considered as carcinogenic agents in POM.

Application to the Skin of Mice

There are several reasons for choosing mouse skin as a test organ, including the relatively low cost of the bioassay and the relatively low demands for pure chemical compounds, especially if one uses inbred mouse strains with high susceptibility to these carcinogens.[830]

Several review articles[80,88,830] and books [17,227,403,812,828] have discussed details of the experimental conditions for mouse skin bioassay. In brief, the more important factors are mouse strain and sex, solvent, dosage and mode of application, and method of observation.

MOUSE STRAINS FOR CARCINOGENICITY TESTING

Genetically homogeneous strains of mice have been developed by selective inbreeding. The advantage of inbred strains is that, unlike random-bred animals of the same species, they show a constant biologic response. Some of the commonly used strains of mice are A, BALB, C3H, C57BL, DBA, and Swiss. Some of these strains are susceptible to spontaneous tumors in particular organs. Strain A mice, for example, develop spontaneous lung tumors and have often been used for lung tumor induction by chemicals. A skin-tumor-susceptible strain, STS, is also available, but it has a high incidence of spontaneous mammary tumors, which makes it less desirable for long-term testing. When using inbred strains for carcinogenicity assay, it is important that no-treatment control groups be used to assess the spontaneous tumor incidence. For the widely used inbred strains, the spontaneous tumor incidences are well established.[190]

Some workers still prefer to use random-bred mice when evaluating new compounds for carcinogenicity, because the induction of tumors in a given inbred strain of mice may be open to question. The use of random-bred mice is probably also more relevant to human exposure to environmental agents. With random-bred strains, it is desirable to use larger groups of animals when testing new compounds for carcinogenicity.

A few years ago, Boutwell developed the STS strain of mice, which appears to be especially suitable for the assay of weak carcinogens, tumor-initiators, and tumor-promoters. However, data have not been available to demonstrate whether reproducible results can be obtained with this strain. To obtain statistically significant data in air pollution carcinogenesis, a given sample must be tested on the skin of at least 20 mice. Because male mice are known to fight, most investigators use only female mice for the skin tests.

SOLVENT

In general, the highest activity on mouse skin is obtained when acetone is used as a solvent. However, mixtures of benzene in acetone must occasionally be used to dissolve the organic pollutants. Dioxane, *n*-hexane, and other solvents are not recommended for the mouse skin test.

CONCENTRATION

A standard rule for the concentration of organic pollutants that must be applied to mouse skin for a carcinogenic response cannot be specified. However, it appears appropriate to apply a dose that contains at least 20% of the amount of benzo[a]pyrene required to induce a low tumor yield in a given strain of mice. In the case of female Ha/ICR/Mil (Swiss random-bred albino) mice, this requires a 0.001–0.002% acetone solution of benzo[a]pyrene. In the past, the test solutions were often painted onto the shaven intrascapular area of the back (about 2 X 2 cm) with a No. 5 camel's-hair brush. However, application of a 50-μl solution should produce more reproducible data than those obtained with skin painting. This can be done manually with a micropipette or, preferably, with an automatic applicator.[828]

Environmental respiratory carcinogens are often derived from combustion. They contain traces of carcinogens, tumor-initiators, and tumor-promoters. The most reproducible data are obtained if one starts the application during the second telogen (resting) phase of the hair cycle, when the mice are about 6–8 weeks old.[47,48] In the past, test solutions with organic pollutants have been applied three times a week to the skin of mice. Under these conditions, one has to test the material for some 8–12 months to obtain a significant number of papilloma- and epithelioma-bearing mice. However, solutions of up to 20% organic pollutants were reported not to induce acute toxic effects in mice, and significant tumor yields might be obtained earlier if the solution is applied five times a week.

MODE OF OBSERVATION

The weight of the mice should be recorded at the outset of the experiment and again every 2 weeks (or at least every 4 weeks). Benign skin tumors should be recorded when they have attained a diameter of 1 mm. They enlarge by nodular growth (papilloma) or by lateral invasion (carcinoma). Some may not enlarge; some will regress. Tumors that remain 1 mm or grow larger for 21 consecutive days are counted and become the raw tumor-yield data for tumor-bearing mice. Microscopic lateral invasion of a tumor into adjacent skin is considered as transformation into a carcinoma. Continued growth of such lesions, however, is required before they can be recorded as macroscopically observed carcinomas.

It is recommended that, on termination of the experiments, in chemical carcinogenesis, the animals be examined for the histologic nature of the skin lesion, for internal pathologic conditions, and especially for adenomas of the lung. The tumors on the test mice are then compared with the incidence of spontaneous tumors in the control group.

The factors outlined above are specifically important in the bioassay of organic pollutants. Most other details are given in standard reviews and books of chemical carcinogenesis[17,227,403,812,828] and in handbooks of breeding and management of laboratory animals.[472]

Subcutaneous Administration

Subcutaneous injection into mice and especially rats is the most widely used form of parenteral administration in chemical carcinogenesis testing. It can be chosen for a great variety of substances, including the ones that are highly reactive, such as some alkylating agents, and the ones that may decompose if administered orally or applied to the skin. For testing a pure chemical, only a single injection into the subcutaneous tissue of the nape of the neck is necessary. Therefore, only small amounts of material are required—for example, 100 mg per mouse for organic pollutants. Because a single monthly injection suffices for the screening of air pollutants, this method saves time and personnel, compared with the mouse skin test.

Homburger and Hsueh[393,394] have demonstrated that the subcutaneous tissue of Syrian hamsters is susceptible to carcinogenic aromatic hydrocarbons. One inbred line gave the most rapid tumor production yet described in any rodent. With 0.5 mg 7,12-dimethylbenz[a]anthracene, the first tumor appeared after 5 weeks, and the mean latency was 9 weeks. If the high susceptibility of connective tissue to carcinogens is confirmed with other hydrocarbons, this Syrian hamster inbred line may become a valuable test animal in air pollution carcinogenesis testing.

Oral Administration

For testing by oral administration, the test substance can be incorporated into the diet, dissolved in the drinking water, or force-fed by a stomach tube. In general, the oral administration of polycyclic aromatic hydrocarbons, especially those present in the respiratory environment, is a relatively poor way of determining their carcinogenicity. However, Shay *et al.*[696] and Huggins *et al.*[406] reported that some strains of rats, especially the Sprague-Dawley, are highly susceptible to some types of carcinogens (polycyclic aromatic hydrocarbons and aromatic amines) when the substances are force-fed in an oily vehicle. Even a single large dose induces mammary carcinoma rapidly (2–6 months). If a week carcinogen is used, the test may require repeated feeding. The agent to be tested should have low water and high lipid solubility. This feeding method with Sprague-Dawley rats merits special consideration for a rapid screening test of polycyclic aromatic hydrocarbons.

Inhalation

The induction of a significant number of lung adenomas in mice (C57BL) has already been demonstrated with artificial smog (ozonized gasoline) by Kotin *et al.*[451,456] and Nettesheim *et al.*[560] The latter group also reported that male mice have a significantly higher lung adenoma incidence than female mice when exposed to ozonized gasoline or an insoluble chromium oxide dust. Gardner[289] reported a slight increase of lung adenomas in mice exposed to Los Angeles ambient atmosphere.

Bladder Implantation

During the last decade, several studies have been reported on the implantation of paraffin or cholesterol pellets containing potential carcinogens into the bladders of mice.[10,509] Recently, the technique has been further developed and standardized by Jull,[419] Boyland and Watson,[86] and Bonser *et al.*[74] With this technique, the sensitivity of the bladder epithelium to tumorigenic stimuli can be used for routine testing of some potential carcinogens. However, bladder implantation is clearly not an appropriate primary route for testing air pollutants. In view of recent epidemiologic evidence that the incidence of bladder cancer is higher in areas of heavy air pollution, bladders should be examined by appropriate techniques[96] after administration of air pollutants by conventional routes.

Factors Influencing Hydrocarbon Distribution in the Host

PARTICLE SIZE

In polluted air, particles occur in sizes ranging from 0.001 to 10,000 μm in diameter. In general, controlled combustion, the major source of carcinogenic POM in polluted air, produces particles 0.1–10 μm in diameter.[388] In several studies, "lung-damaging" components are considered to be particles 0.25–10 μm in diameter.[388] Particles less than 0.25 μm in diameter are retained to a low degree in the lung, and particles with diameters greater than 10 μm lodge in the upper respiratory tract and thus do not reach the bronchi. Although air particles can be separated according to size, separated particles have thus far not been properly analyzed for polycyclic aromatic hydrocarbons. Because practically every combustion leads to the formation of traces of carcinogenic polycyclic hydrocarbons and the particles generated by combustion are primarily 0.1–10 μm in diameter, it can be assumed that POM will be partially deposited in the bronchial tree, especially at the bifurcation.

RETENTION AND ELUTION OF PARTICLES

During normal breathing, the lung retains particles 0.25–5 μm in diameter, with maximal retention of 80% of 1-μm particles and less than 5% retention of particles smaller than 0.1 μm and larger than 5 μm. Falk *et al.*[258] demonstrated experimentally that benzo[a]pyrene and other polycyclic aromatic hydrocarbons are readily eluted from soot samples recovered from the human lung. Carcinogenic hydrocarbons in polluted air appear to be adsorbed primarily on particles that, according to their size range, are compatible with deposition of and retention of a portion of the carcinogenic particles in our urban respiratory environment.

Asbestos fiber has received considerable attention in recent years as an air pollutant.[59] Oil containing carcinogenic hydrocarbon is absorbed by these fibers, especially the γ-chrysotile form. These hydrocarbons are first retained with the particles, readily eluted, and then retained in the lung tissue. This process of retention of carcinogens in lung tissue is accelerated if the respiratory air contains irritants that impinge on the bronchial epithelium.

CHANGES IN CILIARY MOVEMENT AND MUCOUS VISCOSITY

Several theories have been proposed for the importance of cilia in toxicity in respiratory carcinogenesis. In air pollution carcinogenesis, the concept of Hilding[375] appears most relevant. This author emphasizes that, in the pathogenesis of bronchogenic cancer, in both man and animal, without ciliary stasis and concomitant mucous stagnation, subsequent metaplasia from ciliated to squamous epithelium and then to epithelial cancer is not likely to occur. These changes in mucous and ciliary movement, changes in mucous viscosity, and changes in the underlying basal layer of the epithelium are caused by irritants in the respiratory air. The irritant effect seems to be essentially nonspecific, in that chemical, physical, and viral agents are capable of inducing the changes. The large spectrum of irritants in our respiratory environment originates primarily from polluted air and cigarette smoke. These two inhalants also contain carcinogenic hydrocarbons and therefore increase the likelihood of combining biologic action through either simultaneous or sequential inhalation.

The most widely found and more important volatile irritants in our environment are formaldehyde, acrolein, formic acid, acetic acid, peroxy acids, volatile phenols, ozone, nitrogen oxides, sulfur dioxide, and hydrogen cyanide.

Conclusions and Recommendations

Three major groups of agents carcinogenic to experimental animals have been identified or suggested in urban pollutants: polycyclic aromatic hydrocarbons, to a minor extent N-heterocyclic hydrocarbons, and oxygenated neutral compounds of unknown structure. Until now, the most valuable data from carcinogenic bioassays of organic pollutants were obtained with tests on mouse skin and connective tissue. To gain statistically significant bioassay data, at least 20 g of organic particulate matter is needed. That amount of material can be filtered from air in a reasonable time only with special and expensive equipment. Other methods of assay requiring less material should be explored.

The urban environment is known to contain traces of carcinogenic agents, which should impinge directly on the bronchial epithelium of the experimental animal.

The application of artificial atmospheres in model systems, such as those using benzo[a]pyrene and sulfur dioxide,[474] requires detailed exploration. However, direct inhalation studies in animals with ambient atmospheres may not provide useful data because of the high dosage and enormous animal populations that would be required.

It is suggested that the carcinogenicity of POM be assayed by direct intratracheal instillation in hamsters and rats in a model system recently developed by Saffiotti.[648] Direct intratracheal instillation in rats may also be used to test a wide range of fractions and subfractions of POM with Freund's adjuvant.

COCARCINOGENESIS, ANTICARCINOGENESIS, AND RELATED ASPECTS

Research on organic air pollutants and their possible contribution to lung cancer in man is probably far behind current thoughts in tobacco use and its relation to lung cancer in man. Some years ago, there was a single aim and purpose in tobacco research: to quantitate benzo[a]pyrene in cigarette tars so that methods could be devised for removing it from the smoke. During the last decade, it has become evident that, if benzo[a]pyrene has any role at all in tobacco carcinogenesis, it is not necessarily the most significant and that other carcinogens, and particularly cocarcinogens, must play an important role. The same is probably true of organic air pollutants.

Definitions

"Cocarcinogenesis" is the process whereby cancer is induced in animal or man by the combined action of two or more agents, either by a single exposure or, as is more common in both the laboratory and the environment, by repeated exposures. "Tumor-initiating agents" are agents (which may or may not be carcinogenic by themselves) that, when given as a single exposure—and in the case of carcinogens at a subcarcinogenic dose—and followed by a low exposure to a tumor-promoting agent, result in tumors either at the site of application of the initiating agent or at a distant site. "Tumor-promoting agents" by themselves, when given in repeated doses, are usually noncarcinogenic or at most very weakly carcinogenic. The sequential process of induction of tumors by a single dose of an initiating agent followed by repeated low-level exposure to tumor-promoting agents is referred to as "two-stage carcinogenesis." The latter term is somewhat misleading, in that mechanistically multiple steps are most probably involved in this process of tumor induction, and the term refers more specifically to the two stages of treatment with chemicals. In cocarcinogenesis experiments, two or more agents are applied simultaneously or alternately in single doses and repeatedly. "Anticarcinogenic agents" are sometimes carcinogenic but more commonly noncarcinogenic; when they are given in single or multiple doses before, during, or after treatment with a carcinogen, they partially or completely inhibit the induction of tumors at the site of application or at distant sites.

Two-Stage Carcinogenesis

The process of two-stage carcinogenesis as defined above is, as an experimental model, limited to one test system—i.e., mouse skin[784]—although some studies suggest that it may also occur in other systems.[635] Nevertheless, it merits serious consideration in relation to the induction of human lung cancer as caused by air pollutants and environmental exposures to other chemical agents; the example of cigarette smoke condensate as a tumor-promoting agent in two-stage carcinogenesis will be given below. A typical two-stage carcinogenesis experiment on mouse skin is summarized in Table 8-1 with the usual experimental observations. The agent used in these experiments as a tumor-promoter is phorbol myristate acetate, a complex tetracyclic terpenic lipophilic–hydrophilic ester derived from the plant product croton oil.[784,788] The notable features here are the rapid rate of papilloma induction, the high multiplicity of papillomas, and the fact that each agent alone induces very few or no papillomas. Other, less potent tumor-promoting agents are known, e.g., phenol, anthralin, Tweens,

TABLE 8-1 Two-Stage Tumor Induction on Mouse Skin[a]

Primary treatment

Single subcarcinogenic dose of initiating agent (7,12-dimethylbenz[a]anthracene, benzo[a]pyrene, or urethan)

Secondary treatment

Between 3 and 380 days after primary treatment, repeated application of phorbol myristate acetate, 0.5–25 μg, three times a week

Observations

1. First papillomas 30–70 days from beginning of secondary treatment
2. 50–100% of animals bear papillomas by 250 days
3. Number of papillomas per tumor bearer: 1–12 (some even higher)
4. 40–60% of animals bear squamous carcinomas after a year or more on test
5. Control groups with secondary treatment alone usually develop no tumors; some develop late tumors
6. Groups with primary treatment alone usually develop no tumors
7. Few, if any, tumor regressions

[a] Derived from Van Duuren.[784]

Spans, and dodecane. The most noteworthy of these is anthralin; compounds of similar structure may occur as air pollutants.

One of the most important characteristics of two-stage carcinogenesis is its potentially insidious character. An organ or tissue may receive a single subcarcinogenic exposure to a carcinogen or one exposure to a noncarcinogenic initiating agent and later undergo repeated low-level exposures to environmental (in this case airborne) tumor-promoting agents, and the sequence may result in tumors. This persistence of the initiating effect has been demonstrated several times in laboratory experiments on mouse skin, as shown in Table 8-2. It is clear from these experiments that an interval of more than a year between initiation and promotion resulted in rapid induction of papillomas. It must be remembered that this interval represents for the mouse approximately half its life-span.

There is at present little or no information to link the process of two-stage carcinogenesis (as demonstrated on mouse skin) to cancer of the lung in man as induced by organic air pollutants. Recent laboratory experiments by inhalation in rats have shown that combined treatment with benzo[a]pyrene and sulfur dioxide induces squamous carcinoma of the lung; neither agent alone has resulted in these tumors.[468] Convincing evidence of two-stage carcinogenesis—even using the mouse skin model—is not available for air pollutant chemicals, although it has been clearly demonstrated[69, 792] for cigarette smoke

TABLE 8-2 Persistence of Initiating Effect[a]

Primary Treatment and Route	Interval between Primary and Secondary Treatment, days	Secondary Treatment (skin)	Interval from Beginning of Secondary Treatment to First Papilloma, days	No. Mice with Papillomas/ Total No. Mice	Duration of Test, days
7,12-dimethylbenz[a] anthracene, skin	301	Croton oil, two times a week	–	9/22	420
7,12-dimethylbenz[a] anthracene, 150 μg, skin	380	Croton resin, 25 μg, three times a week	37	6/11[b]	500
Urethan, 20 mg, subcutaneous	246	Croton resin, 25 μg, three times a week	32	7/20	456
Urethan, 20 mg, intraperitoneal	246	Croton resin, 25 μg, three times a week	42	8/20	456

[a] Derived from Van Duuren.[784]
[b] Two animals with squamous carcinoma.

condensate. From these findings, it is clear that cigarette smoke condensate is at best a weak tumorigen, but it is a moderately active tumor-promoting agent. In a recent experiment, 30 of 60 mice exhibited papillomas; eight of the 30 also bore squamous carcinomas of the skin if the tar was applied after a single subcarcinogenic (50-μg) dose of 7,12-dimethylbenz[a]anthracene. When the tar was applied alone at 50 mg per application, five times a week, only five of 60 mice bore papillomas; no animals in this group bore carcinomas after 390 days on test (B. L. Van Duuren, unpublished data).

Some agents are not by themselves carcinogenic, but result in substantial tumor yields when followed by repeated exposure to tumor-promoting agents. Urethan is the classic compound in this category; it is not carcinogenic for mouse skin (although it does result in lung adenomas in mice), but it is a potent initiating agent for mouse skin, whether applied topically or given systemically. Urethan and related compounds are not known to be air pollutants, but several noncarcinogenic initiating agents are possible air pollutants. A number of known aromatic hydrocarbon air pollutants—previously considered noncarcinogenic or borderline carcinogens—were recently shown to be active tumor-initiating agents. Compounds in this class include dibenz[a,c]anthracene, chrysene, and benz[a]anthracene.[789]

Cocarcinogenesis and Anticarcinogenic Agents

The problem of cocarcinogenesis is difficult to deal with from the point of view of environmental carcinogenesis, even in a discussion of laboratory models. Whenever two or more agents are applied simultaneously or sequentially, they may interact in a variety of ways so as to alter the effect of each other. This modification of each other's effects results at times in an increase in the biologic effect, such as tumor induction; at times, it results in a decreased effect, owing to competitive reactions at sites concerned with biologic activity, or owing to induction or inhibition of enzymes, which results in the detoxification of an otherwise carcinogenic agent. The agent β-naphthoflavone has recently been shown to inhibit lung adenoma induction by benzo[a]pyrene enzymes.[807] Several other examples are known.[787]

Polycyclic aromatic hydrocarbons of weak or mild carcinogenicity are present in polluted air and cigarette smoke. These have been shown by Falk *et al.*[261] to inhibit subcutaneous sarcoma production by benzo[a]pyrene. The significance of this finding is that a given mixture of POM in polluted air may have a smaller net carcinogenic effect than would be expected from the action of the potent carcinogens present in the mixture.

Airborne Alkylating Agents

Several reports have discussed the presence of "neutral-nonaromatic" fractions of air pollutant concentrates,[448] but none of the reported studies has proceeded with the fractionation, isolation, and chemical identification of these materials. Because olefin hydrocarbons constitute a considerable portion of organic air pollutants, it is reasonable to expect that their oxidation products, whether spontaneous or photochemically induced, may be present. The main source of these olefins is automobile exhaust. Several schematic pathways have been proposed, suggesting possible products of olefin oxidation and peroxidation. Laboratory studies on olefin oxidation and ozonization have resulted in the isolation of epoxides, peroxides, aldehydes, and ketones.[485] Some of the products and possibly reactive intermediate species are shown in Figure 8-1. It has become apparent, also, that oxides of nitrogen and sulfur dioxide interact with olefins so that a great variety of these products can be expected. Some of these compounds are most probably carcinogenic, cocarcinogenic, or both. Extensive studies have been carried out on the carcinogenic activity of many epoxides (mono-, di-, and poly-), β-lactones, hydroperoxides, and peroxides in a variety of test systems, including skin application in mice, subcutaneous injection in mice and rats, intraperitoneal injection in mice, intratracheal instillation in rats, and intragastric intubation in rats. These studies have made it possible to derive some structure–carcinogenicity correlations and predictions of the possible

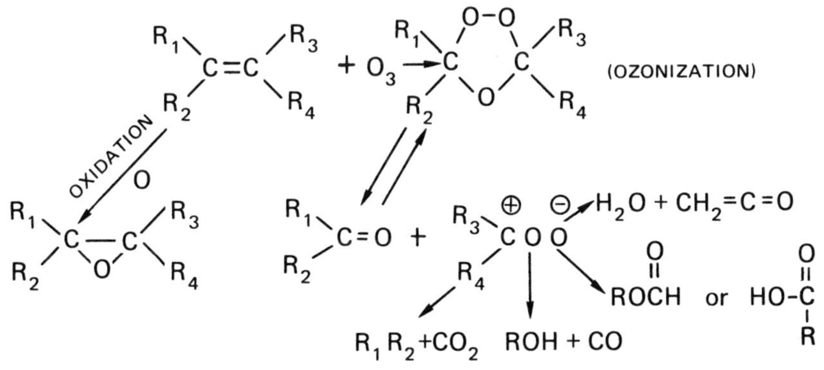

FIGURE 8-1 Olefin oxidation in air. (After Leighton.[485])

carcinogenicity of compounds in this series not yet tested by long-term bioassay.[780] The chemical reactivity of these compounds has been correlated with carcinogenic activity; stereochemical factors, such as molecular flexibility, and the importance of interatomic distance as related to carcinogenic activity and possible intracellular target sites were also considered.

The epoxides, hydroperoxides, and β-lactones are the most active carcinogens within this series, but very few peroxides have been tested for carcinogenic activity, so it is not possible to draw conclusions about them. It is difficult to visualize the formation of β-lactones from olefins; however, some β-lactones can be formed by spontaneous dimerization of such pyrolytic products as ketene and its analogues, as shown in Figure 8-2. The parent compound diketene (β-methylene-β-propiolactone) is inactive as a carcinogen, probably owing to its high chemical reactivity; but its less reactive saturated analogue, β-methyl-β-propiolactone, is carcinogenic.[780] Thus, the possibility that β-lactones are carcinogenic air pollutants cannot be ruled out.

It should be pointed out that, in the extensive network of stations for the collection of air pollutants, apparently only particulate matter is collected. It is expected that most of the potentially deleterious agents just discussed will pass through the particulate-matter filters in vapor form.

Aromatic Hydrocarbon Oxidation Products

The metabolism of carcinogenic aromatic hydrocarbons has been extensively studied in a variety of animal species,[142] but relatively little has been done in explorations on the oxidation products, mostly

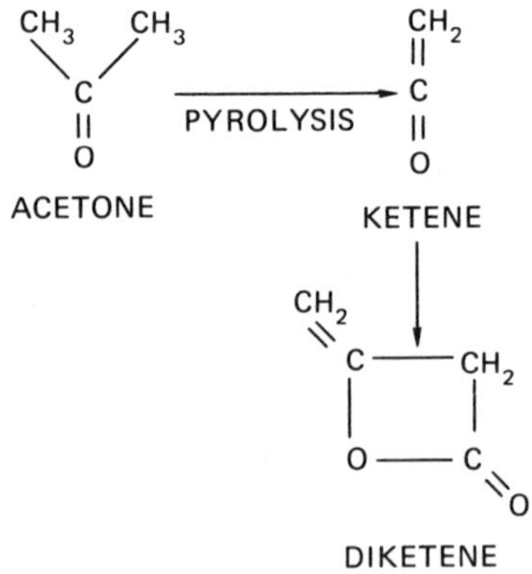

FIGURE 8-2 Formation of diketene from acetone.

photochemical, of aromatic hydrocarbons in air. This is an important area, because it has been demonstrated that some phenolic compounds–e.g., phenol and 1,9-dihydroxy-8-anthrone–are tumor-promoting agents. Further knowledge of these oxidation products is pertinent also with respect to the detoxification of carcinogenic aromatic hydrocarbons in the air, depending on climatic and meteorologic conditions. In addition, some of the conceivable oxidation products, such as hydroperoxides and peroxides, formed photochemically, may be carcinogenic even if the parent hydrocarbons are noncarcinogenic. Benz[a]-anthracene, for example, is a weak carcinogen. It is conceivable that its hydroperoxide or peroxide, shown in Figure 8-3, is carcinogenic. However, neither of these compounds is a known air pollutant. A large variety of phenols and quinones of aromatic hydrocarbons are expected as photochemical oxidation products. Some of these compounds have been tested for carcinogenic activity.[351] The quinones are usually not carcinogenic, but some of the phenols are weakly carcinogenic. It has been suggested that aromatic hydrocarbon epoxides are proximal carcinogens in aromatic hydrocarbon carcinogenesis.[780] Some have been synthesized, e.g., 5,6-dihydro-5,6-epoxydibenz-[a,h]anthracene.[563] However, these compounds are highly reactive and hence unstable, and they have not been isolated as metabolic products of aromatic hydrocarbons *in vivo* or from air pollutants. It is likely that they occur in both.

Conclusions and Recommendations

The polycyclic aromatic hydrocarbons and heterocyclics constitute a group of known carcinogens that are present in the particulate phase of polluted air. However, the extent of their contribution to the inci-

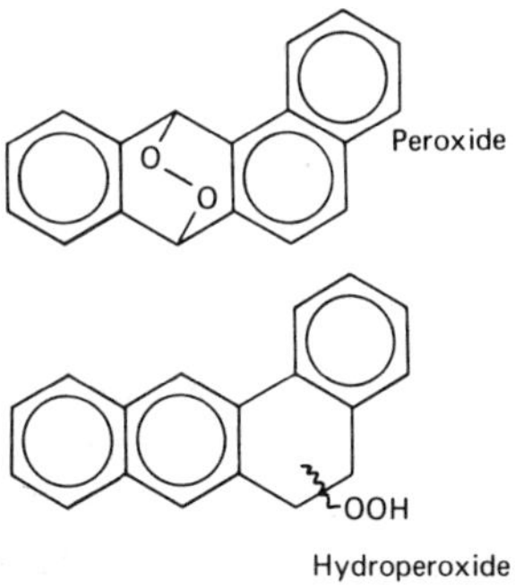

Peroxide

OOH

Hydroperoxide

FIGURE 8-3 Peroxy (peroxide and hydroperoxide) compounds of benz[a]anthracene.

dence of human lung cancer is unknown. A variety of other agents probably contribute to the human health hazard. They include tumor-promoting agents, cocarcinogens, noncarcinogenic tumor-initiating agents, and carcinogens other than aromatic hydrocarbons. The unknown carcinogens include direct-acting alkylating agents (such as epoxides and lactones), peroxides and hydroperoxides of olefins, and aromatic hydrocarbons.

The role played by tumor-inhibiting agents or anticarcinogenic agents in the health effects of air pollutants is at present poorly understood. Research is needed on the chemistry and biologic activity of air pollutant cocarcinogens and tumor-promoting agents, such as polyphenols and paraffin hydrocarbons, and on the oxidation products of airborne olefins and aromatic hydrocarbons, including the nature of the epoxides, hydroperoxides, peroxides, and lactones formed and their biologic properties.

EXPOSURE OF THE LUNG TO POLYCYCLIC HYDROCARBONS

There are relatively few reports of induction of tumors by administration of carcinogenic polycyclic compounds via the airway. Because of the attendant difficulty, there have been few studies in which exposure of the respiratory tract has been carried out in a manner relevant to the problem of air pollution.[148,449,734]

Exposure by Inhalation of Crude Material

Exposure of mice by inhalation of road sweepings, chimney soot, air dusts, etc., produced pulmonary adenomas. One study was carried out with sweepings from an asphalt road. Not only did the pulmonary adenoma incidence rate increase but skin cancer was noted in exposed animals.[108] Other investigators produced increased numbers of pulmonary adenomas with road dust[524] and chimney soot.[686,687]

Mice have been exposed to aerosols of more purified materials consisting of the neutral fraction of coal tars and the same fraction with the addition of the acidic and phenolic fractions. No tumors occurred in the C3H control mice; the incidence of both squamous metaplasia and pulmonary adenomas was significantly greater in the animals that were given both fractions than in those receiving only the neutral fraction.[760]

Inhalation of purified asphalt did not produce lung tumors in guinea pigs or mice,[702] whereas skin painting of similar material led to skin cancer.[703]

Exposure to Ozonized Gasoline Vapor

A number of experiments have been performed in which mice were exposed to "simulated auto smog." This atmosphere was produced by the passage of nitrogen through leaded gasoline, reaction of the vapor with ozone, and introduction of the effluent into exposure chambers.[451] Exposure to this atmosphere increased the incidence of tumors in A strain mice and pulmonary tumors in C57B mice.[454]

Intratracheal Instillation of Carcinogens

An appreciable incidence of lung tumors was produced by the intratracheal administration of 7,12-dimethylbenz[a]anthracene suspended in a balanced saline solution containing 4% casein and India ink powder.[614] Studies were carried out with other agents, including benzo[a]pyrene adsorbed on various types of purified carbon particles.[613,615–618,691,692,694] The method was successful for the production of malignancies of the lung with 7,12-dimethylbenz[a]anthracene; but with benzo[a]pyrene, 30% of these rats had adenocarcinomas and 67% had squamous cell carcinomas.

The adsorption of benzo[a]pyrene on hematite (F_2O_3) particles has produced tumors successfully when equal parts of both were triturated in a mortar, suspended in saline, and injected via the trachea into Syrian hamsters.[649–656] The carcinogen/hematite/Syrian hamster model has been successful in inducing a large number of cancers of the tracheobronchial tree and lung parenchyma, which mimic those occurring naturally in human beings.[654] Furthermore, dosage effects are apparent from both the quantity administered and the number of weekly intratracheal instillations. The hamster is remarkably well suited for this model, in that it is uniquely free of inflammation and spontaneous tumors of the lung. The hamster model has been extended to a primate, indicating that the intratracheal instillation method is effective in producing squamous carcinoma of the lung in another order of mammal. A cancer incidence as high as 76% was reported with benzo[a]pyrene and hematite. Methods using the addition of Tween 60 with benzo[a]pyrene have yielded as many as 50% tracheobronchial tumors. This method has also been confirmed for a combination of asbestos and benzo[a]pyrene.[366,534] The addition of Tween to the carcinogen may be objected to on the grounds that Tween is a promoting agent and may be carcinogenic itself.[191] Other methods of induction of lung cancer by intratracheal implantation of pellets impregnated with 3-methylcholanthrene or benzo[a]pyrene have yielded 45 and 30% cancers, respectively.[474] Experiments have been carried out in which pure hydrocarbon carcinogens in colloidal suspension in gelatin, unaccompanied by other agents, were instilled into the trachea in hamsters. These studies yielded only metaplasia with benzo[a]pyrene but a very high incidence of poorly differentiated adenocarcinomas and other tumors with 3-methylcholanthrene.[474]

Another promising model was developed by Yasuhira.[833] With or without pretreatment with complete Freund's adjuvant, a carcinogen (3-methylcholanthrene) was instilled once into the tracheas of Sprague-Dawley and Wistar rats. More than 66% of the rats developed squamous cell carcinomas and fibrosarcomas of the lung 50–400 days after treatment.

The successful induction of tumors by the intratracheal injection of benzo[a]pyrene appears to be related to the addition of some other physical factor, which in all probability prolongs the residence time of the carcinogen at the target site. Thus, no tumors resulted from injection of benzo[a]pyrene alone,[474] whereas its adsorption on carbon particles,[613–618,691,692] hematite,[649–656] or asbestos[534] produced a striking carcinogenic effect. It has been postulated that the carcinogenic effect was made possible by the adherence of the hydrocarbon to the surface of the particle, from which it was slowly released into the tissue, thereby achieving a more prolonged and constant action. It has been shown that the retention of the benzo[a]pyrene in the lung is proportional to the amount of hematite used, and the rate of elimination suggests a prolongation of retention owing to the presence of the inert particle.[649]

The elimination of benzo[a]pyrene from the lungs appears related to the size of the carbon particle on which it is adsorbed, with elimination progressing more slowly from smaller particles.[694] In experiments using tritiated benzo[a]pyrene in hamsters, a significant slowing of the clearance of radioactivity from the lung was found after a 14-day period when the benzo[a]pyrene was incorporated on carbon or asbestos, compared with the clearance after administration of benzo[a]pyrene alone. The adsorption on the inert particles also resulted in a more pronounced and prolonged increase in the number of pulmonary alveolar macrophages.[618] Similar prolongation of retention of the carcinogen might also be expected with the intratracheal implantation of impregnated pellets, although the role of trauma cannot be ruled out in this method of application of the carcinogen.[474]

Exposure to a Combination of Carcinogens and Gaseous Pollutants

The study perhaps most relevant to air pollution inhalation used a combination of sulfur dioxide and benzo[a]pyrene aerosol in rats and hamsters.[474] In these experiments, animals were exposed to 10 ppm of sulfur dioxide for 6 hr/day, 5 days a week, plus a combination of 10 mg/m^3 of benzo[a]pyrene and 3.5 ppm of sulfur dioxide for 1 hr/day, 5 days a week. Appropriate controls were used. No significant alterations beyond cellular metaplasia were found in hamsters. In rats, however, malignancies of the lung were induced. Two of 21 rats receiving only the benzo[a]pyrene–sulfur dioxide mixture for 1 hr/day developed squamous cell tumors of the lung, whereas five of the 21 rats receiving the same mixture plus the 6-hr/day exposure to 10 ppm of sulfur dioxide developed lung tumors. These data indicate that an atmosphere containing a benzo[a]pyrene–sulfur dioxide mixture is carcinogenic to the lung of the rat and suggest that additional exposure to sulfur dioxide causes an increment in such cancers.

Conclusions

Purified polycyclic compounds, such as benzo[a]pyrene, have produced tumors of the tracheobronchiolar tree or lung parenchyma only when adsorbed on particles and delivered below the larynx. In inhalation experiments, the addition of an irritant, such as sulfur dioxide, to an aerosol of benzo[a]pyrene has induced lung carcinomas in rats. Because many potentially interacting influences—including solid particles, irritant chemicals, and gases—are ubiquitous in polluted air, these may be cofactors as important for the induction of pulmonary cancer as the polycyclic hydrocarbons themselves.

TESTING FOR CARCINOGENICITY IN PRIMATES

Primates as natural hosts for tumors and as experimental animals in carcinogenesis assays have the putative advantage of a phylogenetic relation to man.[302] That advantage is tempered by the initial cost and the expense of maintenance of primates and by their long lifespan, which may require many years of observation for study of a disease with long latent periods. Data on primates as laboratory animals, breeding in captivity, with natural and acquired infections, and as subjects for physiologic studies are available.[302]

Members of the suborder Prosimii appear to be closer to a rodent-like precursor, whereas the suborder Anthropoidea—which includes simians (both Old and New World monkeys), the apes, and man—encompasses the more highly evolved primates. This notion would raise the possibility that the suborder Prosimii is not as good a potential surrogate for man as the suborder Anthropoidea.[7,168]

Luther[504] and Kent[435] summarized the world literature on carcinogenesis in primates to 1960. Kent[435] reviewed both spontaneous and induced neoplasms of simians and concluded that spontaneous tumors increase with age and closely resemble their human counterparts. Vadova and Gel'shtein[779] observed that the greater proportion of epithelial than of sarcomatous tumors among simians parallels the distribution in man. Petrov[592] reviewed 23 monkeys that survived 2 years after the injection of radium and 22 monkeys that survived 2 years after injection of carcinogenic hydrocarbons into marrow cavities of long bones. The first group yielded eight tumors, the second group, one. All tumors were sarcomas of cartilage, bone, or reticuloendothelial cells. Kent[435] regarded the production of bone neoplasms with ionizing radiation as evidence that monkeys are susceptible to experimental tumor production.

Skin and Subcutaneous Tissues

Pfeiffer and Allen[593] administered methylcholanthrene, dibenzanthracene, and benzo[a]pyrene in oil or saline to adult simians by injection into subcutaneous tissues and mammary glands, by intravenous injection, by mouth, by implanting pellets in abdominal and pelvic viscera, and by painting on skin and cervix uteri. At all sites of injection or implantation, fibrotic reactions, granulomas, and accompanying chronic or acute inflammation were produced without proven neoplasm. When administration was by skin painting or cervical canal instillation, papillomatous masses and hyperkeratinization were observed without neoplasms. After multiple exposures to hydrocarbons by various modes of inoculation and with observations continued for up to 10 years in some animals, the authors concluded that failure to produce cancer suggested greater activity of protective mechanisms than in rodents, and possibly detoxification.

Review of this experiment reveals that very few animals were exposed repeatedly to the same agent at the same site for periods of over 2 years. That is significant, in view of the 3–4 years required by

Sugiura *et al.*[735] to induce squamous carcinomas in three of six rhesus monkeys with high-boiling, catalytically cracked oil applied by repeated skin painting. Moreover, the monkeys used by Pfeiffer and Allen[593] were estimated to be 3–10 years old at the outset of treatment. The work of Pfeiffer and Allen should not be regarded as definitively negative, although the possibility that primates resist the effects of hydrocarbons better than rodents is worthy of confirmatory testing. Kelly *et al.*[433] induced hepatic cell carcinomas within 27 months in macaques and cebus monkeys by giving them *N*-nitrosodiethylamine orally and in cercopithecus monkeys by giving it intraperitoneally. The macaques were exposed from birth onward; the other species began exposure before they were a year old.

Levy[493] produced a fibrosarcoma in a marmoset 10 months after the subcutaneous injection of 2 mg of methylcholanthrene. Adamson *et al.*[7] produced fibrosarcomas in three of six tree shrews and one of 12 galagos inoculated subcutaneously as young adults or as newborns with benzo[a]pyrene or methylcholanthrene. Noyes reported induction of fibrosarcoma and rhabdomyosarcoma in a marmoset with benzo[a]pyrene or dimethylbenzanthracene inoculated at different sites in the same animal[565] and fibrosarcomas in three tree shrews inoculated with benzo[a]pyrene.[566] The relative frequency of reports of sarcomagenesis with polycyclic organic hydrocarbons in prosimians, compared with simians, may reflect a true difference in susceptibility to tumor induction or a lack of comparability in experimental plans used in the two suborders.

The foregoing review deals with polycyclic aromatic hydrocarbons and crude materials known to contain them. Chemical carcinogens of other classes have been even less widely studied.

Pulmonary Tissues

Pulmonary carcinogenesis in simian primates was reported by Vorwald,[796] who administered beryllium dust repeatedly to young adult *Macaca mulatta* monkeys over a period of more than 5 years. The earliest lung cancer was observed at 4.5 years after one intramural injection followed by numerous intrabronchial instillations of beryllium oxide as a powder suspended in a saline solution. Bronchogenic neoplasms were also developed after repeated inhalation of beryllium sulfate aerosol.

Squamous carcinoma of the lung and/or bronchi reportedly occurred in 50% (three of six) of the galagos after weekly tracheal instillation of benzo[a]pyrene and ferric oxide dust. These agents were administered for 67–69 weeks beginning at the time of weaning (12 weeks old).[168] Hamsters developed squamous or anaplastic carcinomas of lung after exposure to the same carcinogen-particulate preparation by the same investigators[168] following the schedule reported by Saffiotti *et al.*[650]

Nonneoplastic alveolar lesions in galago and hamster lungs were histologically similar. This raised the question of whether the galago (and possibly other prosimians) may resemble rodents in terms of pathogenetic response.[7,168]

Conclusions and Recommendations

Experimental carcinogenesis caused by polycyclic aromatic hydrocarbons or crude products known to contain them has been achieved in simian and prosimian primates. Nitrosamines have been shown to be carcinogenic in simians. Subhuman primates are therefore susceptible to experimental chemical carcinogenesis.

Not all attempts to produce tumors in simians with polycyclic aromatic hydrocarbons have succeeded, raising questions of resistance due to metabolic or immune characteristics of the suborder Anthropoidea and of appropriate choice of age and method of application. Primates of the suborder Prosimii appear more susceptible to carcinogenesis by this class of compounds and have shorter latent periods.

Pulmonary carcinoma in simians has been produced by particles of beryllium salt but not by polycyclic aromatic hydrocarbons. Pulmonary cancer has been produced in rodents and prosimian primates by intratracheal instillation of particulate preparations of polycyclic hydrocarbons.

Domestically bred simian primates should be tested with methods identical with those which succeeded in prosimian primates and rodents to provide data on the relative susceptibility to systemic, skin, and bronchial carcinogenesis by polycyclic hydrocarbons in the suborder Anthropoidea, which includes man.

EXPOSURE OF OUTDOOR ZOO ANIMALS TO ATMOSPHERIC POLLUTANTS

The only available records on which to base an evaluation of the possible effects of atmospheric pollutants on outdoor zoo animals over long periods are the necropsy reports of the Philadelphia Zoological Gardens from November 1901 to December 1969. These were made available to the Panel by Dr. Herbert L. Ratcliffe, Director Emeritus of the Penrose Research Laboratory of that institution. He and Dr. Herbert Fox were the pathologists in charge during the 68-year period.

The data from necropsies on a total of 21,000 animals suggested a recent increase in the incidence of lung tumors in avian species. There were 350 malignant neoplasms among 19,000 animals in the necropsy series, of which 27 were tumors of the lung. Cancer of the lung in birds, especially ducks and geese, had increased significantly ($p<0.05$) in the second half of the period of observation, compared with the first, although this was not the case in mammals. Birds represented approximately 60% of the animals that came to necropsy.

These results represent an analysis of raw data and must be interpreted with caution, especially because the total numbers of pulmonary neoplasms were small.

The data from the Philadelphia Zoological Gardens provide suggestive evidence of the existence of environmental pulmonary cancer in birds. A controlled experiment is therefore indicated. It could consist of the exposure of a flock of a particular species (for example, sparrows) to an urban atmosphere that is highly polluted, and in a specific geographic area where the prevalence of lung cancer in man is unduly high. The control group would consist of a sheltered flock of the same origin and age distribution that would breathe purified air, but would otherwise be subjected to identical dietary and other environmental conditions. A comparative study of city sparrows with country sparrows of the same species—i.e., from contrasting states of atmospheric pollution—would also be satisfactory, although less well controlled.

9

Modification of Host Factors in *in vivo* Carcinogenesis Tests

HOST IMMUNE STATUS IN CHEMICAL CARCINOGENESIS

The topic of host immune status in chemical carcinogenesis can be divided into two distinct aspects: the role of immunity or hypersensitivity to a carcinogenic chemical itself and the role of immunity to cancer cells as a surveillance mechanism.

Hypersensitivity to chemical carcinogens themselves has not been extensively investigated, and no definite conclusions can be reached on the basis of current evidence. There is one unconfirmed report to the effect that small, repeated, subcarcinogenic inoculations of a polycyclic hydrocarbon may produce resistance to the carcinogenic effects of larger later doses. Preliminary unpublished data by H. Peck and co-workers suggest that the carcinogenic action of a hydrocarbon may be decreased in rats immunized with a hydrocarbon–protein conjugate. Guinea pigs are relatively resistant to chemical carcinogenesis, and this may be related to the ease with which skin hypersensitivity can be induced in them. Unfortunately, this entire area is characterized by contradictory reports and a lack of solid information. Further investigation is definitely needed.

Immunity to cancer cells themselves and its effect on carcinogenesis have been extensively investigated. Although much information has been accumulated, the importance of immunologic surveillance in chemical oncogenesis is still debated. There is much evidence that such a mechanism does exist and can modify the course of chemical oncogenesis, but there is also evidence that immunosurveillance in many systems is weak and largely ineffectual.

The evidence of an immunologic surveillance mechanism in chemical carcinogenesis is of three main types. No one of these types of evidence is by itself conclusive, but in the aggregate they constitute a strong case.

1. It is now generally believed that most, and perhaps all, neoplasms have tumor-specific antigens that can arouse an immune response. The universality of such antigens has become almost a religious dogma in some circles. It is obvious that a nearly universal immunogenicity on the part of tumor cells is prerequisite for effective immunosurveillance.[437]

2. The second type of evidence of immunosurveillance is histologic. It is well established that several types of early neoplasms are characterized by extensive lymphocytic infiltration. Such infiltration, by analogy with the well-known homograft reaction, can be interpreted as evidence of an immune resistance to the developing neoplasm; it is particularly remarkable in early squamous cell neoplasms of the skin and in early malignant melanomas. The kind of breast carcinoma in the human that is heavily infiltrated with lymphocytes (medullary carcinoma) has an unusually favorable prognosis.

3. The third type of evidence of immunosurveillance is the apparent correlation between clinical or experimental conditions of altered immunocompetence and the incidence of neoplasia. This correlative evidence can be further subdivided into two broad categories: experiments of nature and experiments of man. A number of experiments of nature are known in which deficits in immunologic reactivity occur. These include various congenital thymic abnormalities and Down's syndrome, as well as that most widespread of all clinical syndromes, aging. In all these conditions, there appears to be a decreased ability to mount an effective delayed-hypersensitivity type of immune response to an antigenic challenge.[305,437] An increased likelihood of neoplasia is most strikingly manifested in aging. One unconfirmed study suggests that people endowed with a hyperactive immune mechanism may have a low incidence of cancer;[273] it was a retrospective study and showed that allergic disorders—such as

eczema, hay fever, and asthma—were less prevalent in the histories of cancer patients than in a control population. The most widely used man-made experimental modality to decrease immunocompetence has been thymectomy in newborn mice. It has been shown in several studies that neonatal thymectomy renders mice more sensitive to the action of carcinogens administered later. Induction of sarcomas, papillomas, and hepatomas has been influenced by this means.[437] Early returns suggest that patients undergoing immunosuppressive therapy for the purpose of kidney transplantation have a higher than expected incidence of cancer. Immunocompetence, with respect to lymphocyte-mediated immunity, can sometimes be artificially augmented by the administration of BCG vaccine. This procedure may lower tumor incidence in both man and mouse. It is probably more than coincidental that, with perhaps one exception, all the known chemical carcinogens are profoundly immunodepressive.[437] Thus, skin allografts can be induced to survive in mice treated with carcinogenic dosages of a polycyclic hydrocarbon carcinogen. There are claims that dosages too low to be immunosuppressive may still be carcinogenic; this point needs further investigation.

Although the three types of evidence cited make, in the aggregate, a strong case for the importance of immunologic surveillance in chemical carcinogenesis, there are several arguments against this thesis. These arguments are not decisive, but they do suggest that immunosurveillance is a relatively unimportant mechanism of homeostasis in many systems. The first of these lines of evidence concerns the immunogenicity of tumor cells. Obviously, an efficient immunosurveillance mechanism would require that most tumor cells be highly immunogenic. As pointed out, it is now commonly held that all tumor cells have tumor-distinctive surface antigens and that these are immunogenic in the host. The evidence is not entirely convincing. For example, adenomas induced by urethan in mice appear to have relatively little immunogenicity.[603] Baldwin has not been able to detect evidence of tumor antigens in some of the rat tumors he has induced with 2-acetylaminofluorene.[30] Many spontaneous tumors in both rats and mice have very little, if any, immunogenicity.[29,571] Other examples of at least very weak immunogenicity could be cited. This relative lack of immunogenicity may or may not reflect a paucity of tumor-distinctive surface antigens. Perhaps enhancing antibodies or other masking mechanisms prevent or diminish immunogenicity. Whatever the mechanism, many and possibly most so-called spontaneous neoplasms seem to have little functional immunogenic capacity.

Despite this apparent lack of widespread potent immunogenicity, one could still argue that immunosurveillance is a very effective mechanism; indeed, the very lack of immunogenicity in many neoplasms may suggest its effectiveness. Immunoselection may bring about a situation in which the tumors that reach a clinically detectable size represent a small surviving minority of the real total, most of which were suppressed by an efficient immune reaction. If that is so, neoplasms that arise in a host owing to a deficient immunosurveillance system should usually be highly antigenic and immunogenic. However, this is not the case, as is shown by studies of tumors arising in tissue cultures or in diffusion chambers.

Although it has been shown that there is an immunologic surveillance mechanism, some evidence suggests that it is relatively weak as a defense against the development of neoplasia. Many tumors have little or no immunogenicity, and this cannot be attributed entirely to immunoselection. Even if potentially immunogenic, a neoplasm may fail to immunize the host until late in its course of growth. Immunodepression does not regularly result in an increment in the development of epithelial neoplasia, which would be highly suggestive of depression of a major and critical defense mechanism against tumor development. Other types of surveillance having little to do with specific acquired immunity may also constitute a major defense against incipient, chemically induced neoplasia.

AGE AND CARCINOGENESIS TESTING

Although many studies have been carried out in newborn and young adult mice, not many studies have been carried out in aged animals. There have been even fewer comparative studies on animals of different ages. The study of chemical carcinogenesis in aged laboratory animals is complicated by a number of factors: Aged animals are not always readily available; the life-span is too short for chronic carcinogenicity studies; and the incidence of spontaneous neoplasms and other disease states increases with age and interferes with the interpretation of experimental results. Some studies suggest that aged animals are more susceptible to skin carcinogenesis,[790] whereas others suggest little difference between young adult and aged animals in tumor induction by aromatic hydrocarbons.[529] In one study, 7,12-dimethylbenz[a]-anthracene was given by intragastric intubation to rats less than 2 weeks old, 5–8 weeks old, and 26 weeks old. The overall tumor incidences were very similar in the two older groups but considerably higher in the group treated when they were less than 2 weeks old.

In some kinds of carcinogenesis studies, age is vitally important in determining the results, in that it is related to such factors as hormonal status and immune response. It has been shown, with a variety of carcinogens and routes of administration, that newborn mice (usually 12–24 hr old) are more susceptible to carcinogenesis than young adult mice.[190] A noteworthy case is liver carcinogenesis using aromatic hydrocarbons: Adult mice are normally resistant to liver carcinogenesis, but hepatomas developed in 21 of 25 8-day-old male suckling mice that were fed 3-methylcholanthrene.[439]

The importance of hormonal status in chemical carcinogenesis is apparent from experiments on mammary tumor induction with 7,12-dimethylbenz[a]anthracene in rats 50–55 days old.[407] With rats 75–100 days old in the same experimental protocol, the time to induction of mammary tumors is much longer and the tumor incidence is lowered.

The effect of age on 3-methylcholanthrene carcinogenesis has been examined in guinea pigs. The carcinogen was injected subcutaneously, and the animals were examined for sarcoma induction. The animals were in four age groups: young, mature, old, and senile. There was no significant difference among the first three groups, but in the senile animals there was a 50% reduction in sarcoma induction. The frequency of metastases from induced tumors also declined with age.[65]

In a number of additional studies, newborn mice have been found to be more susceptible than adults to chemical carcinogenesis, particularly in organs distal to the site of administration. Klein[441] produced a high incidence of hepatomas in week-old male mice with 2-acetylaminofluorene at a dosage lower than had been reported to be hepatocarcinogenic. Kelly and O'Gara[432] carried out an extensive study of carcinogenesis after a single subcutaneous injection of 0.06 mg of dibenz[a,b]anthracene and 0.1 mg of 3-methylcholanthrene into day-old mice; they found a high incidence of a great variety of lung tumors, fibrosarcomas, leukemias, sebaceous gland adenomas, and hepatomas that appeared in 8–32 weeks. Toth *et al.*[755] compared the oral administration of dimethylnitrosamine in adult and its subcutaneous administration in newborn BALB/c mice; in addition to a number of other types of tumors, hepatomas were induced only in the newborn mice.

Epstein *et al.*[241] tested the carcinogenic activity in newborn mice of air pollutants collected from six cities in the United States. The mice received 15–25 mg of material subcutaneously. Some toxicity was observed, and a number of mice developed hepatomas and pulmonary adenomas. Epstein and Mantel[243] found that the herbicide maleic anhydride produced hepatomas after subcutaneous injection into newborn mice, and Epstein *et al.*[234] found similar results with the fungicide griseofulvin.

The differences in the carcinogenicity of compounds between newborn and adult mice appear in some cases to be paralleled by differences in metabolism. Domsky *et al.*[210] showed that 7,12-dimethylbenz[a]-anthracene injected subcutaneously in olive oil was retained at the site of injection for a much longer time in newborn than in adult mice. The same thing was found by Mirvish *et al.*[535] when urethan was injected intraperitoneally.

Experimental chemical carcinogenesis in newborn animals has been reviewed by Toth,[754] who cautions that the issue of the greater susceptibility of newborn animals is a complex question still open to controversy.

It appears, then, that newborn mice in many cases are more susceptible to chemical carcinogenesis, particularly in organs distal to the site of injection, than are adult mice. Increased sensitivity to chemical agents, with the requirement of smaller doses, suggests that greater use of newborn mice for carcinogenicity tests of air pollution fractions would be advisable.

NUTRITION AND CARCINOGENESIS

During the 1930's and 1940's, there was considerable optimism over the possibility that a solution to the cancer problem might be achieved through the science of nutrition. It seemed reasonable that, just as the normal organism is responsive to its nutritional state, cancer cells might be found to require specific nutrients. However, extensive studies *in vivo* and investigations in cell and organ culture showed that malignant tissue has no unique nutritional requirement that can be exploited.[741]

Several generalizations about nutrition and carcinogenesis seem valid. As was the case with established cancer, no unique and specific nutritional factor could be demonstrated whose absence prevented the formation of tumors. Nor could specific nutritional factors be found whose presence protected the host against tumor formation.[741]

Relatively minor inhibition of tumor formation has been achieved by manipulating the level of known nutritional elements, but none of these procedures has developed to the point of practical application.[741] Current research suggests that ample vitamin A intake imparts some protection to epithelial tissues against chemical carcinogens, as demonstrated by the protection afforded by vitamin A in hamsters against lung cancer produced by instillation of hydrocarbons and other particles.[655]

In the case of liver cancer induced by azo dye in rats, a specific nutritional effect was discovered. Diets low in riboflavin resulted in the full carcinogenic potential of these dyes. In contrast, diets adequate in riboflavin facilitated efficient degradation of the azo linkage by liver enzymes, thereby producing metabolites of the dye incapable of inducing liver cancer.[533]

Although the protective effect of adequate riboflavin nutrition on azo dye carcinogenesis is specific, it serves as a model for possible nutritional effects on other carcinogens. The existence of dietary factors specific for the metabolic activation or inactivation of most classes of carcinogens—especially carcinogenic hydrocarbons—has been thoroughly investigated, and none has been found. However, in view of the ubiquity of carcinogens in our environment and the fact that an adequate nutritional state generally facilitates the detoxification of foreign molecules, adequate nutrition for all should be an important goal in cancer prevention.

A role in cancer formation has been determined for caloric intake.[741] If food is always present in a cage, it is characteristic for animals to become obese. Food intake may be reduced to 65 or 75% of the *ad libitum* level with generally beneficial results, particularly as measured by longevity. In addition to fewer deaths from infection and degenerative diseases in general, the incidence of spontaneous cancer is lowered. In some experiments, deaths from spontaneous mammary tumors occurred in two thirds of the fully fed female mice, whereas none of the calorically restricted mice developed mammary cancer. Inhibition of formation of every type of spontaneous and chemically induced tumor that has been tested has been achieved by caloric restriction. The inhibition appears to be inversely related to the strength of the carcinogenic stimulus and directly related to the degree of caloric restriction.

The effect of caloric restriction may also be observed in man. Evaluation of insurance statistical studies indicates that persons who are overweight when past middle age are more likely to die of cancer than are persons of average weight or less.[742] Insurance statistics then support the conclusion that a considerable portion of potential cancers in man might be prevented or substantially delayed by avoiding overeating. It should be emphasized that this conclusion is justified with respect only to the formation of cancer; there is no evidence that caloric restriction is a practical way to affect the growth of an established tumor.

Aryl hydrocarbon hydroxylase activity is not detectable in the small intestine and lungs of rats fed purified diets but is measurable in these tissues in animals fed crude diets. Therefore, dietary induction of enzyme is necessary. There is evidence that some vegetables may be responsible (L. Wattenberg, personal communication).

RESPIRATORY INFECTION AND PULMONARY CARCINOGENESIS

Experimental data on the possible role of infection in respiratory carcinogenesis are extremely meager. There have been only a few attempts to study the problem, the first by Campbell,[109] who infected mice with influenza virus and later exposed them to dust containing tar. Kotin[447] and Nettesheim *et al.*[560] tested the effect of influenza viral infection on the incidence of lung tumors in mice chronically exposed to ozonized gasoline. Leuchtenberger and Leuchtenberger[491] and Harris and Negroni[344] infected mice with influenza virus and exposed them to cigarette smoke. The effect of influenza viral infection on the incidence of spontaneous lung tumors in mice was studied by Steiner and Loosli,[723] and Imagawa *et al.*[410] explored the effects of influenza virus on urethan-induced lung tumor formation in mice. The results obtained in these studies do not permit any definite conclusions. Except in the experiments of Nettesheim *et al.*, no measures were taken to exclude spontaneous respiratory infections, and it is well known that conventional mice are heavily infected with a number of respiratory agents. The data reported by the various investigators are contradictory. Campbell found a reduction in the incidence of tar-induced lung tumors after viral infection, but the number of mice included in the study was small; Nettesheim *et al.* found that influenza virus decreased the incidence of gasoline-fume-induced lung tumors; and Steiner and Loosli found that the virus decreased the incidence of spontaneous lung tumors. In contrast, Kotin and Imagawa *et al.* found that viral infection increased the incidence of lung tumors induced by gasoline fumes and urethan, respectively. The findings of the Leuchtenbergers, using tobacco smoke, and Harris and Negroni, using tobacco smoke and benzo[a]pyrene, are ambiguous. The mouse lung-tumor system is less than ideal, in that, with few exceptions (e.g., Kotin *et al.*), squamous hyperplasia and alveologenic adenomas were produced in the mice, and no laboratory has yet developed a method that will consistently produce tumors topographically and morphologically similar to those most often seen in man.

However, the inconclusiveness of the available data can at this time be regarded only as a consequence of insufficient experimental efforts. In fact, the seemingly contradictory results of Kotin and Nettesheim *et al.* suggest that respiratory agents might play an important role in the pathogenesis of lung cancer. The designs of the experiments of Nettesheim *et al.* and Kotin (with ozonized gasoline fumes) were almost identical, with two exceptions: Kotin used repeated viral infection, whereas the animals of Nettesheim *et al.* were infected only once. In the study of Kotin, control animals, as well as animals exposed only to ozonized gasoline, showed a high incidence of pneumonitis, and it is not unreasonable to assume that the mice used for the adaptation of the three types of viruses used in the study were also contaminated with various unidentified respiratory agents. Viral and bacterial agents isolated from commercially raised mice and rats have caused not only chronic bronchial pneumonia, but also extensive squamous metaplasia of bronchiolar and alveolar epithelium. The mice used in the studies of Nettesheim *et al.*, however, were derived from a germfree colony and were kept free of pathogens throughout the experiment. No squamous metaplasia or squamous cell tumors developed, and the incidence of pulmonary adenomas and adenocarcinomas was reduced by influenza viral infection. It is therefore conceivable that the squamous cell tumors observed by Kotin *et al.* after viral infection and smog exposure developed from chronic pulmonary lesions caused by adventitious microbial agents.

A cocarcinogenic effect of respiratory infection appears to be an attractive hypothesis. The respiratory system has a number of very efficient protective devices: a protective layer of mucus, protective layers of nonproliferating superficial cells, mucociliary clearance, and alveolar clearance by phagocytes. In addition, the tracheobronchial tree and the alveoli are lined with an epithelium with a low proliferative rate, and lymphatic tissue is closely associated with various parts of the respiratory tract. Acute and chronic respiratory infections disturb the mucus production, the integrity of superficial cell layers, the ciliary action, the deep-lung clearance by pulmonary macrophages, the normal regeneration and differentiation of epithelial cells, and the local immunologic surveillance mechanisms. Respiratory infection could result in easier penetration of the carcinogen to susceptible basal cell layers (disruption of mucous blanket and the superficial cell layers) and in protracted residence and accumulation of carcinogenic particles in various parts of the respiratory tract (perturbation of the mucociliary clearance mechanism and the deep-lung clearance). Cell necrosis and vigorous subsequent regeneration, often with disturbed cellular differentiation (metaplasia), after respiratory infection could render the target tissues more susceptible to malignant transformation. Increase in cell proliferation has been shown to be an effective "promoter" in a number of tumor systems.

Respiratory infections may suppress local and systemic immunocompetence, and thus the immunologic surveillance that would normally suppress the growth of malignant clones could be rendered nonfunctional (antigen competition). This hypothesis is supported by some recent experiments in which the incidence of either spontaneous or chemically induced tumors was compared in germfree and conventional animals. These experiments suggest that both the type and the number of tumors developing in experimental animals are affected by the bacteriologic status of the animals. The tumor systems used in these kinds of studies are the mouse lung tumor[103] and liver tumor,[634] leukemia,[799] and myeloma.[527] Although such data are open to different interpretations, one reasonable explanation is that it is the difference in immunologic status of the germfree animal that is responsible for the lower tumor response.

In summary, because of the lack of sufficiently controlled experimentation, no conclusive experimental evidence is available to support or refute the hypothesis of a cocarcinogenic effect of respiratory infection. However, because respiratory infections are detrimental to a number of local and systemic defense systems and have a profound effect on cell proliferation and differentiation, the hypothesis of cocarcinogenicity of respiratory infections is very attractive and needs extensive study.

INTERACTION BETWEEN PHYSICAL AND CHEMICAL CARCINOGENESIS

Of the several physical factors that have been associated with the carcinogenic process *in vivo*, ionizing radiation is the only one on which there is a significant body of experimental data. Radiation alone is a potent carcinogen, capable of inducing tumors in nearly all tissues of most species.[121,283] Because radiation produces chemical alterations in cellular DNA that may lead to chromosome damage and mutational changes, it has been thought to initiate the carcinogenic process. The interactions between ionizing radiation and chemical carcinogens have been studied primarily in three tissues—skin, mammary gland, and lung.

A cocarcinogenic effect in skin has been observed when irradiation has been preceded or followed by topical applications of various chemical agents, including 3-methylcholanthrene, benzo[a]pyrene, dibenzanthracene, croton oil, and cigarette-smoke condensate. In the case of the first three chemicals, there is no convincing evidence that the effect is more than additive. In the induction of skin tumors by locally acting carcinogens, however, systemic factors may play an important role. In one investigation,[68] for example, the carcinogenic effect of topically applied cigarette-smoke condensate was considerably increased by local irradiation of the skin at points some distance from the site of exposure to the chemical.

Shellabarger[697] has investigated the effects of systemically administered 3-methylcholanthrene and external x irradiation, singly or in combination, on mammary carcinogenesis in rats. The agents were given 10 days apart and in either order. Both 3-methylcholanthrene and x irradiation given singly led to a significant incidence of tumors, but the effect of combined administration was clearly additive. A similar effect of combined treatment had been observed in earlier studies on the leukemogenic action of these agents in mice.[283]

The interactions of radiation and chemical agents in respiratory carcinogenesis are particularly important in air pollution control, considering the possibility of simultaneous human exposure by inhalation to radioactive particles and chemical agents. A number of problems have been encountered in attempts to produce pulmonary cancer in experimental animals by ionizing radiation alone, and the induction of tumors by inhalation exposure has proved to be especially difficult; all these studies have recently been reviewed by Bair.[27] Many of the successful experiments have involved the implantation of radioactive wires or pellets into the lung parenchyma to deliver an intense local dose of radiation (about 10^4–10^6 rads) or the use of microcurie amounts of radioactivity sufficient to deliver high doses of radiation to large volumes of lung tissue. Although these studies have generally shown a clear relation between radiation dose and tumor incidence, the results have been complicated by the surgical trauma to the lung or, more important, when large lung volumes were exposed, by the direct radiation injury of functional lung tissue (including inflammatory changes, necrosis, and fibrosis) caused by the high radiation doses.

In three recent studies, significant numbers of lung cancers were produced with relatively low doses, in the range to which man might be exposed. Gross *et al.*[318] produced lung cancers in 43% of rats and 2% of hamsters exposed to 3,000 or 4,000 rads of external chest x irradiation. Little *et al.*,[500] using 15 weekly intratracheal injections of the alpha-emitting radionuclide polonium-210 adsorbed onto hematite particles, produced lung cancers in 48% of hamsters that received lifetime doses of 225 rads to the whole lungs. Yuile and co-workers[834] found primary lung tumors in 3–13% of rats that received whole-lung doses of 71–538 rads after a single inhalation exposure to polonium-210 in a sodium chloride aerosol. A continuing study of the importance of various physical and biologic factors in lung carcinogenesis from inhaled radioactivity is being carried out by Bair and his colleagues in several species of animals.[27,658] On the basis of their results and the other available experimental data, these workers concluded that, for deposited particles, alpha radiation is more carcinogenic than beta or gamma radiation, that nonuniform irradiation of the lung by radioactive particles is more carcinogenic than external irradiation, and that the local radiation doses required for a substantial tumor incidence must be very high near the radioactive particle or source. Although these conclusions appear generally valid for the high-incidence portion of the dose–response curve, the relative importance of "hot-spot" or point-source irradiation of relatively small tissue volumes—compared with a lower dose more uniformly distributed to all critical cells in the lung—is not entirely clear at present (the results of Gross *et al.*[318] are a case in point). The available data suggest, however, that, in the case of human lung exposure, the most hazardous source of radiation exposure consists of highly active alpha-emitting particles small enough to be inhaled and deposited in the lower respiratory tract.

As far as the interactions of radiation and chemical agents in respiratory carcinogenesis are concerned, the experimental data from animals are limited to two investigations. Gross and co-workers[318] exposed rats and hamsters to 3,000 or 4,000 rads of external radiation to the chest and, beginning 8 weeks later, a series of intratracheal injections of 7,12-dimethylbenz[a]anthracene. They found that the lung cancer prevalence in irradiated animals that received injections of 7,12-dimethylbenz[a]anthracene was no higher than in irradiated animals that did not receive injections. Although the injected dose when administered by itself gave rise to no tumors in the rats, it exhibited a 6% tumor prevalence in the hamsters. The authors suggest that the failure of 7,12-dimethylbenz[a]anthracene to increase the yield of radiation-induced tumors substantially may be because each agent exerts its primary carcinogenic action on a different tissue within the lungs. A similar conclusion may be drawn from the studies by Little *et al.*,[500] who administered either polonium-210 or benzo[a]pyrene adsorbed onto hematite carrier particles to hamsters by intratracheal injection. Either agent given in sufficient dosages led to a high incidence of bronchogenic cancer, but the benzo[a]pyrene-induced tumors varied in histologic type and arose primarily from the trachea and large bronchi, whereas the radiation-induced tumors were mostly peripheral and were all mixed adenocarcinomas and squamous cell carcinomas. In other experiments (B. N. Grossman, J. B. Little, and W. F. O'Toole, unpublished observations), low doses of the two carcinogens adsorbed onto the same carrier particles have been administered simultaneously (Table 9-1). Although the data are limited, the effects of the two agents administered together appear to be additive, rather than synergistic.

Thus, although these two investigations do not show a synergistic effect between radiation and chemical agents in lung cancer, the lack of an effect may be due to technical factors, such as dosage of the specific agents, or to the differences between the target tissues for the carcinogens within the lung. Anatomically, the origin of the carcinogenic process initiated by external radiation 8 weeks before may not be exposed to intratracheally injected 7,12-dimethylbenz[a]-anthracene. However, owing to their differing chemical natures and solubilities, the deposition and clearance patterns of benzo[a]pyrene and polonium-210 after combined intratracheal administration in hamsters may be very different, and the two agents may thus act at differing sites in the lung. Another unknown variable is the impor-

TABLE 9-1 Comparison of Two Carcinogens Administered Singly and Simultaneously

Carcinogen	No. Hamsters Developing Tumors	Time of Appearance of First Tumor	Tumor Incidence (After First Tumor), %
Polonium-210	30	40th week	48
Benzo[a]pyrene	26	64th week	9
Polonium-210 + benzo[a]pyrene	24	38th week	73

tance of the chronic inflammatory process often associated with high doses of radiation to the lung. Further experimental work is needed for the relation between physical and chemical carcinogenesis in the lung to be established.

A final piece of evidence bearing on this subject comes from the epidemiologic studies of lung cancer among the Colorado Plateau uranium miners exposed by inhalation to alpha radiation in the air of the mines.[27,265] The data indicate that the very high incidence of lung cancer in these miners, which is related to the degree of exposure to radioactivity in the mines, is associated primarily with the group who are also cigarette smokers. Only two cancers have occurred thus far among nonsmoking miners in whom 11 would be expected if cigarette smoking played no role in the excess cancer deaths; the excess respiratory cancer deaths per 10,000 person-years of observation were 10 times greater for cigarette smokers than for nonsmokers. Furthermore, among the group of 761 underground uranium miners who are American Indians (most of whom do not smoke), the incidence of lung cancer was not significantly increased. These observations suggest a synergistic effect in man between exposure to alpha radiation from inhaled radioactive particles and components of cigarette smoke.

In summary, it has been found that ionizing radiation and polycyclic aromatic hydrocarbons combined produce an additive carcinogenic effect at various sites. It has been difficult to induce pulmonary tumors with irradiation because of difficulties in delivering the radiation to the lungs. Progress has been made by the use of polonium-210 adsorbed on hematite particles, and a combination of this and benzo[a]pyrene (also adsorbed on hematite) delivered by intratracheal instillation produced an additive carcinogenic effect.

10

Distribution, Excretion, and Metabolism of Polycyclic Hydrocarbons

DISTRIBUTION AND EXCRETION OF POLYCYCLIC HYDROCARBONS

Before the use of isotopically labeled carcinogenic hydrocarbons, fluorescence was applied to the study of the tissue distribution and excretion of these compounds. One of the earliest such studies was that of Peacock,[588] who injected colloidal suspensions of anthracene, dibenz[a,h]anthracene, and benzo[a]pyrene intravenously into fowls and rabbits and found that fluorescent material was rapidly cleared from the blood and excreted into the bile. Some unidentified fluorescent material was also seen in the livers. Doniach *et al.*[212] found that, after inoculation of benzo[a]pyrene into mice and rabbits, the fluorescence of the kidneys, lungs, and liver changed from the characteristic violet of the hydrocarbon to blue. Blue fluorescence also persisted for several weeks in the skin after a topical application of the benzo[a]pyrene. Chemical characterization of the blue-fluorescing hydrocarbons was unsuccessful.

The distribution of radioactivity in mice after administration of the first sample of dibenz[a,h]anthracene that was ever labeled with carbon-14 was reported in 1948 by Heidelberger and Jones.[354] They measured the radioactivity in a number of tissues after intravenous injection of a colloidal suspension of the hydrocarbon and found that the material was first concentrated in the liver and then excreted via the bile and intestinal tract into the feces. Similar studies were done with other routes of administration, and some radioactivity was detected in subcutaneous sarcomas induced in mice by the labeled carcinogen. Heidelberger *et al.*[355] carried out various fractionations of the tissues and excreta from the mice they studied.[354] Relatively small amounts of unchanged hydrocarbon were found in tissues, along with larger quantities of acidic and phenolic compounds, which undoubtedly represented detoxication products.[355] Heidelberger and Weiss[356] studied the rates of disappearance of radioactivity from the sites of subcutaneous injection in tricaprylin of three labeled hydrocarbons; the disappearance rates increased in this order: dibenz[a,h]anthracene (half-life, 12 weeks), methylcholanthrene (half-life, 3½ weeks), and benzo[a]pyrene (half-life, 1¾ weeks). The relative carcinogenic activities of these three compounds at low doses injected subcutaneously were directly proportional to the durations of their retention at the site.[356]

Kotin *et al.*[452] studied the elimination of [^{14}C]benzo[a]pyrene from rats and mice and confirmed the biliary excretion seen in the earlier cases. Only radioactivity was measured and no attempt was made to fractionate the tissues. The solvent used for subcutaneous injection was found to affect the retention of the radioactivity at the site. Similar excretory patterns were demonstrated after intratracheal instillation in rats.[452]

The biliary excretion in rats of hydroxylated derivatives of benzo[a]pyrene was decreased after administration of piperonylbutoxide, a methylenedioxyphenyl derivative that is a potent inhibitor of microsomal enzyme mixed-function oxidases.[257]

The distribution of radioactivity in rats after administration by stomach tube of dibenz[a,h]anthracene, methylcholanthrene, and 7,12-dimethylbenz[a]anthracene was comprehensively studied by Daniel *et al.*[179] They also found biliary excretion into the feces and a rather prolonged retention of the radioactivity in body fat, ovaries, and adrenals. Shabad has reviewed in English some of the spectrofluorescence research in the Soviet Union on the distribution of carcinogenic hydrocarbons; elimination from the skin and elimination after intratracheal instillation have been measured.[693] Dontenwill *et al.*[213] have carried out a thorough study of the elimination of benzo[a]pyrene after intratracheal instillation into Syrian hamsters. The rate of resorption from the lungs was measured when the compound was dissolved in solvents or adsorbed on hematite.

It is evident that no definitive study on the metabolism, tissue distribution, and excretion of carcinogenic hydrocarbons has yet been carried out. Measurement only of radioactivity of a compound yields no information on the chemical nature of the radioactive material. Fractionation of the radioactive material has been carried out, but no characterization of the compounds has been achieved. Studies aimed at the complete characterization of metabolites and excretion products, including tissue distribution and binding to macromolecules, have not yet been attempted but should be carried out.

METABOLISM OF POLYCYCLIC AROMATIC HYDROCARBONS

The metabolism of carcinogenic and noncarcinogenic hydrocarbons has been investigated for a number of years in intact animals, in various types of cells in culture, and in cell-free systems. The original studies were done with fluorescence spectra as the primary analytic tool, but recently the use of radioactive labeled hydrocarbons has provided much more precise quantitative data, particularly when combined with thin-layer chromatography. It has been known for a long time that simple aromatic compounds are metabolically hydroxylated to phenols, which are then conjugated with glucuronic acid or sulfate to yield water-soluble, easily excreted products.

Although mouse skin is the tissue most commonly used in testing for carcinogenic activity, little work has been done on the metabolism of hydrocarbons in mouse skin. Heidelberger *et al.*[352] studied the metabolic conversions of dibenz[a,h]anthracene injected into mice and identified several quinonoid metabolites. Boyland and Sims[83] investigated the metabolism of benz[a]anthracene given intraperitoneally to rats, mice, and rabbits; they identified several phenols and dihydrodiols and their conjugates. Later, they studied the metabolism of a number of polycyclic hydrocarbons in rat liver slices and homogenates. Although these carcinogens do not normally induce tumors in liver, that organ is convenient for metabolic studies. The liver microsomal system will be considered in detail below.

Several more recent studies have been carried out with cells in culture. Diamond *et al.*[199] and Duncan *et al.*[220] have measured the conversion of polycyclic aromatic compounds to unidentified water-soluble metabolites in various types of cells. Sims[707] has characterized a number of metabolites produced from several hydrocarbons in mouse embryo cells and has found them to be similar to those previously identified in liver homogenates. In no case, however, have all the metabolites of any hydrocarbon yet been identified.

An example of what is known about the metabolism of an important carcinogenic hydrocarbon is shown in Figure 10-1, derived largely from the work of Sims.[706] Benzo[a]pyrene gives rise to the following metabolites: the monophenols (3- and 6-hydroxy-), the

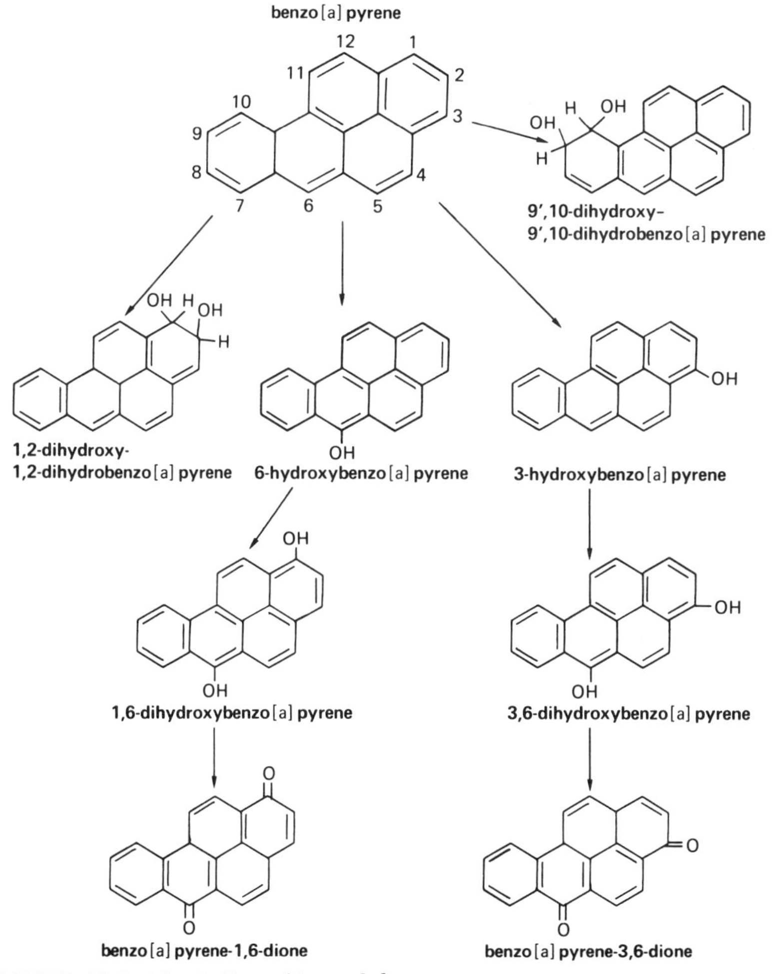

FIGURE 10-1 Metabolism of benzo[a]pyrene.

diphenols (3,6- and 1,6-dihydroxy-), the corresponding quinones (3,6-dione and 1,6-dione), and two dihydrodiols (1,2- and 9′,10-). None of these metabolites is carcinogenic. A number of unidentified conjugated products were also present. It is of interest that the so-called K regions (4,5 and 11,12 bonds), which the Pullmans have postulated to be the critically reactive sites[610] and are most reactive toward some chemical reagents, appear to be metabolically inert; Sims[707] found no evidence of the presence of either the 4,5- or the 11,12-dihydrodiol. The details of the metabolism of other polycyclic hydrocarbons are beyond the scope of this document.

In the case of almost every chemical carcinogen that has been properly examined, there is evidence of covalent binding of some derivative of the carcinogen to DNA, RNA, and protein. This work has been thoroughly reviewed by Miller.[532] In the case of polycyclic carcinogens, the covalent binding to mouse skin DNA and RNA has been studied by Brookes and Lawley,[93] Goshman and Heidelberger,[307] and Brookes and Heidelberger.[92] In contrast with the aromatic amines, the chemical nature of whose binding to nucleic acids and proteins Miller has established,[532] the chemistry of binding of hydrocarbons to nucleic acid is not yet known, but it is under active investigation. The binding of aromatic hydrocarbons to proteins has been studied most extensively by Abell and Heidelberger[4] and by Tasseron *et al.*[744] These authors have identified and partially purified a soluble protein fraction from mouse skin to which the polycyclic hydrocarbons are bound in direct proportion to their carcinogenic activities. The function of this protein fraction is being investigated. It is not now known which, if any, of these macromolecules is the primary cellular target of carcinogenic action.

A number of carcinogens are chemically reactive, including alkylating agents, aliphatic epoxides, lactones, and methylnitrosoureas. These compounds react directly with nucleophilic sites in DNA, RNA, proteins, and probably other molecules; the chemistry of the interactions has been reviewed by Miller.[532] By contrast, many of the more important carcinogens, including polycyclic aromatic hydrocarbons, are not chemically reactive. Therefore, it seems highly probable that they must be converted metabolically into chemically reactive compounds that are capable of reacting directly with the macromolecules of the cell. This process, referred to as "metabolic activation," is likely to be obligatory if polycyclic hydrocarbons are to induce cancer.

Three types of metabolic intermediates have been proposed as the active metabolic form of carcinogenic hydrocarbons. Boyland and

Sims in 1964[83] proposed that epoxides are formed from hydrocarbons and are the reactive form. It is known that dihydrodiols, such as have been identified as benzo[a]pyrene metabolites (Figure 10-1), can be formed from epoxides. Thus, it seems likely that epoxides can be visualized as the metabolic precursors of dihydrodiols and probably also of phenols. Very recently, Grover and Sims[321] have shown that the K-region epoxides of phenanthrene and dibenz[a,h]anthracene can react covalently in the test tube with DNA and histones. This, plus other evidence that is accumulating–largely from work in cell-free systems–suggests strongly that epoxides are the activated form of carcinogenic hydrocarbons.

Dipple *et al.*[203] have proposed, alternatively, that the active form is a cation resulting from attack on an electronically delocalized part of the molecule, whereas Wilk and Girke[818] suggest that the active metabolite may be a radical cation. Clearly, more work is needed to establish firmly the nature of the ultimate carcinogenic form of polycyclic hydrocarbons.

This question is of more than purely academic interest. It is now almost generally accepted that polycyclic aromatic hydrocarbons must be metabolically activated if they are to induce cancer. Hence, surveys of human tissues for their ability to carry out this reaction may be of practical value in assessing the contribution of this class of compound, found so extensively in air pollution, to human cancer. The enzymic nature of this conversion has been studied most fruitfully in the rat liver microsomal system. It has been possible to alter the *in vivo* metabolism and carcinogenicity of polycyclic aromatic hydrocarbons by treating the animals with various agents. The alteration is most likely due to effects on the microsomal enzyme system to be discussed below.

An interesting effect of metabolism on carcinogenicity and toxicity was discovered by Boyland *et al.*[85] It had been known that 7,12-dimethylbenz[a]anthracene, in addition to being highly carcinogenic, also produced severe adrenal necrosis. It had also been known that it was metabolized to two hydroxymethyl derivatives: 7-hydroxymethyl-12-methylbenz[a]anthracene and 12-hydroxymethyl-7-methylbenz[a]anthracene. Although the former derivative was much less carcinogenic than 7,12-dimethylbenz[a]anthracene, it produced adrenal necrosis at a much lower dose; the latter derivative was inactive in both respects.

It has been demonstrated that K-region epoxides of several polycyclic hydrocarbons are much more active than the parent hydrocarbon and the corresponding dihydrodiols and phenols in producing malignant transformation *in vitro* in hamster embryonic and mouse prostate cells.[322] This constitutes strong evidence that epoxides are the metabolically activated ultimate carcinogenic form of the polycyclic hydrocarbons. It has also been suggested, on the basis of nuclear magnetic resonance considerations, that activation of benzo-[a]pyrene could theoretically occur at carbon-6.[124]

ARYL HYDROCARBON HYDROXYLASE: AN ENZYME SYSTEM

The primary enzyme system responsible for the metabolism of polycyclic hydrocarbons is a multicomponent complex generally localized in the microsomal fraction of the cell.[157,292,297] This enzyme system has been called "benzpyrene hydroxylase" but is more suitably named "aryl hydrocarbon hydroxylase," in that it metabolizes a variety of polycyclic hydrocarbons. It also metabolizes steroids and chemicals of exogenous origin, such as drugs, pesticides, and preservatives.

The enzyme system contains reduced nicotinamide adenine dinucleotide phosphate–cytochrome C reductase, cytochrome B5, cytochrome P_{450}, cytochrome P_{450} reductase, and other unknown components. It converts polycyclic hydrocarbons to epoxides,[416] phenols, dihydrodiols, and quinones.[84,706] Some phenols and dihydrodiols are conjugated to yield glucuronides or sulfates by conjugating enzymes. Little is known about the mechanism of enzyme action. The enzyme system is found in about 90% of the tissues of the monkey, mouse, hamster, and rat.

An important feature of the system is its inducibility. Enzyme level depends on exposure to polycyclic hydrocarbons, a variety of such drugs as phenobarbital, various steroids, flavones, and nutritional and hormonal conditions.[157,292,297,557] Polycyclic hydrocarbons induce the hydroxylase in the liver, lung, gastrointestinal tract, kidney, and skin. The level and inducibility are genetically determined and vary in different strains of mice, the hepatic enzyme being inducible in the Swiss, C57, C3H, and A strains, but not in AKR/N or DBA strains (according to Nebert and Gelboin[557] and S. H. Yuspa *et al.*, unpublished data). The enzyme system is also induced transplacentally in hamsters and rats: Exposure of the mother to polycyclic hydrocarbons increases enzyme concentration in fetal tissues and placenta. There is a high correlation between the placental content and the smoking habits of pregnant women.[559,813] The enzyme is also present and inducible in cells grown in culture derived from whole embryos of hamster, rat, and mouse; in cells from individual hamster fetal tissues, such as liver, lung, small intestine, and kidney; and in several cell lines, such as mouse 3T3, HeLa cells, and mouse epidermal cells. The enzyme is absent or very sparse in most established cell lines. The mechanism of induction and various characteristics of the enzyme system—such as cofactor requirements, kinetic behavior, and half-life—have been studied.[555,556,558]

The microsomal enzyme catalyzes reactions *in vitro* that cause a covalent binding between benzopyrene and DNA or protein.[291,320] In cell culture, the enzyme converts polycyclic hydrocarbons to toxic products.[293] The level of enzyme activity correlates positively with the susceptibility of the cells to the cytotoxicity of benzo[a]pyrene, suggesting that the enzyme converts benzo[a]pyrene to cytotoxic metabolites. One of the metabolites of benzo[a]pyrene, 3-hydroxybenzo[a]pyrene, is cytotoxic to cells that are either susceptible or resistant to benzo[a]pyrene.[293] A compound, 7,8-benzoflavone, that inhibits the metabolism of benzo[a]pyrene and 7,12-dimethylbenz[a]anthracene in hamster embryo cell cultures, liver microsomes, and skin homogenates also protects the cells against the toxic effects of these carcinogens.[198] This inhibitor markedly reduces mouse skin tumorigenesis caused by repeated treatment with 7,12-dimethylbenz[a]anthracene or by a single treatment followed by weekly administration of croton oil.[294,807] These findings indicate that this enzyme system is responsible for the activation of 7,12-dimethylbenz[a]anthracene to its carcinogenic form.[293] These results with the inhibitor of 7,12-dimethylbenz[a]anthracene tumorigenesis may be related to an activation step involving hydroxymethyl formation, rather than ring hydroxylation. Thus, it is possible that inhibition of some types of hydroxylation of the polycyclic hydrocarbons results in an increased tumorigenicity, whereas an increase in other types of hydroxylation, perhaps ring hydroxylation at some positions, may result in reduced tumorigenicity.[294]

Tumorigenesis is markedly influenced by preinduction of the enzyme. For example, Huggins *et al.*[405] found that pretreatment of rats with small amounts of polycyclic hydrocarbons decreases 7,12-dimethylbenz[a]anthracene-induced tumor formation in the mammary gland. Wattenberg and Leong[808] reported similarly that the inducer 5,6-benzoflavone inhibits 7,12-dimethylbenz[a]anthracene-induced tumorigenesis in the lung and mammary gland of rodents. The protective effects of pretreatment with inducers may be due to increased enzyme contents in the target tissue, which may more rapidly eliminate the carcinogen or convert it to a less carcinogenic form. Inducers also cause an increase in enzyme content in the liver, the major site of metabolism, and this may lower the concentration of the carcinogen in the target tissue.

The enzyme plays a central role in polycyclic hydrocarbon carcinogenesis and, because it is found in most tissues and its activity is affected by environmental conditions, the nature of its role in hydrocarbon metabolism and activation needs to be clarified. The enzyme system converts hydrocarbons to products that are largely noncarcinogenic, but in the process of conversion it may be responsible for carcinogen activation. The most difficult and important question to answer is that of the relation between the possible enzymic activation of the carcinogen and its enzymic detoxification to inactive products. The role of related enzymes, such as the epoxide hydrase and the conjugating enzymes, also needs clarification. Although there is evidence that polycyclic hydrocarbons require activation if they are to be active, it is not certain whether the microsomal enzyme system is responsible for the activation or whether all or only some of the carcinogenic hydrocarbons require activation. In addition to inducers of the enzyme, there are compounds, like 7,8-benzoflavone, that inhibit the enzyme.

Thus, it seems that the tools are at hand to examine more closely the role of the enzyme in polycyclic hydrocarbon tumorigenesis. When this role is clearly illustrated, a most important second step will be to clarify the nature of the environmental factors—such as pesticides, drugs, nutritional level, and hormonal states—that enhance the detoxification activity of the enzyme system, reduce the activation of the carcinogen, or both. Recently, an epoxide has been isolated as an intermediate in the microsomal hydroxylation of dibenz[a,h]anthracene.[20,690]

In summary, the metabolism of polycyclic aromatic hydrocarbons has been studied in various tissues of test animals, in cells in culture, and in cell-free systems. The following types of metabolites have been characterized *in vivo:* phenols, dihydrodiols, quinones, and various water-soluble conjugated products. This metabolism is carried out primarily by the drug-metabolizing enzyme system of the microsomes. This complex system is found in many tissues of many species and is inducible by polycyclic hydrocarbons and other types of compounds, such as pesticides and drugs. The enzyme system can also be inhibited by several compounds. It can increase or decrease the toxicity of hydrocarbons and is probably responsible for their metabolic activation to chemically reactive carcinogens.[293] Assay of this enzyme system in various human cells and tissues could provide valuable information on their susceptibility or resistance to hydrocarbon carcinogenesis. There is a need to determine the complete metabolic profile of several carcinogens in various tissues. There is also a great need for a reliable source of purified metabolites and potentially reactive intermediates, as well as specific inducers and inhibitors of the enzyme system.

11

In vitro Approaches to Carcinogenesis

CHEMICAL CARCINOGENESIS

It has been known for many years that normal cells in culture can be transformed into cancer cells by a number of DNA and RNA oncogenic viruses. The first report of chemical transformation *in vitro* was by Berwald and Sachs in 1963.[52] They described the production with polycyclic hydrocarbons of characteristic morphologic alterations in the growth patterns of hamster embryonic fibroblasts in culture. This alteration involved a disorientation of the arrangement of the cells and the formation of criss-crossed piled-up colonies. The hydrocarbons were more toxic to the normal fibroblasts than to transformed cells. Berwald and Sachs later demonstrated that mass cultures of the morphologically transformed cells gave rise to tumors in hamster cheek pouches, whereas the control hamster cells did not.[53] Quantitative cloning techniques were developed, and it was found that there was a proportionality between the number of transformed clones produced by a given hydrocarbon and its carcinogenic activity. Huberman and Sachs,[398] using benzo[a]-pyrene, developed quantitative dose–response curves, which were interpreted as indicating that transformation was a one-hit process and was separate from the toxicity exerted by the compounds.

DiPaolo *et al.*[201] confirmed the results of the Sachs group and also demonstrated that individual clones of transformed cells could give rise to tumors (all fibrosarcomas) in hamsters.[202] Kuroki and Sato[467] have obtained malignant transformation in cultures of embryonic hamster fibroblasts with 4-nitroquinoline-*N*-oxide and its derivatives.

It is well known[659] that embryonic mouse fibroblasts readily undergo "spontaneous" malignant transformation in culture, as Berwald and Sachs also found. However, spontaneous transformation in the embryonic hamster fibroblasts was not observed. In spite of this, it seemed desirable for various reasons to try to develop another system for chemical carcinogenesis *in vitro.* Lasnitzki[475] found that polycyclic hydrocarbons produced profound histologic alterations in organ cultures of mouse ventral prostate. Röller and Heidelberger[638] obtained even more striking morphologic changes suggestive of malignancy; but, on inoculation of these altered pieces of prostate into isologous mice in a variety of conditions, no tumors were produced. Thus, carcinogenesis *in vitro* had not been accomplished.

Although the organ cultures of mouse ventral prostate treated with carcinogenic hydrocarbons were not malignant, Heidelberger and Iype[353] succeeded in obtaining from them lines of cells that grew in a disoriented fashion and produced tumors on inoculation into isologous mice. This indication of successful carcinogenesis *in vitro* led to an effort to cultivate the mouse prostate cells directly in cell culture. Chen and Heidelberger[131] succeeded in obtaining a number of cell lines from dispersed untreated organ cultures of adult C3H mouse ventral prostates. Unlike most other mouse cell lines, these only rarely underwent spontaneous transformation. These cells grew exponentially until they reached a monolayer, at which point no further growth occurred. When 10^7 of these control cells were injected subcutaneously into irradiated C3H mice, no tumors were obtained. When these cells were treated in culture with 3-methylcholanthrene, their growth rate increased, they did not reach a saturation density, they piled up in a disordered array, and they produced a 100% incidence of fibrosarcomas on subcutaneous injection of 1,000 cells into nonirradiated C3H mice.[132]

These results led to the development of a quantitative system with the prostate cells in which transformation and toxicity were both measured in sparsely plated cells; the cells gave rise to individual colonies that coalesced into a monolayer and then formed individual piled-up colonies that were very easy to score.[133] It was shown that each piled-up colony was capable of yielding tumors in mice, thus fully justifying the scoring of these morphologically transformed colonies as truly malignant. With this quantitative system, it was shown that there was an excellent correlation between the number of piled-up, malignant, transformed colonies and the carcinogenic activities of nine polycyclic hydrocarbons. Furthermore, there was no dose–response relation between the processes of transformation and toxicity. It appears that systems like these might be used to screen for potentially carcinogenic activities in fractions of all sorts obtained from air pollution.[133]

This system of carcinogenesis *in vitro* with hydrocarbons has already furnished considerable information on fundamental cellular mechanisms of hydrocarbon carcinogenesis. As mentioned before, it has been possible to produce 100% transformation of clones derived from single prostate cells, thereby proving (at least in this system) that the carcinogen directly transforms nonmalignant into malignant cells and eliminating the possibility that the hydrocarbon selects for pre-existing malignant cells.[537] Such experiments would not be possible in whole animals.

It has been demonstrated by Mondal *et al.*[538] that individual transformed clones obtained from the same dishes have surface transplantation antigens that are non-cross-reactive, as is the case with hydrocarbon-induced sarcomas *in vivo.* This provides added confidence that this system is a valid model for polycyclic hydrocarbon carcinogenesis.

In vitro systems have been used productively by Inbar and Sachs[411] to detect differences in the surface properties of normal and transformed cells by the use of Concanavalin A—a protein that binds to glycoproteins on cell surfaces. When it is used quantitatively during the process of carcinogenesis *in vitro* in embryonic hamster embryo cells, there is an alteration in the topology of the membrane such that masked binding sites are released.

It has been found in many of the studies alluded to here that polycyclic hydrocarbons exert more toxicity to normal than to transformed cells.[52,53,133,353,398] Diamond has also shown that this is true in rodent cells, whether transformed by chemicals or by viruses.[197] Although it does not seem to be true in the case of primate cells in cell culture,[197] monkey and human respiratory epithelium and mucosal connective tissues in organ culture are as sensitive as rodent tissues to toxicity and induction of metaplasia by polycyclic aromatic hydrocarbons.[167]

The embryonic fibroblast and the prostate cell systems for *in vitro* chemical carcinogenesis have interesting differences and similarities and complement one another. Inasmuch as both systems are transformed by carcinogenic hydrocarbons, they must contain the enzymes to activate the hydrocarbons. Therefore, intensive studies are now under way in several laboratories to study the metabolism of hydrocarbons and their binding to macromolecules in direct relation to the process of malignant transformation. It is clear that these systems will be used very fruitfully in the future for the investigation of many theoretical and practical problems in chemical carcinogenesis that cannot be studied in whole animals.

ORGAN CULTURE

Easty[225] describes the organ culture approach to studying carcinogenesis, its applications thus far in viral and chemical carcinogenesis, and its uses in studying cell differentiation and organogenesis. Other reviews of the organ culture system in a variety of applications are available in Willmer's three-volume text.[820] Moscona, Trowell, and Willmer,[547] Grobstein,[315] Wolff,[826] Lasnitzki,[477] Fell and Rinaldini,[266] Bang,[31] and Rapp and Melnick[623] discuss special applications of organ and cell cultures.

Organ culture is a technique for maintaining organs or organ pieces *in vitro* so as to retain normal histologic associations among cell types and to preserve cell differentiation and growth at rates resembling those *in vivo.* Adult organ pieces can be maintained for 2-4 weeks. The cells of an organ in culture are regarded as responsive to test materials in the same way as cells of that organ would be *in vivo,* but adequate proof that this is so is not commonly obtained. As a useful reference point, histologic, histochemical, and cytologic alterations in organ culture explants can be compared with the alterations produced *in vivo* after comparable times of exposure to test materials.

Dose-response analysis of the relative biologic activity of test materials can be undertaken in organ culture, with better control of concentration and duration of exposure than *in vivo.* Human tissues can be exposed directly to compounds that could not be safely given to living human subjects, thus permitting assessment of human tissue reactions.

Metabolism of native and environmental chemicals by a target organ can be observed. Metabolic products can be measured in culture medium or in homogenates of explanted tissue. Penetration of substrate and binding of substrate or its metabolites can be recorded by cell type by using autoradiography with radioactive substrates.

Most applications of this system to study the biologic effects of atmospheric pollutants have been made with crude materials or pure compounds already studied in animals. Results in organ culture have demonstrated the parallelism between organ responses *in vivo* and *in vitro*. For example, respiratory[170,480] and prostatic[482,638] tissues undergo epithelial metaplasia, pleomorphism, or devitalization in approximate proportion to the toxic or carcinogenic activity of polycyclic organic air pollutant materials, as established in animals.[186,241,404,633,662] Environmental materials that have been tested in organ culture include cigarette smoke and cigarette-smoke condensates,[476,478,479] benzene-soluble organic materials extracted from particles trapped by air filters, fractions of these extracts, and a number of carcinogenic, weakly carcinogenic, or noncarcinogenic polycyclic aromatic hydrocarbons.[169,204]

Carcinogenic polycyclic aromatic hydrocarbons applied to such target tissues as the mouse prostate[482,638] or rodent airways[170,479,481,574] produce similar epithelial metaplasias *in vivo* and in organ culture. In further examples, neoplastic lesions that followed exposure to benzo[a]pyrene were inhibited by vitamin A in living animals[138,655] and in organ culture,[171] whereas similar inhibition of prostatic lesions produced by methylcholanthrene occurred *in vitro* with vitamin A.[482] This series of studies established the comparability of *in vivo* and *in vitro* responses and demonstrated the value of integration of *in vitro* and *in vivo* methods in defining factors in lung and prostate carcinogenesis.

The value of the organ culture system in studies of polycyclic aromatic air pollutants lies in the production of histologically typical early lesions by direct action of the test agent. Such lesions indicate that the agents act proximately; hence, enzymes or other cellular components necessary for metabolic activation of polycyclic aromatic materials are present in the target tissue *in vitro* and therefore *in vivo*. The toxic hazard of such pollutants can be assessed, and potential carcinogenic hazard can be regarded as worthy of study. The production of lesions permits dose–response estimates of early tissue responses in man and lower animals by criteria of cell physiology and histopathology, in which abnormalities more subtle than cell death can be identified. Moreover, the organ culture system may offer the only direct means of comparing human and animal target tissue responses to toxic or carcinogenic POM.

Evidence of neoplastic transformation of epithelia has not been tested often in organ culture with chemical carcinogens. Failure in tests so far attempted may have resulted from the use of unfavorable methods. In any event, failure has led to reduced confidence in this system for *in vitro* study of epithelial carcinogenesis.[148,256]

The value of *in vitro* production of neoplastic transformation or other cellular changes in detecting environmental health hazards or in setting air quality criteria cannot be stated definitively, but no single method of testing is perfect. Animal tests are time-consuming and require large amounts of material. *In vitro* methods take less time and material and can give evidence of toxicity, carcinogenicity, and mutagenesis, but they may be too sensitive. For the present, *in vivo* and *in vitro* methods need to be compared as to their relative utility.

Cultured cells have been used as "first-line" preliminary screening devices for detecting biologic activity and toxicity of air pollutants.[643,644] The highly developed cell culture methods for identifying neoplastic transformation of fibroblasts[52,53,201,202,353,398,467,537] by carcinogenic chemicals already place this *in vitro* system in a position to detect carcinogenic potential of air pollutants. Animal tissues *in vitro* have been compared with tissues in intact animals in comparable conditions of acute exposure[169] and provide evidence that *in vitro* tests give close short-term parallels with *in vivo* reactions.

RECOMMENDATIONS

Animals and cultures should be compared by use of one exposure system for evaluation of biologic response to atmospheric pollutants. If *in vitro* methods are tested in parallel with animal models, decisions as to usefulness of *in vitro* methods can be made, or improved *in vitro* methods can be developed for use in evaluation of environmental hazards.

Animal studies of air pollutants have provided important data, but carryover to man is lacking. *In vitro* comparisons of human and animal tissues should be conducted to attempt to bridge this gap. Organ cultures may be more useful for this type of comparison than cultured cells, because the organized differentiated tissue resembles the tissue of the intact animal.

12

Indirect Tests for Determining the Potential Carcinogenicity of Polycyclic Aromatic Hydrocarbons

SEBACEOUS GLAND SUPPRESSION

In 1954, W. E. Smith *et al.*[709] reported the efficacy of the sebaceous gland suppression test in determining the carcinogenicity of some petroleum fractions. The disappearance of the sebaceous glands could be correlated directly with the carcinogenicity of these fractions. Pullinger[609] was the first to note that sebaceous glands disappear from mouse skin within a few days after application of some carcinogens. Simpson and co-workers[704,705] found that, if a carcinogen, such as 3-methylcholanthrene, were dissolved in lanolin, it would be noncarcinogenic and would not suppress sebaceous glands. On the basis of these observations, the test was used extensively as a guide to the fractionation of cigarette-smoke condensate by a number of investigators[458, 539,737] and most recently by Chouroulinkov *et al.*[134] as a screening method for various types of cigarette-smoke condensate. A close correlation between the results of long-term carcinogenicity testing and sebaceous gland suppression was reported.[134]

Bock and Mund[71] used whole mounts of mouse skin and found that the sebaceous gland suppression effect was pronounced for polycyclic hydrocarbon carcinogens and was parallel to the carcinogenic activity of this group of compounds. They applied test solution by pipette twice daily for 3 days on the dorsal area of Swiss mice 55–65 days old. The mice were sacrificed 4 days after the last application. Potent carcinogens like 3-methylcholanthrene and 7,12-dimethylbenz[a]anthracene showed the highest suppression index. The noncarcinogens, phenanthrene and 1,9-benzanthrone, were inactive. However, chrysene and benz[a]anthracene also had pronounced suppression effects, even though their carcinogenicity with regard to mouse skin is weak. Furthermore, other carcinogens, such as β-naphthylamine, had no effect. Bock and Mund later extended their studies to 103 compounds[70] and found high levels of sebaceous gland suppressor activity associated with the benz[a]anthracene structure. It was emphasized, however, that sebaceous gland suppression was not always associated with carcinogenic activity. The potent carcinogen, 7,9-dimethylbenz[c]acridine, was demonstrated to be a weak sebaceous gland suppressor, whereas colchicine, which is not a skin carcinogen, was a moderately active suppressor. The subject of sebaceous gland suppression by aromatic hydrocarbons has been extensively reviewed by Bock.[67]

In summary, the sebaceous gland suppression test is not a reliable indicator of carcinogenicity but may have limited use in predicting the carcinogenicity of some groups of compounds, such as substituted benz[a]anthracenes.

PHOTODYNAMIC ASSAY

Photodynamic activity is demonstrated by the immobilization and death of *Paramecium caudatum,* a ciliate, when exposed to otherwise harmless long-wave ultraviolet radiation after incubation with photosensitizing polycyclic compounds in a pure state or in crude organic mixtures.[238,239,246–248] The time required for immobilization of 90% of *P. caudatum* is considered as the end point and reflects concentrations of the photosensitizing agents.

The photodynamic activities of 240 polycyclic aromatic compounds, determined with *P. caudatum,* have been compared with their *in vivo* induction of zoxazolamine hydroxylase activity in rats.[240] A highly significant association was demonstrated between photodynamic and enzyme-inducing activities. A significant statistical association between photodynamic activity and carcinogenicity of polycyclic compounds of wide structural range has been demonstrated, but the photodynamic assay cannot identify a particular polycyclic compound as being carcinogenic or noncarcinogenic.[246]

The photodynamic assay permits differentiation of pollutants from different sources and of various fractions of pollutant extracts derived from any one source.[247,249] Relative photodynamic potency, expressed as apparent micrograms of benzo[a]pyrene per 1000 m^3 of air, bears no relation to atmospheric concentrations of particles, organic compounds, or derived fractions. For the aromatic fraction, which contains nearly all of whatever benzo[a]pyrene is present in the parent organic extract, photodynamic potencies are strongly and positively correlated with benzo[a]pyrene concentrations. The assay has been applied to organic atmospheric pollutants and six fractions thereof from more than 100 different sources in the United States, exemplifying a wide spectrum of urban and rural pollutant characteristics. Pollutants were assayed over a range of 1–100 μg/ml, using benzo[a]pyrene concentrations of 0.001–100 μg/ml as a standard.

Very high photodynamic activity and steep dose–response slopes have been demonstrated in basic fractions of organic particulate pollutants from more than 50 U.S. cities. This is of particular interest in light of the presumptive isolation of dialkylated azaheterocyclic carcinogens from basic fractions (E. Sawicki, personal communication) and the very high photodynamic activities of such carcinogens.[246]

More recently, a composite neutral fraction of organic extracts of particulate atmospheric pollutants has been separated chromatographically into 217 subfractions; nine polycyclic aromatic hydrocarbons and five polycyclic carbonyl compounds were identified in these subfractions. The distribution of photosensitizing polycyclic compounds in these subfractions has been determined with a photodynamic bioassay using *P. caudatum*.[244]

The economy, rapidity, and simplicity of the photodynamic bioassay, which can be conducted on less than 1-mg amounts of organic extracts, are attractive. The data suggest that the bioassay provides a biologic index of potential carcinogenic hazard attributable to polycyclic compounds. Evaluation of this concept, however, demands correlated photodynamic carcinogenic and chemical studies on numerous samples and fractions of organic atmospheric pollutants collected from sources exemplifying a wide epidemiologic spectrum of respiratory tract cancer incidence. Such studies are in progress.

13

Teratogenesis and Mutagenesis

TERATOGENESIS

Teratology is the study of congenital malformations. These are generally defined as structural abnormalities that can be recognized at or shortly after birth and can cause disability or death.[241] Generically, teratology also includes microscopic, biochemical, and functional abnormalities of prenatal origin. The incidence of human congenital malformations is unknown in the absence of a comprehensive national registry; it has been variously estimated at about 3–4% of total live births. Three major categories of human teratogens have so far been identified: viral infections, x irradiation, and such chemicals as mercurials and thalidomide. Although the teratogenicity of various chemicals has been experimentally recognized for several decades, only after the thalidomide disaster of 1962 were legislative requirements for teratogenicity testing established.

Teratogenic effects of chemicals and other agents should, of course, be identified in experimental animals, rather than in human beings after accidental or unrecognized exposure. Test agents should be administered to pregnant animals during active embryonic organogenesis. Shortly before anticipated birth, embryos should be removed by cesarean section and examined. Characteristics to be considered in test and concurrent control animals include the incidences of abnormal litters, of abnormal fetuses per litter, of specific congenital abnormalities, and of fetal mortality; maternal weight gains in pregnancy; and maternal and fetal organ : body weight ratios. Additionally, some pregnant animals should be allowed to give birth in order to identify abnormalities that may be manifest only in the perinatal period.[241,454,456]

Agents to be tested for teratogenic effects and their known metabolites should be administered singly and repeatedly to two or more mammalian species of more than one order, in various nutritional conditions, during active organogenesis, by a variety of routes, and at dosages reflecting possible human exposure. Of interest in this connection is the total lack of data in the available literature on teratogenicity testing by the respiratory route; respiratory exposure is particularly important for pesticide aerosols and vapors, besides being the obvious route for testing air pollutants.

To date, there are no available data on teratogenicity testing of air pollutants by any route. Therefore, community atmospheric pollutants and defined components thereof should be tested in at least two mammalian species by inhalation and by parenteral administration. Test materials should be administered acutely, subacutely, and chronically.

MUTAGENESIS

Mutagenicity Testing: *In vivo* Methods

Recent recognition of genetic hazards due to chemicals has been paralleled by the development of a variety of methods for testing mutagenicity. Submammalian tests—in bacteria, bacteriophage, *Neurospora,* plants, and *Drosophila*—help to elucidate basic mechanisms. However, in view of the wide range of metabolic and biochemical differences between these systems and man, submammalian tests should be used to provide data ancillary to more relevant test systems. Of these, three *in vivo* mammalian tests are practical and sensitive: dominant lethal assay, host-mediated assay, and *in vivo* cytogenetics. Results from such tests may be extrapolated to man with a relatively high degree of confidence.[172,766]

DOMINANT LETHAL ASSAY

Dominant lethal mutants are convenient indicators of major genetic damage that have been used in mammals for measuring effects of x rays[34] and, more recently, chemical mutagens.[35,123,229,235,236,245,295,636] Data on induction of dominant lethal mutants in mammals may be appropriately extrapolated to man, especially inasmuch as most recognizable human mutations are due to dominant autosomal traits.[763] The genetic basis for dominant lethality is the induction of chromosomal damage and rearrangements, such as translocations and aneuploidies, resulting in nonviable zygotes. Evidence of zygote lethality induced in mammals by x rays and by chemical mutagens has been obtained embryologically[368,712,713] and cytogenetically[242,295,647] (A. J. Bateman, personal communication), respectively. Additional evidence of the genetic basis of dominant lethality is derived from the associated induction of sterility and heritable semisterility in F_1 progeny of males exposed to x irradiation[444,712] and to chemical mutagens;[122,237,255] translocations have been cytologically demonstrated in such semisterile lines in mice[123,237,443,708] and in hamsters (K. S. LaVappa and G. Yerganian, personal communication).

The induction of dominant lethal mutations in animals can be assayed with a high degree of sensitivity and practicality after acute, subacute, or chronic administration of test materials, either orally or by any parenteral route, including respiratory.[232] After administration of a drug, male rodents are mated sequentially with groups of untreated females over the duration of the spermatogenic cycle. For mice, the entire duration of spermatogenesis is approximately 42 days, comprising the following stages: spermatogonial mitoses, 6 days; spermatocytes, 14 days; spermatids, 9 days; testicular sperm, 5.5 days; and epididymal sperm, 7.5 days. Thus, matings within 3 weeks after single drug administration represent samplings of sperm exposed during postmeiotic stages, and matings 4-8 weeks later, samplings of sperm exposed during premeiotic and stem cell stages.

The classic form of the dominant lethal assay involves autopsy of females approximately 13 days after timed matings, as determined by vaginal plugs in mice and vaginal cytology in rats and enumeration of corpora lutea and total implants (as evidenced by living fetuses and early and late fetal deaths).

A modified test in mice allows the determination of effects of drugs on pregnancy rates. Corpora lutea counts—which are not only notoriously difficult to carry out but inaccurate in mice—can be omitted. Numbers of total implants in test animals can be related to those in controls, yielding a simple measure of preimplantation losses. With such modified procedures and computerized data handling, large numbers of test agents can be simply and rapidly tested for mutagenic activity.[8]

Dominant lethal mutations are directly measured by enumeration of early fetal deaths and indirectly by preimplantation losses. Results are best expressed as early fetal deaths per pregnant female, rather than the more conventional mutagenic index (i.e., early fetal deaths $\times$ 100 divided by total implants). The latter index can be markedly altered by variation in the number of total implants. Preimplantation losses offer a presumptive index of mutagenic effects, but there is no precise parallelism between preimplantation losses and early fetal deaths; these should be regarded as concomitant and not alternate measures.

HOST-MEDIATED ASSAY

In spite of the universality of the genetic code, some compounds are actively mutagenic in animals but not mutagenic in microorganisms; conversely, some compounds are mutagenic in microorganisms but are detoxified in mammalian systems. The host-mediated assay was developed to determine the influence of *in vivo* mammalian factors in activating or detoxifying chemical mutagens. In this assay, the chemical under test is administered to the mammal, which then receives an injection (by another route) of indicator microorganisms, thus simplifying measurement of mutation frequencies. The microorganisms are later recovered and scored for induction of mutants. Comparison of the mutagenic action of the test agent in the microorganism directly and in the host-mediated assay indicates the influence of host biochemical metabolism in activating or detoxifying the potential mutagen.[284-287] The formation of mutagenic metabolic products from dimethylnitrosamine and cycasin with this procedure has been reported. Although this is an indirect test, it is the only practical method for detecting point mutations *in vivo*.

In vivo CYTOGENETICS: CHROMOSOMAL ABERRATIONS

Experimental animals and man, exposed acutely, subacutely, or chronically to pollutants by any route, can be investigated cytogenetically for structural and numerical chromosomal aberrations. Chinese hamsters are favored in cytogenetic studies, although rats are more commonly used. Cytogenetic effects on metaphase or anaphase preparations of somatic (bone marrow, spleen, and embryo homogenates) or germinal cells can be studied singly or serially after recovery periods. High quality of standardized preparations and the use of coded slides are critical. Distinctions between chromosome and chromatid breaks and between gaps and open breaks are probably less important than hitherto assumed, inasmuch as these effects are generally parallel. Chromosomal aberrations are regarded as indicators of induced genetic instability and correlate well with mutational frequencies.

Data on mutagenicity testing of air pollutants in *in vivo* mammalian systems—the dominant lethal assay, the host-mediated assay, and *in vivo* cytogenetics—are scanty. There are no published data on mutagenicity testing by inhalation. High concentrations of benzo[a]pyrene administered parenterally to male mice induced dominant lethal mutations in F_1 embryos; however, an organic extract of particulate atmospheric pollutants and three derived fractions were not found to be mutagenic.[245] Trimethylphosphate, used as a fuel additive in gasoline at a concentration of approximately 250 mg/gal, was mutagenic in mice after oral or parenteral administration;[3] cumulative effects were also demonstrated. Evaluation of potential human hazards requires data, as yet unavailable, on the concentration of unreacted trimethylphosphate and of any biologically active pyrolysis products in automobile exhaust.

Consideration of potential biologic hazards due to environmental contaminants like air pollutants extends to chronic toxic effects, including mutagenesis, carcinogenesis, and teratogenesis. Such effects may be induced directly by components of air pollutants themselves, or indirectly after interactions between air pollutants and other environmental pollutants, irrespective of route of exposure.

Test systems must be designed to reflect the role of microsomal enzyme function in activation and detoxification and the role of possible interactions between test agents (administered by any route) and between dietary factors and other chemicals, such as unintentional and intentional food additives, drugs, and air pollutants. In testing air pollutants by inhalation, it is necessary to investigate the effects of defined components by themselves and in combination with other defined and undefined components, including their reaction products.

Pollutants must be tested at higher concentrations than those of general human exposure;[766,770] irrespective of route of administration, maximally tolerated dosages are recommended for this purpose as the highest dosage in dose–response studies. Testing at high doses is essential to the attempt to reduce the gross insensitivity imposed on animal tests by the routinely small sample groups—e.g., 50 or so rats or mice per dosage per chemical, compared with the millions of humans at presumptive risk.

Mutagenicity Testing: *In vitro* Methods

A method of detecting point mutations in mammalian somatic cells is to use *in vitro* tissue-culture systems. The potential of using mammalian somatic cells *in vitro* for genetic studies has long been recognized, but substantial progress was not made until improved and simplified techniques for mammalian cells were developed by Puck and associates.[607,608] These methods made possible quantitative analysis of genetic variations in cell populations via the plating technique for mammalian cells.

It was demonstrated[137,423] that gene mutations are induced by treatment of Chinese hamster cells in cultures with alkylating agents. In addition, physical agents, such as x rays and ultraviolet radiation, and other chemical agents, such as carcinogens, have been shown to induce forward and back mutations at several genetic loci in these cells.[90,91,135,136,424] Thus, the *in vitro* cell culture offers a new system for testing the mutagenicity of chemicals in the human environment. The question whether somatic mutation may cause cancer can now be re-examined more critically, because both carcinogenesis and mutagenesis have been shown to occur experimentally in the same target cell system *in vitro*. Furthermore, human somatic cells from normal and neoplastic tissues can also be tested directly.

Chu and co-workers[136] have tested, in Chinese hamster cell cultures, the mutagenicity of a few selected groups of chemical carcinogens and their related compounds and derivatives. Table 13-1 lists the compounds tested so far and their relative carcinogenicity (based on animal studies) and mutagenicity. The genetic marker assayed in the hamster cells was the change from 8-azaguanine sensitivity to resistance. The results obtained thus far indicate that there is a direct relation between the degree of carcinogenicity and mutagenicity and that metabolically activated derivatives of the test compounds often play important roles in mutagenic action. It has recently been demonstrated that epoxides of polycyclic hydrocarbons are much more mutagenic to mammalian cells than are the corresponding hydrocarbons, dihydrodiols, and phenols (C. Heidelberger, personal communication).

TABLE 13-1 Relative Carcinogenicity and Mutagenicity of Selected Compounds[a]

Test Compound	Carcinogenicity[b]	Mutagenicity
Benzo[e]pyrene	–	–
Benzo[a]pyrene	+	–
3-Hydroxybenzo[a]pyrene	–	±
Dibenz[a,c]anthracene	–	–
Dibenz[a,h]anthracene	±	±
7,12-Dimethylbenz[a]anthracene	+++	+++
2-Acetylaminofluorene	+	–
N-Hydroxy-2-acetylaminofluorene	+	–
N-Acetoxy-2-acetylaminofluorene	+++	+++

[a]Derived from Chu *et al.*[136]
[b]Key:
– = not carcinogenic (or mutagenic)
± = uncertain or weakly carcinogenic (or mutagenic)
+ = carcinogenic (or mutagenic)
+++ = strongly carcinogenic (or mutagenic)

Parallel results have been obtained in the induction of mutations with the same series of compounds at the adenine-3 region of *Neurospora* (H. V. Malling, personal communication). Similarly, *N*-acetoxy-2-acetylaminofluorene has been shown to be mutagenic in T4 bacteriophage,[163] transforming DNA in *Bacillus subtilis*[507] and *Escherichia coli*.[551]

Clearly, these results are promising, but more data using more representative compounds and additional genetic loci will be needed before a more definitive conclusion may be drawn. The use of mammalian cells *in vitro* for a combined and coordinated test for chemical mutagenesis and carcinogenesis may be expected to yield significant information on cellular mechanisms of cancer formation. However, data on mutagenesis derived from somatic cells *in vitro* are limited by the present inability to identify the factors involved by conventional genetic techniques.

ASSOCIATIONS BETWEEN MUTAGENICITY AND CARCINOGENICITY

It is now generally accepted that mutagenesis involves a change in the structure of DNA. Whether such a change is essential for chemical carcinogenesis remains unknown. Most chemical substances that react with nucleic acids also react with proteins. That has often made it difficult to identify the significant cellular receptors responsible for the biologic effects of carcinogens. Because of the uncertainty, two general molecular mechanisms of chemical carcinogenesis have been proposed: somatic mutation resulting from the binding of a chemical to DNA and alteration of its structure; and modification of gene expression, which could occur in several ways, including derepression.[596] There is no compelling evidence for or against the nonmutational theory of chemical carcinogenesis, but recent evidence from mammalian cell systems that lends some support to the somatic-mutation theory is considered in the paragraphs that follow.

The somatic-mutation theory of carcinogenesis, which is generally quoted as originating with Boveri,[81] has received intermittent support from various authors up to the present. However, a review of the evidence pertinent to this concept led Burdette[101] to conclude that a general correlation between mutagenicity and carcinogenicity could not be established. The principal objection to the theory was that a number of chemical carcinogens had not been found to be mutagens and well-established mutagens had not been shown to be carcinogens. It was, nevertheless, pointed out that these arguments were not conclusive, inasmuch as some chemicals may be demonstrably mutagenic only after metabolism and may differ in their ability to yield particular types of mutations. Furthermore, the prolonged testing necessary to eliminate the possibility that known mutagens are carcinogenic has not been carried out in many cases.

If any association between mutagenicity and carcinogenicity is sought, it seems desirable that the experimental tests be carried out in the same species, preferably mammals. Various test systems now available for mutagenicity in mammals have been discussed. The ensuing discussions deal primarily with the mutagenicity tests with chemical carcinogens in mammalian systems.

Mutations may be classified into chromosomal alterations, point mutations of nuclear genes, and mutations of extranuclear genes. It has been shown that most, if not all, chemical carcinogens have the capacity to induce mitotic and chromosomal abnormalities.[56] Variations in chromosome number and structure have been reported in several neoplasms induced by carcinogenic hydrocarbons in several species of rodents. Nevertheless, there is no established correlation between karyotypic abnormalities and the initiation of neoplasia. More recently, it has been shown that some carcinogenic hydrocarbons, but not the structurally related noncarcinogenic compounds, can induce chromosome aberrations in mammalian cells both *in vivo* and *in vitro.*[426,427] The latter type of study is promising, but more extensive tests using a larger selection of components would be desirable. It is clear that the cytogenetic effects of environmental chemicals could provide a convenient and valid indication of genetic damage in cells and organisms. But it must be pointed out that many chemicals can induce point mutations but not chromosome mutations.

In 1948, Darlington[180] proposed a plasmagene theory of the origin of cancer. He believed that the cancer determinants that arise in the cytoplasm are due to mutations in either hereditary plasmagenes, infectious viruses, or proviruses. More recent evidence indicates not only that some viruses are tumorigenic, but also that chemical agents might activate or modify the infectious viruses or proviruses (oncogenes?) that are present in the cell.[399] In addition, in view of the demonstrations of extrachromosomal inheritance in protozoa and fungi, epinuclear hereditary factors may be present in mammalian cells. Efforts in this area, particularly with respect to the possible alterations of these epinuclear factors in relation to cancer, may turn out to be rewarding.

Chu[135] has recently devised a modified version of host-mediated assay by placing Chinese hamster cells in dialysis bags that are then implanted surgically into the peritoneal cavities of rats. After various periods, the hamster cells are removed for mutagenic studies *in vitro.* Both point and chromosomal mutations can be assayed this way. Although the procedures are still being improved, this test system may be useful for the study of mutagenicity of carcinogens.

Point mutations in the germ cells of mammals have been experimentally induced and quantitatively analyzed. A great majority of dominant lethal mutations probably involve chromosomal aberrations. The test procedure designed for the analysis of radiation-induced, specific-locus mutations in mice has been readily adapted for mutagenicity testing of chemicals (e.g., see Ehling[228]). Although this specific-locus method can be extended to include chemical carcinogens, the costs for mammalian breeding experiments are so prohibitive as to make routine testing impractical, at least at the initial stages of screening for chemical hazards. However, somatic mutations in intact mammals may be more relevant to carcinogenesis, and methods for their detection are urgently needed.

14

Vegetation and Polycyclic Organic Matter

Polycyclic organic matter is widely distributed in water, soil, air, and plants. Widespread occurrence of POM has been reported in many plants and plant products, such as tobacco smoke,[156,160-162,792] snuff,[111,113] peat,[298] wood soot,[463] charred biscuits,[463] stack gases from pulp mills,[66] nonurban soil,[9,66] roasted coffee beans,[130,465] plant tissues,[77,312,324,325] pyrolyzed cellulose, lignin, pectin,[296] tobacco leaves,[114,792] wood-smoked foods,[496,736] incinerator effluents,[343] and marine fauna and flora.[107,111,510] Combustion of almost any organic material contributes polycyclic organic compounds and their partial oxidation products in trace amounts to the environment for probable contamination of every receptive surface; and POM is produced by baking, barbecuing, broiling, or frying of many foods.[325]

Although several polycyclic compounds generated by burning of vegetation are known to be carcinogenic when applied externally to particularly susceptible animal tissue, few have been adequately tested by ingestion by experimental animals.[325] Tolerance or resistance to low concentrations of the carcinogenic materials has undoubtedly developed through the process of natural selection in animal species that consume smoke-contaminated foods or are exposed repeatedly to the by-products of combustion. Man has apparently acquired some degree of tolerance to POM through continual exposure to the external environment and to foods. If the limits of their tolerance are exceeded by additional exposures, the carcinogenic threshold of a particular compound or combination of compounds may also be exceeded, and that will result in the development of cancer. It is important to examine the possible contribution of vegetation to the total burden of carcinogenic or potentially carcinogenic compounds in the environment.

POM apparently does not induce cancer-like tumors in plant tissue, although a variety of tumors do occur naturally. The most common of these unusual growths is crown gall, which is initiated by a microorganism.[17,88] The ability of bacteria to produce crown gall is related to their production of β-indoleacetic acid. Cultivated mushrooms exposed to fumes emitted from coal tar, diesel oil, and a component fraction of tar acids developed tumorous growths.[254] Tumors were also produced on mushrooms by incorporating soot and diesel oil in the nutrient medium. Attempts to produce similar tumors by direct application of carcinogenic polycyclic aromatic hydrocarbons to the mycelium of the mushroom failed.

Both the carcinogenic benzo[a]pyrene and its inactive isomer benzo[e]pyrene have been found to be fairly abundant, even in rural soils remote from major highways and industries.[76] These benzopyrenes are among the pyrolytic products of wood, and they also occur in the transformation of plant organic matter to peat and lignite.[298] The POM content of soils is also increased by exposure to industrial effluents, products of oxygen-deficient burning of vegetative matter, deposits of petroleum products, and exhaust gases from the automobile. Mallet and Héros[511] detected benzo[a]pyrene in tree leaves and in decaying organic matter under the same trees. They suggested that the benzo[a]pyrene was absorbed through the tree roots and was translocated through the transpiration stream to the leaves. These observations and the report by Guddal[323] that several hydrocarbons may be absorbed from contaminated soils by plants were substantiated by Dörr's experiments,[215] which showed that benzo[a]pyrene was absorbed from soil and water cultures by barley roots and was translocated to the shoots.

In polluted atmospheres,[155,156,218,324,662,664,752] POM may contaminate plants used for food by settling on surfaces or by absorption into tissues. Grimmer and Hildebrandt[313] found that grain samples from the Ruhr district contained 10 times more carcinogenic polycyclic matter than samples from nonindustrial areas. Grimmer[312] reported increased concentrations of phenanthrene, anthracene, pyrene, anthanthrene, fluoranthene, benz[a]anthracene, chrysene, benzo[a]-pyrene, benzo[e]pyrene, perylene, benzo[ghi]perylene, dibenz[a,h]-anthracene, and coronene in samples of lettuce, kale, spinach, leeks, and tomato collected from the field. The concentrations varied widely between fields and between locations in the same field. The maximal concentration of benzo[a]pyrene in lettuce was 12.8 μg/kg of tissue, whereas the maximal concentration in kale was twice that. Spinach, leeks, and tomato samples contained benzo[a]pyrene at 7.4, 6.6, and 0.2 μg/kg, respectively. It was possible to remove approximately 10% of the benzopyrene from the vegetables by washing them in cold water, but Grimmer[312] reported that the remainder of soot film could not be washed off with water. Plants with the smallest amount of surface area exposed to the atmosphere, such as tomatoes and leeks, had the least benzo[a]pyrene. Howard and Fazio[397] indicated that little information was available on the extent of contamination of our food supply via air pollutants. Gunther *et al.*[325] found anthracene (25 ppm) and five unidentified polycyclic compounds in orange rinds obtained from an area adjacent to a heavily traveled highway. Bolling[73] reported higher concentrations of POM in wheat, corn, oats, and barley grown in industrial surroundings than in those grown in more remote areas; and Gräf and Diehl[309] identified eight polycyclic compounds in various plant leaves, with concentrations of benzo[a]pyrene as high as 40 parts per billion (ppb) in some leaves.

Borneff *et al.*[77] conclusively demonstrated the biosynthesis of POM in plants. Algal cultures were grown in nonlabeled and ^{14}C-labeled acetate as the sole carbon source. Data showed that POM was synthesized and that plants have a normal concentration of about 10 μg of benzo[a]pyrene per kilogram; the total amount may in some cases be greater than 100 μg/kg. According to Borneff *et al.*,[77] the health risk in orally introduced polycyclic compounds has not been completely clarified. Wynder *et al.*[832] indicated that the consistent uptake of polycyclic compounds by vegetation may exert an influence on the incidence of human intestinal neoplastic disease. Hakama and Saxén[334] found a significant correlation between the consumption of cereals and the occurrence of gastric cancer. Grimmer[312] found that benzo[a]pyrene in 23 samples of cereal grain varied from 0.2 to 0.4 μg/kg, and flour and bread made from some of these samples appeared to retain most of the benzo[a]pyrene. Grimmer also found that the water extract from 1 kg of tea contained about 4 μg of benzo[a] pyrene; however, 11 coffee samples contained hardly any hydrocarbons.

In addition to the polycyclic compounds synthesized in plant tissue or deposited on plants by polluted air, the preparation of foods may increase the total burden of carcinogenic compounds. Davies and Wilmshurst[185] reported the formation of 0.7 μg of benzopyrene per kilogram of starch heated to 370–390 C and suggested that temperatures reached during the toasting of bread (390–400 C) may be capable of producing polycyclic compounds. Chassevent and Héros[130] found little benzopyrene in commercially roasted coffee, but considerably more was found in "home"-roasted coffee, which included the endosperm of the seed. Kuratsune and Hueper[465] reported that polycyclic aromatic hydrocarbons were found in roasted coffee but did not specify whether the samples tested contained the endosperm.

Benzopyrene is apparently present in a wide variety of both cooked and uncooked foods. According to Raven and Roe,[625] 0.3–2.1 μg/kg of Icelandic smoked meat and fish was found by Bailey and Dungal;[23] Lijinsky and Shubik[496] reported 8 μg/kg of smoked salmon; and Gorelova and Dikun[306] found up to 10.5 μg/kg of home-smoked sausages. Much of this benzopyrene was no doubt produced by pyrolysis of fat in the meat, but condensed smoke from the oxygen-deficient combustion of organic fuel used to produce smoke as a meat preservative was also a contributor.

Wynder and Hoffmann[831] found 5.3, 4.4, 2.4, and 1.4 μg of benzo[a] pyrene per 100 cigarettes made from Virginia, Turkish, burley, and Maryland tobaccos, respectively. The quantity of benzo[a] pyrene isolated by various workers around the world varies widely. A review of these results by Wynder and Hoffmann[831] shows a range from 0.2 μg/100 cigarettes reported in Denmark to a maximum of 12.25 μg/100 cigarettes for one sample in the United States.

As might be expected, smoke from tobacco products other than cigarettes also contains polycyclic compounds. A significantly higher benzo[a] pyrene content for the smoke of pipe tobacco (especially when compared with cigarette tobacco smoked in a pipe) suggested that additives for pipe tobacco, particularly sugars, may become precursors of benzo[a] pyrene on pyrolysis.[390] Extensive studies of tobacco pyrolysis have been made in an effort to determine whether specific compounds in tobacco can be considered precursors of POM in tobacco smoke.

Polycyclic compounds in processed tobacco itself may be derived from polluted air or from tobacco processing (curing, aging, etc.).

Studies by Lyons[505] and Bentley and Burgan[46] showed that traces of polycyclic compounds found in tobacco do not contribute an appreciable amount to the total aromatic hydrocarbons in tobacco smoke. However, Bentley and Burgan[45] and Wynder and Hoffman[827] reported up to 12 ppb and 20 ppb of benzo[a]pyrene, respectively, in tobacco. Van Duuren *et al.*[792] found tumor-promoting agents in tobacco leaf and in the smoke condensate. More recent work by Van Duuren *et al.*[789,791] also pointed out that aromatic hydrocarbons may be involved in a two-stage process of carcinogenesis. Several noncarcinogenic polycyclic compounds were found to function as initiating agents with croton seed oil as a promoter of tumor formation. In addition to the indicated tumor-promoting property of croton seed oil, Hecker[348] reported that the oil may contain a carcinogenic principle. The significance of polycyclic content of tobacco on the emission of carcinogenic hydrocarbons from cigarette smoke has been questioned by several researchers because of the small amounts detected in the tobacco. Campbell and Lindsey[113] used the analysis of cherry laurel leaves for polycyclic compounds as a check against results obtained with tobacco to show that there was essentially no difference between them.

Food may become contaminated with polycyclic hydrocarbons and other carcinogens when crops are sprayed for pest control. To test this assumption, Gunther *et al.*[325] selected oranges growing in southern California, because fruits remain on the trees almost all year and orchards are planted both in heavily polluted regions and in areas relatively free of pollutants. The studies were designed to determine residue persistence of five selected polycyclic hydrocarbons (3-methylcholanthrene, dibenz[a,h]anthracene, benzo[a]pyrene, dibenzo[a,i]pyrene, and anthracene) in and on Valencia orange rind. The first four of these occur in agricultural environments from air pollution, from industrialization, and from petroleum oil pest-control operations. With the exception of dibenzo[a,i]pyrene (persistence half-life, 12 days), the compounds that had penetrated into the rind to the extent of 1–12% were considered to be long-lived, with persistence half-lives of 120–200 days in the field. Degradation half-lives were uniformly 1–2 days. Apparently, 85% of the degradation losses, presumably by volatilization and oxidation, occurred in a few days with little penetration into the fruit pulp, and there was no evidence of translocation into twig tissue.

Anthracene and five unidentified fluorescing materials were found in Valencia oranges grown near a major highway where there was relatively high air pollution.[325] Surface contamination of the fruits was apparently degraded rapidly, but any portion of the polycyclic compounds that became incorporated into cuticular oils and waxes persisted for long periods.

In summary, no information was found to indicate that carcinogenic polycyclic hydrocarbons affect vegetation. Polycyclic compounds absorbed by roots from contaminated solutions, by foliage from polluted atmospheres, and by aquatic plants from contaminated bodies of water are added to the traces of these compounds produced metabolically. One researcher has reported abnormal growths on mushrooms grown on contaminated media. Burning of vegetation and some plant products may produce significant quantities of several carcinogenic hydrocarbons. The increased concentrations of these materials in organic soils and in sediments in large bodies of water suggest that many of the polycyclic compounds are produced in decayed organic matter.

The following recommendations are appropriate:

1. Encourage research to determine the contribution of pesticidal sprays, herbicides, and polluted atmosphere to accumulation of polycyclic hydrocarbons in and on vegetation.
2. Determine the influence of traces of carcinogenic materials in vegetable foods on the incidence of cancer in man and animals.
3. Produce a reasonably accurate estimate of the amounts of carcinogens generated by wild fire and by the combustion of solid waste.
4. Investigate the effect of long-term exposure and massive dosages of polycyclic compounds on plant growth, development, and reproduction.

15

Introduction to Appraisal of Human Effects

A historical review of man's reaction to airborne pollutants containing what are now recognized as polycyclic aromatic hydrocarbons reveals that occupational incidents provided the first evidence of cause–effect relations. The degree of exposure to such materials is often much greater in occupational settings than that encountered in community air pollution. The first recorded description of an occupational disease related to the burning of fossil fuel appeared in 1775, when Percivall Pott, a surgeon at St. Bartholomews Hospital in London, published a paper[600] about cancer of the scrotum in chimney sweeps and related the disease to their constant exposure to soot. Since then, and particularly since the large-scale use of fossil fuels has expanded, components and degradation products of such fuels after burning, refining, distilling, or cracking have been demonstrated to have a close association with a high incidence of skin cancer affecting the scrotum and other heavily exposed skin areas. For example, a high incidence of cancer of the skin has been observed among workmen in coal-tar industries and gas plants (particularly in operations in oil and shale refineries) and among machine operators using lubricating oils (from particular sources) in the textile industries and in machine shops. Clinical incidents of this kind have been well described and reviewed by Henry.[359,360] In recent years, polycyclic organic materials proved to be carcinogenic in experimental animals have been found in the derivatives of fossil fuel associated clinically with skin cancer. It seems logical for skin problems to be described first, inasmuch as the skin is the most vulnerable of organ systems and its diseases are easy to identify.

Descriptions of lung disease related to dust or airborne particles were made in the sixteenth century. More than 300 years later, Härting and Hesse[345] demonstrated that lung cancer was prominent among the pulmonary diseases from which miners were suffering. The ores from the Joachimsthal and Schneeberg mine areas were eventually shown to contain radioactive dusts; it was from the pitchblende from this area that Marie Curie first extracted radium. Although there is no evidence that POM played a role in the lung cancer of the miners, this is an excellent historical example of the devastating biologic effect of an airborne pollutant in man.

It was not until the twentieth century that cancer of the lung was shown to be associated with coal processing operations, such as coking, and the manufacture of illuminating gas. Substantial increases in lung-cancer mortality rates over those of the general population have been noted by Kawai *et al.*[430] in gas-generator workers and by Alwens *et al.*,[13] Kuroda,[466] the Kennaways,[434] and Doll[205] in coke-oven workers and gas-retort workers. In most instances, polycyclic aromatic compounds, such as benzo[a]pyrene, have been recovered from the airborne contaminants.

There has been strong suspicion for some time that the smoking of tobacco is related to human disease, particularly lung cancer. From 1939 to 1964, at least 29 retrospective epidemiologic studies of lung cancer were published.[769] It was studies like those of Hammond and Horn[341] that provided the evidence that lung cancer could be correlated positively with cigarette smoking. Since 1964, several highly significant reports have related the frequency of illness to smoking.[54,207,208,339,422]

ENVIRONMENTAL POLLUTION: THE PUBLIC HEALTH PROBLEM

Consideration of the importance of physical, chemical, and thus environmental factors in disease first emerged from recognition that disease patterns in Americans had changed and were different from those affecting people in underdeveloped countries. In the latter, deaths appear to be related largely to infection, with tuberculosis, pneumonia, and diarrheal diseases still the major killers. In the United States, the major causes of death include cardiovascular diseases, cancer, and stroke,[116] as well as chronic bronchitis and emphysema.

In 1969, lung cancer was the greatest single cause of cancer deaths in men, killing almost 50,000. A rare cause of death in women 30 years ago, it was responsible for 10,000 deaths in women in 1969. Such increases have occurred in almost every industrial nation in the world, and Clemmesen[144] predicts that lung cancer will achieve epidemic proportions within the next decade.

The relation between cigarette smoking and lung cancer[771,772] has stimulated concern over the role of air pollution in cancer, because some urban air pollutants, including POM, are similar to those found in cigarette smoke. Epidemiologic studies of occupations in which similar substances were present in large quantities also revealed a sharp increase in skin and lung cancer in persons with long exposures. The rising incidence of and poor prognosis in lung cancer, even with early diagnosis, has led to recognition of the necessity for defining all significant etiologic factors. Primary prevention is the only effective means of control.

DATA SOURCES AND PROBLEMS IN INTERPRETATION OF HEALTH EFFECTS DATA

The difficulty in defining causal and dose–effect relations arises from the peculiar natural history and characteristics of lung cancer. It has a long latent period—possibly as long as 30 years—and its peak incidence occurs after the age of 50. The long interval between initial exposure to a cancer-inducing agent and the appearance of detectable disease makes etiologic analysis difficult. This is particularly true in view of the many changes that may occur in the person in terms of occupation, residence, nutrition, habits, and socioeconomic status. For example, in the United States in the last 30 years, there has been a remarkable migration from rural town and farm areas to large cities and, more recently, from cities to suburbs. In addition, the nature and amount of environmental contaminants have changed, with the development of new industries and the closing down of old ones and changes in modes of transportation and types of fuels. Many of these factors apply as well to the pathogenesis of cancers in other body systems.

The use of laboratory animals is a valuable source of material. It makes it possible to study short- and long-term biologic effects of daily doses of single and multiple substances under different conditions in a controlled environment not analogous to the human condition. Animal studies with airborne pollutants, particularly polycyclic aromatic hydrocarbons, have indicated that pathologic changes, such as neoplasms and inflammatory responses, occur through the painting, injecting, or implanting of the material on or in the skin and through intratracheal instillation, implantation, or inhalation.

In vitro methods have also provided methodologies for studying the induction of neoplastic or toxic changes in single cells, small cell populations, or whole organ cultures.

Lung cancer is associated with many etiologic factors, some of which, like cigarette smoking and occupational exposure to asbestos, are well known. There are a number of methodologic approaches to the study of the etiology and pathogenesis of lung cancer. Data are available from epidemiologic studies of the disease in workers exposed to high concentrations of known carcinogenic substances in different industries. Some insight into cause–effect relations may be obtained by reviewing available data on occupational exposures. But the populations under observation are biased, in that they consist mainly of young, healthy persons exposed for only a portion of each day or week. They do not include those who, because of increased sensitivity to the pollutant or development of related diseases, are forced out of the industry. In addition, lung cancer affects mostly older people, many of whom might have left the industry that was causally related to the disease and would thus be lost to follow-up.

The epidemiologic method of studying the effect of air pollution on the incidence of lung cancer involves the comparison of lung cancer death rates in communities that have demonstrably different levels of pollution. The largest environmental differences are found in contrasts between different countries, but the interpretation of such contrasts is made difficult by virtue of the wide differences between countries in smoking habits and other characteristics. Comparisons between urban and rural areas within a country are attractive, in that they tend to maximize differences in levels of pollution. Contrasts between urban areas offer the important advantage that direct measures of air pollution are generally available only for urban centers.

Each such comparison requires a suitable method of adjustment for the major known extraneous variables. Because the lung-cancer death rate depends heavily on age and sex and because age distributions vary markedly, only age-specific or age-standardized and sex-specific lung cancer death rates should be compared. This is generally possible. Cigarette smoking has been shown to be correlated with lung cancer. However, adjustment for amount of cigarette smoking is difficult and in any case uncertain. Detailed cigarette-consumption statistics of specific areas are often difficult to obtain, and consumption specificity by age and sex is virtually undocumented. Details of smoking practice—in particular, butt length—may be important. Unfortunately, no usable documentation of such differences is available.

Despite difficulties, some types of comparison strongly support the proposition that urban pollution is related to an increased lung cancer death rate. It is much more difficult to relate the increment in deaths directly to specific contaminants, and efforts to do so have led to variable results.

The problem of assessing the effect of POM on the incidence of lung cancer is compounded further by other factors that make it more difficult to carry out and evaluate epidemiologic studies. Areas to be considered are listed below.

1. *The adequacy of measurements of POM*
In the past and even today, measurements were carried out in only a few areas of large cities. In Chicago, for example, Carnow[117] found that wind patterns often create conditions in which levels of pollutants may be remarkably different in various areas on different occasions, so limited measurements might not truly represent the degree of exposure of individuals, particularly if values are expressed as citywide or annual averages.

2. *Whether the increase in incidence rates is real or apparent*
The extensive routine use of x rays by physicians and in mass surveys for tuberculosis in the last 25 years and physicians' increased awareness of this disease are undoubtedly factors in the reporting of disease incidence. Greater use of methods for making tissue diagnosis—such as bronchoscopy, bronchial and pleural biopsy, and exfoliative cytology—and greater reliance on autopsy also increase the number of detected cases of carcinoma, which in the past might have been considered tuberculosis, pneumonia, or other diseases. Greater access to physicians in urban and rural areas increased the case-finding potential. Aging of the population naturally contributes additional cases in persons who, in the past, might have died earlier in life from other diseases. Kotin,[448] Lew,[494] Clemmesen,[144] Kreyberg,[461] and others have

reviewed and carefully considered these issues. There seems little room for doubt that a substantial portion of the observed increase is real.

3. *Whether a rural–urban difference can be ascribed to persons who die in large cities to which they have come from rural areas for diagnosis and treatment*
Studies by Stocks,[728] Haenszel,[329] and others have shown that migration to cities for diagnosis and treatment is not a factor in urban–rural differences.

4. *Whether the apparently significant higher incidence in lower socioeconomic groups is related to factors other than the urban factor, i.e., inadequate medical care, poor nutrition, type of heating, or greater occupational exposure to carcinogens*
This question has been studied by Manos and Fisher,[515] Cohart,[149] and others; the urban–rural differences persist after adjustment for these factors.

5. *Whether differences can be ascribed to ethnic factors*
Graham *et al.*[310] documented high rates of lung, prostate, and gastric carcinoma in Poles. Religious differences may also be significant, as suggested by the work of MacMahon,[506] who found upper respiratory cancer rates in Catholics and Protestants to be 3 or 4 times those in Jews but found little difference between foreign- and native-born groups. This may be related to cultural, dietary, occupational, or genetic factors. Racial differences were also studied but require further investigation. Duchen[219] found no increased incidence in Caucasians contrasted with Bantu natives that could not be explained by differences in longevity, whereas Hoffman and Gilliam[383] found a higher lung cancer rate in Caucasian than in Negro males.

16

Characteristics of Human Disease Related to Polycyclic Organic Matter

LUNG CANCER

In no instance has exposure to a specific polycyclic aromatic hydrocarbon been proved to have caused a tumor in man. That does not, however, deny the risk of exposure. There is now good evidence of the overwhelming importance of cigarette smoking in the etiology of lung cancer in man, and polycyclic aromatic hydrocarbons in cigarette smoke have been considered as an identifiable group of components in this connection. Although the effects of known dosages of specified substances acting alone should perhaps be assessed first, the possibility of additive or potentiating effects of other factors must also be considered. For example, absence of evidence of carcinogenic effect of atmospheric polycyclic aromatic hydrocarbons in nonsmokers does not preclude an effect in the pathogenesis of lung cancer in smokers.

Carcinogenic polycyclic aromatic hydrocarbons from a variety of sources are known to be present in urban atmospheres. To assess carcinogenic potential, the composition of the atmosphere must be considered. If the ambient air concentration of a supposedly carcinogenic substance, "A," is found to correlate with the higher incidence of lung tumors in a specific human population, but not in all, one of several inferences can be drawn:

1. "A" is not a lung-specific carcinogenic agent and another substance, "X," or substances, may in fact be the responsible agent.
2. "A" is carcinogenic for the lung but only if:
   a. its effect is added to that of another carcinogenic substance or substances;
   b. it follows or accompanies other substances that possess an initiating or cocarcinogenic effect;
   c. a carrier substance, which may itself be inert, keeps "A" in contact with the target cells;
   d. other conditions are favorable for bringing "A" into effective contact with the susceptible tissue; or
   e. there is some combination of the above.

Carcinogenicity of air pollutants has been demonstrated not only by bioassay of crude benzene extracts of deposited soot,[241,487] but also by identification of specific substances known to be carcinogenic in experimental animals, notably benzo[a]pyrene.

Irritant or toxic gases are known to exist in various concentrations in the atmosphere—e.g., sulfur dioxide, oxides of nitrogen, and ozone. Such gases are known to have effects other than simple irritation of the conjunctiva—specifically, a potentiating action on the carcinogenic properties of polycyclic aromatic hydrocarbons.[12,19] This has been demonstrated by the higher incidence in CAF/Jax mice of pulmonary adenomas produced by simultaneous exposure to ozone and carcinogens.[734] Probably more significant is the role of sulfur dioxide in benzo[a]pyrene carcinogenicity in rats, inasmuch as some of the tumors have been squamous.[474]

The interaction of a presumed carcinogenic inhalant and a target tissue—in this case the lung—can be considered in the framework of an "ideal model" for inhalation carcinogenesis in man. An ideal model is one that provides in quantitative terms knowledge of the response of a specific tissue to a defined agent and of the factors governing the response. Such a model requires examination of data from analyses of atmosphere and tissues, from physiologic and biochemical studies, from animal experiments, and from epidemiologic studies. Inferences for man from experimental studies involving animals, as considered here, must be drawn with particular caution.

Components of a model for inhalation carcinogenesis in man are given in Table 16-1.

Determinants of Concentration in Tissue at Risk

According to most investigators,[449] polycyclic aromatic hydrocarbons in the atmosphere are bound to particles that when condensed can be characterized as "soot." The size distribution of the particles (0.125–2.5 μm in diameter) is well within the range likely to be aspirated into the lower respiratory tract.[182,775]

Hatch[347] has stated that the effective dose of an inhaled pollutant is the dose that reaches the critical site in the body. Duration of contact is also important. These are determined by the vector sums of delivery of the pollutant and its clearance or degradation.

Particles larger than 5 μm in diameter for the most part become entrapped within the upper respiratory tract or are removed by ciliary action or coughing from the lower respiratory tract, usually carried within mucin. Some material whose size makes it aspirable into the lower respiratory tract is removed in the same way as the larger particles. Precise details depend on the properties of the material, the physio-

TABLE 16-1 Components of Model for Inhalation Carcinogenesis in Man

A. Composition of polluted atmosphere
   1. Concentration of each pollutant
   2. Physical state of each pollutant
   3. Presence of potentiating agents
   4. Presence of "inert" particles or absorbents

B. Factors determining concentration at target tissue
   1. Physiologic characteristics
   2. Protective mechanisms
      - ciliary function
      - mucous barrier
   3. Clearance mechanisms
      - phagocytosis
      - leaching
      - ciliary action
      - removal by blood and lymph
      - metabolic transformation

C. Mechanism of carcinogenesis

D. Factors modifying reaction at target site
   1. Synergistic or antagonistic substances
   2. Host factors
      - previous disease
      - genetic factors

logic characteristics of the subject, and the subject's state of health. These variables are difficult to establish accurately and represent one of the deviations from the "ideal model."

Impairment of ciliary transport mechanisms by chemicals, including pollutants in the atmosphere or in cigarette smoke, could be important in increasing the concentration of damaging substances within the lower respiratory tract. Such ciliostatic effects have been studied by Hilding[376] and Dalhamn *et al.*[178] Table 16-2 is a partial list of ciliostatic substances whose effects have been investigated. Living pathogenic agents can have a similar effect.[449]

Clearance of fine particles of polycyclic substances can be remarkably rapid. In mice, methylcholanthrene is cleared within 6 hr of inhalation and within 24 hr of intratracheal instillation.[618] In some experimental animals, presumably inert carrier substances have been shown to have an important effect on the concentration–time determinants of the damaging action of inhalants. The experiments of Boren[75] have shown that carbon functioning as an absorbent greatly increases the damaging action of nitrogen dioxide on the lung. When tritiated benzo[a]pyrene is incorporated in carbon or asbestos, clearance from the lungs of hamsters is slowed.[694] Increase in the carcinogenic effect of benzo[a]-

TABLE 16-2 Compounds in or Related to Constituents of Polluted Urban Air and Cigarette Smoke That Can Inhibit Ciliary Activity[a]

Compounds in Air Pollutants	Compounds in Cigarette Smoke
Paraffin	Nicotine
2-Methylpentane	Pyridine
Olefins	Ammonium hydroxide
2-Methylbutene-2	Methylamine
2-Methylpentene-2	Trimethylamine
Aromatic	Acetonitrile
Benzene	Thiocyanic acid
Aldehydes	Methanethiol
Formaldehyde	Phenol
Propionaldehyde	
Acid	
Formic acid	
Peroxides	
Acetyl peroxide	
Peracetic acid	
Epoxides	
Propylene oxide	
Cyclohexene oxide	

[a] Derived from Kotin and Falk.[449]

pyrene by means of carbon and carrier particles[613,616] and hematite[650] has also been shown.

The rapid clearance of fine particles of polycyclic aromatic hydrocarbons from the lower respiratory tract has already been mentioned. In animal experiments, the major mechanism operative shortly after inhalation or intratracheal instillation is probably ciliary action. Phagocytosis by macrophages, however, must play a role, and this process becomes more important when the polycyclic aromatic hydrocarbons are adsorbed on relatively inert carrier particles, such as those of carbon or hematite. The phagocytes, too, can be moved by ciliary action or can travel through the tissues into lymphatics to regional and ultimately more distant lymph nodes or into the bloodstream. The material can also be released if it remains intact and can then become subject to phagocytosis again.

A process of interest is the "leaching" of polycyclic aromatic hydrocarbons. This was inferred from analysis of human tissues by Falk, Kotin, and Markul.[258] They found no benzo[a]pyrene in the residue of soot within lymph nodes that, on the basis of the composition of soot in the atmosphere, would be expected to contain a considerable percentage of this carcinogen. The fate of the leached material is not known. There did not appear to be a correlation between the amount of soot and the presence of metaplastic or neoplastic change. Further studies on this important subject are warranted.

Modifiers of the Reaction at the Target Site

How such gases as ozone and sulfur dioxide potentiate the effects of carcinogenic polycyclic aromatic hydrocarbons in the lung is largely unknown, and it is not clear how tissues altered by disease may enter a "precancerous" state. Metaplastic changes and continuing regenerative activity are common to injured and diseased tissue. One interpretation is that cells in mitosis are more susceptible to carcinogenic agents. A well-known example is the frequent occurrence of carcinoma of the skin in burn scars.

More relevant to the problem under consideration, especially because the observations were made on human lungs, is the demonstration of the relation between chronic interstitial pneumonia, honeycombing, and cancer of the lung. Honeycombing is the revision of pulmonary architecture that takes place in the healing phases of interstitial pneumonia where loss of alveoli and interstitial fibrosis accompanied by hyperplasia of smooth muscle and by cellular infiltration have taken place simultaneously. On the thick walls of the labyrinths that remain, there is ingrowth of epithelium that can be astonishingly hyperplastic and even metaplastic. In a series of 153 consecutive resected lung tumors, some 22% were associated with honeycombing and atypical proliferation.[530] In many of these specimens, transitions could be traced from the latter to obvious neoplastic change. Of the associated tumors, 83% arose in the periphery of the lung, and there was radiographic as well as anatomic evidence of such origin in some cases. A peripheral lung carcinoma has its center of mass clearly beyond a segmental bronchus. All the patients who had both honeycombing and carcinoma were men, and all for whom the relevant information was available were cigarette smokers. Of persons with honeycombing, 58% had prior histories indicative of pneumonia 5 or more years before developing pulmonary carcinoma, whereas only approximately one third of patients with other lung cancers had such a history. Of the other patients in this series with pulmonary reactions, 89% were male and 16% were nonsmokers. In a control necropsy series, the incidence of honeycombing was 4.7%. Among a total of 403 control patients, 86% of patients with this lesion were men. Four of 19 persons with honeycombing (21%) also had cancer of the lung; all were men.

In the relation of lung cancer to honeycombing, cigarette smoking and sex were associated factors. Focal interstitial pneumonia with honeycombing is not invariably present in smokers, nor is it confined to males. Indeed, a high incidence of peripheral tumors has been noted in lungs with honeycombing.[36] These studies illustrate that many factors can enter into pulmonary carcinogenesis in man and may be as relevant to consideration of the effects of atmospheric pollutants as they are to consideration of the effects of smoking.

Problems of Drawing Inferences for Man from Animal Data

Numerous pitfalls are inherent in any attempt to extrapolate data bearing on dose–response relations from organ to organ or from species to species. Thus, dosage data, insofar as they can be calculated from skin painting or injection experiments[241] and applications to cervical epithelium,[72] are inapplicable to the tumorigenic dose for the lung. Some of the external factors that modulate interspecies variation are age, sex, size, hormonal status, state of health, and type and extent of supportive treatment. Rall[620] has cited examples indicating that predictions as to effective dose of drugs are sometimes in proportion to body surface area for a number of diverse species. There is no evidence that this applies to carcinogenesis. Even among individuals of a single species, the presence of disease may alter (usually potentiate) the development of tumors.

Considering the numerous variables that are known and anticipating that some may be totally unknown, it is best to focus attention largely on the target tissue, the lung, and to consider the air as the primary route of entry of carcinogenic agents. This is not to deny the possibility that an additional moiety of the same or a complementary factor might reach the pulmonary tissue through some other route, especially the bloodstream.

The initial reaction occurs between the agent and tissue. But complex interactions involving chemical, physical, physiologic, and pathologic characteristics must also be considered. Cocarcinogenesis[657] compounds the problem.

Some of the problems in the interpretation of experimental data are illustrated by studies of the relation of cigarette smoke to lung cancer, and these are relevant to investigations designed to determine the effects of atmospheric polycyclic aromatic hydrocarbons. Auerbach, Hammond, and associates[18,340] have reported experiments in which beagles were trained to "smoke" through a tracheostomy over periods of approximately 2½ years. Animals exposed to unfiltered smoke in amounts thought to be comparable with heavy human exposure developed not only pulmonary lesions interpreted as emphysema by objective criteria "blindly" applied, but also a high incidence of peripheral bronchiolo–alveolar tumors, some of which were interpreted as invasive. The findings were considered dosage-related, inasmuch as animals smoking filter-tip cigarettes, and therefore estimated to receive approximately half the "tar" and considerably reduced nicotine, developed less emphysema and fewer tumors, of which only two were localized squamous lesions of the bronchi interpreted as locally invasive carcinomas. This is especially remarkable in that, in all of Auerbach's previous work in human smokers, stress was laid on squamous metaplasia, squamous carcinoma *in situ,* and invasive squamous carcinoma of the major bronchi. In this respect, the canine lesions are more like those of the patients with honeycombing in the surgical series studied by Meyer and Liebow,[530] especially because all the patients in that series were male smokers. It is also noteworthy that at least some of the emphysema and fibrosis in the dogs would undoubtedly fulfill the criteria of "honeycombing" as defined in the work on the patients.

The recent work of Auerbach and associates can be criticized on the following grounds:

1. The controls were not subject to the same conditions as the smoking dogs—although they had tracheostomies, they were not made to "smoke" unlighted cigarettes, and they were not made to stand.
2. The smoking dogs had intercurrent infections, and some were given antibiotic treatment for these complications.
3. At least two smoking dogs had evidence of aspiration of food, and another animal had lesions replete with "brown pigment and fat."
4. Two control animals had bronchiolo–alveolar tumors at the age of approximately 5 years, although they must be extremely rare in dogs at this age.

It may therefore be questioned what part of the effect in these experiments can be attributed to smoking and what part to other conditions imposed. Possible factors include the lesser degree of cleanliness of tubing in animals smoking cigarettes without filters and the hypersecretion in the smoking dogs. The sequence might be increased secretion in the smokers, with aspiration leading to infection; pulmonary damage; regenerative changes; and bronchiolo–alveolar tumors.

In support of the importance of smoking is the fact that the effect of smoking given numbers of filter-tip cigarettes was less than that of smoking equal numbers of ordinary cigarettes. It is possible, therefore, that the results of these experiments suggest a potentiating effect of smoking on pulmonary damage related to other causes, as in Meyer and Liebow's observations in man.

The applicability of these observations to problems of air pollution, with its potential for pulmonary damage, and the potentiating effects of cigarette smoking can be considered only suggestive at this time.

It is frustrating, but also perhaps significant, that it has proved more difficult to induce squamous and undifferentiated tumors in animals than peripheral "pulmonary adenomas." The latter are least like tumors in man associated with environmental agents, which are predominantly squamous and undifferentiated. More attention should therefore be paid to the experimental squamous tumors and to the factors in their induction.

Squamous cell carcinomas were produced in rats by the impaction within bronchi of pellets consisting of carcinogen, either pure or diluted within cholesterol in concentrations from 0.1% to close to 100%. Dose–response relations are illustrated in Figure 16-1.[474]

It has not been possible in experiments reported to date to produce epidermoid carcinomas in experimental animals given pure carcinogen

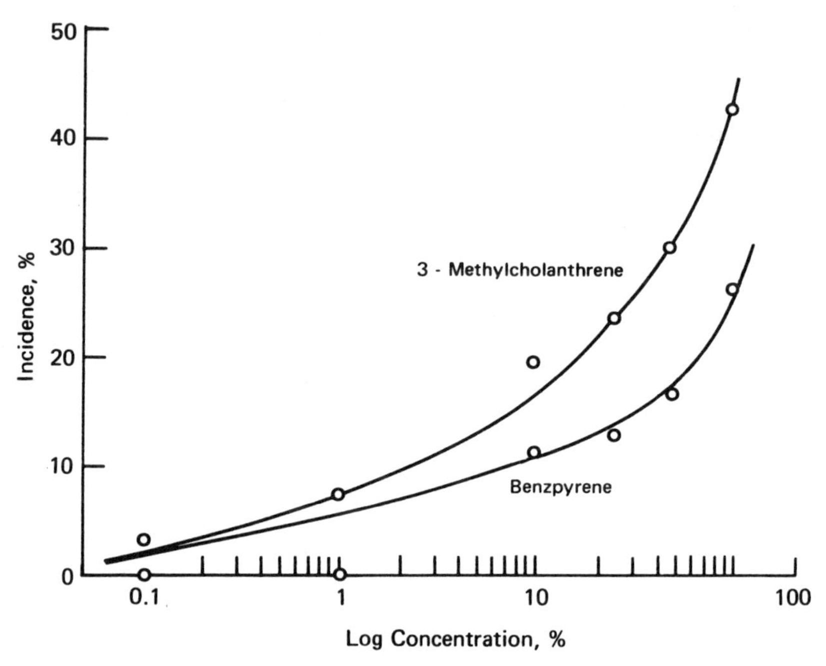

FIGURE 16-1. Dose–response relations after exposure of the lungs of rats to a graded series of concentrations of 3-methylcholanthrene and benzo[a] pyrene in pellets impacted within bronchi. Concentration refers to percentage of carcinogen in cholesterol carrier. Each pellet weighed 3–5 mg. The ordinate indicates the percentage of animals developing bronchogenic carcinomas after correction for early mortality. (Reprinted with permission from Laskin *et al.*[474])

intratracheally, but such tumors developed when benzo[a] pyrene was given with the detergent, Tween 60.[366]

The first squamous lung tumors induced by the intratracheal instillation of a carcinogen were produced by Pylev[614] and Shabad[693] in rats with 9,10-dimethyl-1,2-benzanthracene suspended in balanced saline solution containing 4% casein and India ink powder. Similar results were obtained with benzo[a] pyrene introduced in a mixture containing purified carbon particles.[613,616] The experiments of Saffiotti, Cefis, and Kolb[650] are especially noteworthy; they produced squamous tumors in hamsters, which are relatively free of intercurrent pulmonary disease and spontaneously occurring tumors. Benzo[a] pyrene was administered mixed with equal parts of hematite. The particles were ground to a size range averaging less than 3 μm in diameter, and 6 mg of the dust were instilled once a week for 15 successive weeks. Some

76% of the animals developed tumors, mostly of the squamous type, but including anaplastic carcinomas and adenocarcinomas.

Of great interest are lung tumors produced by inhalation of vapors of ozonized gasoline and influenza virus.[456] Although the exact composition of substances in the ozonized gasoline vapor is unknown, these experiments suggest a potentiating effect of viral infection in pulmonary carcinogenesis. However, similar experiments carried out by Nettesheim *et al.* with influenza virus had an opposite effect.[560]

Of greatest relevance are the studies of Laskin, Kuschner, and Drew[474] on the combined effect of sulfur dioxide and benzo[a]pyrene. Among 21 rats that received 534 exposures to an atmosphere containing 10 ppm of sulfur dioxide and 494 exposures to an atmosphere containing 3.5 ppm of sulfur dioxide plus a benzo[a]pyrene concentration of 10 mg/m^3, five developed squamous cell carcinoma. Of 21 rats that received only the 494 treatments of the combination of the lower concentration of sulfur dioxide plus benzo[a]pyrene, two developed lung carcinoma. Both the cytologic characteristics and the presence of renal metastases confirmed the malignancy of the tumors. One criticism of these experiments is the high incidence of bronchopulmonary lesions to which infection probably contributed.

The difficulties of extrapolation from animals to man are at once obvious when an attempt is made to compare the dosage schedule in the several experiments just summarized with what might be received in the distal pulmonary parenchyma of man breathing a polluted atmosphere. Let it be assumed that the atmosphere contains benzo[a]pyrene in a concentration of 30 $\mu g/1{,}000\ m^3$, which is within the range at unfavorable seasons in Birmingham and Detroit, and that this concentration remains constant through the year.[667] If it is assumed that all the inhaled carcinogen is retained, that the tidal volume is 500 ml, and that the respiratory rate is 14/min, then it would take about 99 days to retain 30 μg and 272 years to retain 30 mg. Especially after applying a correction for the difference in body surface area between rodents and man, it is evident, for example, that the animals in the experiments of Saffiotti *et al.*[650] on hamsters and Laskin *et al.*[474] on rats received enormously greater dosages of benzo[a]pyrene adsorbed on particles and in a much shorter time. Of course, the amount deposited in the various subjects would be determined by particle size; by structural and physiologic factors, including tidal volume and respiratory rate; by efficiency of the clearance mechanism; and possibly by the facilitating effects of viral infections or other lower respiratory tract disease. Additional factors are the effects of particles in delaying clearance, the presence of cocarcinogens or other carcinogens in the atmosphere, and the damaging effects of gases, cigarette smoke, and other substances. The overwhelming importance of cigarette smoke is generally accepted.

Conclusions

The gaps in knowledge concerning inhalation carcinogenesis become evident when it is considered that not even the concentration and physical state of damaging elements at the human receptor site have been adequately established.[1] There is compelling evidence that carrier substances are important, but their exact role can only be surmised.

The assumption that lung tumors are the consequence of a single pollutant is almost certainly wrong, and there is much to support the idea of synergism or cocarcinogenesis, especially with respect to cigarette smoking. Pre-existing pulmonary disease can also be a predisposing factor. Thus, the factors for lung cancer must be more complex than suggested by the simplified schema of Table 16-1.

It is probably significant that the successful experimental production of pulmonary squamous tumors by intratracheal insufflation or by inhalation has required the simultaneous introduction of inert particles or additional pulmonary injury produced by toxic gases or viruses.

Examination of the dosage of carcinogens used in experiments that have been successful in producing squamous cell carcinoma in the lungs of animals makes it obvious that these doses have been much higher than those to which man is likely to be subjected throughout his lifetime with any known specific agent. It is possible, however, that the lungs of rats and hamsters are more resistant to carcinogenesis by polycyclic aromatic hydrocarbons.

On the basis of available evidence, the best approach to assessing the significance of atmospheric pollutants in the etiology of pulmonary carcinoma in man is the epidemiologic one. Urban–rural differences undoubtedly offer clues to the problems of pulmonary carcinogenesis in man. Special attention should be paid to imperfections or contradictions in the correlations of measurements of air pollution or of any specific suspected agent, such as benzo[a]pyrene, with the incidence of lung cancer.[99,100,112,330,332,400,460,662,667,670,731,732] There is general agreement on the increment in the incidence of lung cancer produced by smoking, although interpretations vary. This is of great interest, because urban air and cigarette smoke have carcinogenic substances and some damaging gases in common.[783] However, the complexities of the problem of pulmonary carcinogenesis are compounded, rather than simplified, by that fact.

Problems for Investigation

The epidemiologic approach remains open; although potentially it is highly significant, it must be used with utmost caution to avoid *post hoc ergo propter hoc* reasoning. The well-established urban–rural differences in the prevalence of lung cancer must harbor important etiologic clues and are worthy of the most thorough investigation and the most circumspect analysis. In addition, a possibly fruitful epidemiologic approach would be to compare disease in locations where there are extremes of photochemical pollutants or extremes of polycyclic aromatic hydrocarbons. In all such studies, the most careful attention must be paid to the effects of cigarette smoking in quantitative terms.

Further studies of the composition both of the atmosphere and of cigarette smoke, with a search for similarities and differences, should be carried out. Exposure factors are incompletely known and have great relevance.

Base-line data with respect to some pollutants could be obtained for man by analysis of tissues of human beings of past centuries. Accurate dating is possible, because clothing, coins, etc., often remain intact.

Human physiologic characteristics that have a bearing on pulmonary carcinogenesis have been incompletely studied. An example is the fate and clearance of particles from the respiratory tract. The effects of some gases can also be investigated. Furthermore, advantage should be taken of known industrial exposures and accidents that might have relevance to the problem.

Further investigations of experimental models of pulmonary carcinogenesis that appear to be most relevant to the human disease are indicated. These should be directed mainly toward inhalational exposure to carcinogenic polycyclic aromatic hydrocarbons known to be present in the atmosphere, with due attention to quantitative factors and to adequacy of controls. The following are to be considered in the design of such experiments:

1. The similarity of the experimental species to man,
2. Degrees of exposure in urban atmospheres (this is not to deny the use of greater exposures as a first approach),
3. The high probability that lung cancer is a multifactorial disease

and that the most significant factor is cigarette smoke (experiments designed to investigate interacting factors must be planned in a manner permitting analysis), and

4. The use of germfree animals, once a suggestive model has been established (this is desirable, because the role of infection is problematical and difficult to control in ordinary circumstances).

CHRONIC BRONCHITIS AND EMPHYSEMA

Chronic bronchitis and emphysema are two pulmonary diseases that should be considered when evaluating the human health effect of urban pollution.

The etiology of chronic bronchitis is not known,[269,303,629,645] and the disease has not been produced experimentally in animals by irritants, at least in its severe stages, in which substantial airway obstruction occurs. POM cannot be excluded with certainty as an etiologic factor in chronic bronchitis, particularly because it might be associated with disease when combined with other atmospheric pollutants. The response of humans to complex mixtures of pollutants in which POM is merely one component is not known.

The belief that exposure to POM may be of etiologic importance in chronic pulmonary disease is based on two epidemiologic observations: Workers producing coal gas in England had an increased incidence of both chronic bronchitis and lung cancer, compared with the general population, but the increase in the incidence of chronic bronchitis was considerably greater than that for lung cancer and these workers had enormous exposures to POM, particularly benzo[a]pyrene; and an urban–rural gradient for lung cancer and chronic bronchitis parallels that for POM. Of course, it has been demonstrated that benzo[a]-pyrene, a known carcinogen present in fossil fuels, is not particularly irritating to the normal lungs of experimental animals, except at very high dosages.[269,303,629,645]

Chronic bronchitis and emphysema are different forms of cumulative injury that become overt disease when the damage reaches a level sufficient to cause appreciable disability. The severity of the preclinical stages of a disease like emphysema can be shown by autopsy studies to correlate well with levels of urban pollution.[415] However, the role of POM cannot be ascertained from these data.

Stocks[730] reported on data collected from 26 localities in northern England and Wales. Standardized mortality for chronic bronchitis and pneumonia in 1950–1953 was correlated with annual concentrations of smoke, benzo[a]pyrene, benzo[ghi]perylene, pyrene, and fluoranthene. A statistical process of successive elimination was applied to discover which hydrocarbon was responsible for the demonstrated correlation of mortality rates with smoke concentrations. For lung cancer and chronic bronchitis, benzo[a]pyrene was found to be the substance of prime importance. For pneumonia, benzo[a]pyrene was not important. This seems to cast doubt on the significance of benzo[a]-pyrene in relation to bronchitis mortality.

In a later report, Stocks[729] correlated lung cancer and chronic bronchitis mortality with consumption of cigarettes and solid and liquid fuel in 20 European countries. The consumption of solid or liquid fuel did not appear to be as important as cigarette smoking in bronchitis mortality rates in these countries.

The evidence indicates that POM concentrations in polluted urban air do not significantly influence the pathogenesis or outcome of non-neoplastic lung diseases, such as emphysema and chronic bronchitis. However, the total urban pollution content appears to be a significant factor in disease development.

SKIN DISORDERS

Although it is apparent that, in occupational exposures in man, the most common pathologic response to high dosages of POM is cutaneous neoplasia, various nonneoplastic skin responses have been observed. They include nonallergic dermatitis, cell-mediated hypersensitivity (allergic contact dermatitis), phototoxic and photoallergic reactions, pilosebaceous responses (such as folliculitis and acne), and pigment changes (such as hypermelanosis and hypomelanosis). It should be clearly understood that these pathologic responses in man have not been reported as related to community airborne POM, but to the use of POM in work or at home.

Nonallergic Dermatitis

Nonallergic dermatitis in man caused by materials containing polycyclic aromatic hydrocarbons is reportedly associated with the same kinds of work exposures that may produce skin cancer. The materials include derivatives of fossil fuels, such as coal tar, pitch, creosote, and asphalt from coal; paraffin distillates, high-boiling petroleum residues, asphalt, and lubricating, cutting, and coolant oils from petroleum; and shale oil.[683 (pp. 300–335)] Many polycyclic aromatic carcinogens, such as benzo[a]pyrene and 7,12-dimethylbenz[a]anthracene, are primary irritants for animal skin in concentrations as low as 1% in equal parts of acetone and olive oil or 0.5% in ethanol or in pharmaceutical-grade white mineral oil (R. R. Suskind, personal communication).

The inflammatory response of the skin to irritants and antigenic agents in man is known as an eczematous reaction. It is characterized clinically by erythema, swelling, and vesiculation. One cannot distinguish histologically between skin reactions provoked by a chemical irritant and by a sensitizer. In man, the histologic features of the acute process (acute contact dermatitis) are intercellular and intracellular epidermal edema, which may lead to vesiculation or blister formation; vascular dilatation in the upper dermis; and edema of the dermis. The cellular infiltrate is usually composed of neutrophils and lymphocytes. In chronic contact dermatitis, the epidermis is thickened and there is elongation of the epidermal rete ridges and a marked thickening of the protective horny layer (hyperkeratosis and acanthosis). Some microscopic vesicles may be present in the subacute process but as a rule are absent if the problem is longstanding. The upper dermis may contain a moderate to large number of cells that are predominantly lymphocytes. Histocytes, fibroblasts, and eosinophils may also be found. The infiltrate is usually perivascular. Neutrophils are rare. The number of capillaries seen in sections may be increased and the walls of arterioles thickened.

The skin response in mice and guinea pigs with single or repeated exposure to benzo[a]pyrene is characterized by an inflammatory response in which erythema and edema are primary events. This is followed by epidermal and some dermal necrosis, hair loss, and depigmentation.

Cell-Mediated Hypersensitivity

When the skin of guinea pigs and mice is exposed to a carcinogen, such as benzo[a]pyrene, a significant degree of immunoblast (pyroninophilic cell) response in regional nodes can be induced.[274] The immunoblast response is a characteristic primary event when a mammalian host is exposed to antigenic molecules. When guinea pigs that have been repeatedly exposed to benzo[a]pyrene are challenged with appropriate low concentrations (0.001–0.005% in ethanol), a delayed hypersensitivity reaction of the skin—allergic contact dermatitis—is produced (R. R. Suskind, personal communication). The clinical response in the skin in 24–48 hr is characterized by erythema and edema that persists longer than a primary irritant response. Allergic contact dermatitis in man has been reported from therapeutic coal-tar preparations, but it is rare. The list of polycyclic organic materials known to sensitize after skin contact is sizable, e.g., anthraquinone and its derivatives, bisphenols (such as bithional and hexachlorophene), tetrabromfluorescein, tetraiodofluorescein, mercaptobenzothiazole, β-naphthol, α-naphthylamine, phenothiazines, phthalic anhydride, rhodamine, rotenone, and halogenated salicylanilides (such as tetrachlorsalicylanilide). However, these are rarely if ever found as community air pollutants.

Cutaneous Photosensitization

Exposure to POM in the presence of solar radiation or ultraviolet radiation from other sources may produce an inflammatory skin response in man. There are essentially two types: Phototoxic reactions are dose-dependent; no cell-mediated hypersensitivity state prevails; and clinically they present as exaggerated sunburn. Photoallergic reactions are not dose-dependent; a cell-mediated hypersensitivity mechanism is involved; and clinically they take the form of eczematous allergic contact dermatitis. The antigenic agent may be the original airborne material, an ultraviolet-mediated degradation product, or a physiologic metabolite. Examples of naturally occurring phototoxic agents are furocoumarins like 8-methoxypsoralen, found in celery rot, and 5-methoxypsoralen, found in oil of bergamot,[585] a common ingredient of perfumes and scents. The furocoumarins are the phototoxic agents in such plants as cow parsnip, St. John's wort, mustard, and figs. Contact with these plants may produce photodermatitis. Aerosol solvents, such as methylated naphthalenes, used in insecticide application, are notorious for their phototoxic potential.[221]

In the production and preparation of some drugs in industry, hospitals, clinics, etc., contact with the dust or solutions of the drugs may produce photodermatitis. Polycyclic organic compounds that are photosensitizers include phenothiazine derivatives, such as chlorpromazine and promethazine; hydrochlorthiazide; fluorescent dyes, such as eosin and trypaflavine; and antibiotics, such as demethylchlortetracycline.

In industrial uses, components of pitch, coal tar, and creosote, to which roadbuilders, roofers, gas workers, coke-oven workers, etc., are exposed, may produce a phototoxic reaction that is characterized by a short induction period (a few hours) and the appearance of an exaggerated sunburn. The redness and scaling subside after removal from exposure. After decline of the inflammation, hypermelanosis is often observed. The photoallergic dermatitis cannot be distinguished histologically from allergic contact dermatitis. Photosensitive eruptions are usually limited to the sun- or ultraviolet-exposed areas of the skin—commonly the face and hands.

Pilosebaceous Reactions

Some types of POM and mixtures in which they are found are known to induce changes predominantly in hair follicles and sebaceous glands. When the inflammatory reaction is limited to or around the follicle, the skin problem is called "folliculitis"; when the entire pilosebaceous apparatus is involved, it is known as "acne."

Acne is characterized primarily by lesions involving the pilosebaceous apparatus of the skin, such as comedones, milia, cysts, nodules, follicular inflammation, erythematous papules, excessive oiliness, pustules, and abscesses. Although the problem is usually associated with puberty (acne vulgaris), such lesions may be provoked in adults by a variety of chemical agents. Among the known causative substances are components of crude petroleum, cutting oils, coal tar and some of its products, and chlorinated aromatic compounds, such as chlorinated naphthalenes, chlorinated diphenyls, and chlorinated diphenyloxides like "dioxin." The latter is a common name for a group of potent acnegenic compounds that are chlorinated dibenzodioxins, for example, 2,3,7,8-tetrachlorodibenzo-*p*-dioxin.

Occupational acne can also occur in persons involved in the manufacture of inorganic and organic chlorinated compounds in which acnegenic substances are produced, either inadvertently in small amounts or as intermediate compounds.[738] These may gain access to the skin through the air or by direct handling.

PETROLEUM AND ITS DERIVATIVES

Workers in oil fields and in refineries who may have prolonged skin contact with either the crude oil or the heavier oil fractions may develop acneiform lesions, folliculitis, or both in the exposed areas.[738] Weeks or months of contact with the material in relatively unhygienic conditions are usually necessary. These conditions are now uncommon among oil-field and refinery workers in this country.

By far the most common sources of acneiform skin eruptions and folliculitis are the cutting oils used in machine-tool operations, such as cutting, grinding, milling, boring, and honing. The acnegenic oil is carried to the skin in mist, in aerosol form, or by direct contact.

COAL-TAR PRODUCTS

Persons who maintain unhygienic contact with coal-tar oils and pitch—such as coal-tar plant workers and handlers of roofing, roadbuilding, and construction materials like pitch and creosote—may develop acne, folliculitis, or both.[738] Because some of these materials also photosensitize the skin, exposure to sunlight may provoke photochemical dermatoses, including exaggerated sunburn and melanosis.

Other types of dermatoses that may be provoked by petroleum and coal-tar fractions are contact dermatitis (nonallergic dermatitis or allergic sensitization), papillomas, keratoses, and cancer.

CHLORINATED HYDROCARBON COMPOUNDS

The chlorinated aromatic compounds are among the most potent acnegenic materials. The chlorinated naphthalenes, diphenyls, and diphenyloxides have unusual dielectric and flameproofing properties and are used as electric wire and cable insulations, as well as condenser dielectrics. Mixtures of the chlorinated naphthalenes and diphenyls are called "halowaxes." These hydrocarbons, with three or more chlorine atoms substituted for hydrogen atoms, are acnegenic. Dioxin was found to be responsible for outbreaks of severe acne in U.S. and West German plants manufacturing trichlorophenoxyacetic acid, a widely used weed killer.[738]

PATHOGENESIS AND PATHOLOGY

As in acne vulgaris, the primary cellular response in chemically induced acne is proliferation of the follicular epithelium, which lines the sebaceous duct and follicle pore. Hyperkeratinization of the duct and pore cells results in plugging of the orifice, which prevents normal extrusion of sebum. A modulation of the sebaceous cells occurs. The lipid-bearing cells of the sebaceous gland appear to be replaced by keratinizing cells, and the process later produces a cyst or sac filled with keratin and retained sebaceous lipid. These are the events that occur in the evolution of comedones and keratin cysts.[738]

Any of the three categories of hazardous materials already discussed may produce plugging, cyst formation, folliculitis, furuncles, and abscesses; and it is not possible to distinguish between the reaction patterns of the different causative agents. It is generally agreed, however, that petroleum products that produce comedones and cysts will also provoke more severe and widespread follicular inflammatory reaction and furunculosis than will coal-tar products. In the latter case, increased melanin formation is seen much more often than in response to petroleum products, and the retention cysts are usually smaller than in other types of acne. Melanosis is also seen in acne provoked by chlorinated hydrocarbons.

Pigment Disturbances

Two types of pigment reactions to POM are possible: hyperpigmentation and hypopigmentation.

Most of the agents that induce hyperpigmentation and contain POM are photosensitizing. They include coal-tar products, low-boiling petroleum fractions (such as methylated naphthalenes used in insecticide sprays[221]), essential oils from plant sources containing furocoumarins, and such dyes as tetrabromfluorescein. Again, ultraviolet radiation is a critical factor, and the chemical agent enhances pigment darkening and pigment synthesis. Most of the phototoxic agents will induce hypermelanosis of the skin.

Decrease in the color of the skin may result from damage to the melanocytes or interference with melanin biosynthesis or maintenance. Hence, severe irritation or chemical burns may result in temporary or permanent loss of pigment as a result of cell death. There are no recorded incidents of a polycyclic aromatic hydrocarbon inducing depigmentation or hypopigmentation on a biochemical basis, as in the case of the reaction of skin to the monobenzyl ether of hydroquinone or the reaction of hair to mephenesin carbamate, a muscle relaxant. Persons who have chronic photodermatitis in which the inflammatory component is severe and prolonged may develop irreversible pigment loss (leukoderma). This is the consequence of decreasing the population of melanocytes by cell death.

17

Clinical and Epidemiologic Studies

OCCUPATIONAL SKIN EFFECTS

Most of the data regarding cutaneous effects in man of exposure to POM are found in reports of occupational incidents. There is no documentation that particulate materials containing POM in community air have caused any adverse skin effects.

In the occupational problems that have been described, the hazardous material reaches the skin either by direct contact or as an aerosol, dust, or mist. Studies of industrial exposures in which the skin is affected do not attempt to differentiate exposures by air from exposures by other means. This aspect should be considered carefully in attempting to extrapolate the information derived for occupational problems to the possible hazards of community air pollution.

Numerous materials are recognized as carcinogenic for man, and polycyclic aromatic hydrocarbons have been identified in some of these. It is likely, however, that many of the actual carcinogens contained in combustion and distillation products of carbonaceous substances are either identical or similar. Most of these are polycyclic aromatic hydrocarbons of the benzo[a]pyrene and benzanthracene types. The few studies in which clinical evaluations have been correlated with chemical analyses support the belief that the carcinogens in question are polycyclic aromatic hydrocarbons.

Hendricks *et al.*[358] showed that cancer of the scrotum occurred in wax pressmen but affected only workers with extensive, continuous, or prolonged exposure to slack or crude wax containing high concentrations of polycyclic aromatic hydrocarbons. Workers exposed to finished waxes that were low in aromatics did not develop cancer. W. E. Smith *et al.*[710] confirmed these epidemiologic suspicions when they demonstrated that only the aromatic portions of the crude wax were carcinogenic in mice. A convincing discussion of evidence incriminating benzo[a]pyrene as a carcinogen for man has been presented by Falk *et al.*[259] However, many studies, especially in earlier years, related only such complex chemical mixtures as pitch and tar to the cancer problem, and it is only on circumstantial evidence that the polycyclic aromatics can be suspected of having been the actual carcinogens in those instances.

A variety of factors influence the induction of cutaneous cancer in man, including degree or level of exposure, concentration of carcinogen, and duration of exposure; factors affecting absorption of carcinogen or cocarcinogen, such as vehicle, presence of surfactant, and concomitant or prior physical or chemical injury to epidermal barrier; factors affecting the carcinogenic activity of material on target tissue, such as chemical or physical cocarcinogens, long-chain hydrocarbons, ultraviolet radiation, and ionizing radiation; and genetic pigmentation, the primary factor in determining skin cancer as related to exposure to ultraviolet radiation.

Pitch, Tar, and Asphalt

Cyclic hydrocarbons are found in so-called high-temperature tar distillation, which is carried out above 370 C. The use of horizontal retorts is said to increase the amount of polycyclic aromatic hydrocarbons by a factor of 4. Skin carcinomas are often observed in occupations in which "high-temperature tar" is the exposure material, and they are not considered to be associated with low-temperature tar fractions, which do not contain polycyclic hydrocarbons.[402]

Cancer associated with these products has been recognized since 1876, when Volkmann described skin cancer in a worker employed in a Saxony tar distillery. It has since been the subject of many reports, including those by Bridge and Henry,[89] Haagensen,[326] Carozzi,[118] Staemmler,[719] and Uytdenhoef.[778]

Henry[360] analyzed the official data issued by H. M. Chief Inspector of British Factories from 1920 through 1945. Pitch, tar, and tar products were responsible for 2,229 (59.4%) of the 3,753 cases reported. Shale oil, mineral oil, and bitumen were associated with 1,515 cases (40.3%). The occupations in which skin cancers due to pitch, tar, or tar products were reported were diverse. Most of the cases, however, were among persons occupied in tar distilling (538), briquette manufacture (364), coal-gas manufacture (305), and pitch loading (36).

Neve[562] observed more than 2,000 kangri cancers in Kashmir. The kangri is an earthenware bowl 6 in. or more in diameter held against the skin as a personal heating appliance. The bowl is filled with hot wooden embers and cooled by sprinkling with water. The temperature may reach 150 F. It has been maintained that the cancers are due to repeated burns,[640] but it is likely that the carcinogenic agent is a tarry distillation product of woodcoal whose action may be increased by small burns[402] or continuous exposure to soot. Similar factors may be associated with skin carcinomas related to the use of the kairo in Japan.[401]

Fractionation and Distillation Products of Oils

Occupational cancers due to contact with fractionation and distillation products of oil have been recognized for many years. Some oils and their products appear to be more carcinogenic than others. In general, the data from epidemiologic surveys and animal experiments are correlated with the presence or use of polycyclic aromatics.[58] Other factors play a significant role in carcinogenesis; for example, the concentrations of accelerators and cocarcinogens influence the rate of tumor development in shale-oil refinery exposures.

The hazards to health previously found in such persons as petroleum refinery and shale-oil workers, cotton mule spinners, and machinists exposed to cutting oils can be eliminated by good hygienic practices.

Skin cancer in the shale-oil industry has been studied in Scotland, where Bell[43] first recorded the occurrence of two cases of scrotal cancer among paraffin pressmen. In the previous year, Volkmann had described three skin cancer cases among workers distilling paraffin wax from lignite (brown coal). Scott[684] reported that 89 cases of epithelioma had been observed in a Scottish shale refinery from 1900 to 1928 among workmen employed as wax pressmen or in the distillation of shale oil. He calculated that, inasmuch as the oil company employed 5,000 men, the cancer incidence was 2% over the 28 years surveyed. However, because most of those workers had no prolonged or significant exposures, the actual risk among the exposed was considerably higher.

Henry,[360] in his analysis of cases of occupational skin cancer officially recorded in Great Britain from 1920 to 1945, cited 52 cases of skin cancer in 42 workmen engaged in oil refining. In all but four of these cases, the contact had been with shale oils. Most of the cancers were on the exposed hands and forearms, but 30% occurred on the scrotum.

Twort and Twort[759] treated mice with oils from various sources. They found Scottish shale oil to be the most potent carcinogen, the unfinished lubricating fraction being more carcinogenic than the finished oils or the crudes. They felt that the carcinogenicity could be removed by treatment with sulfuric acid purification. Schwartz *et al.*[683 (pp. 726–737)] refer to the observation that shale oil is the most carcinogenic and cite evidence that the incidence with various oils is inconsistent with the assumption that benzenoid hydrocarbons are carcinogenic and are almost all destroyed by treatment with sulfuric acid, by oxidation and reduction processes, or by refining. Bingham and Horton[57] found that, when mice were painted with various crude or partially refined oil fractions, carcinogenesis was correlated best with the chromatographic mass spectrometric analyses of four- and five-ring aromatic compounds.

MULE SPINNING

Some of the most dramatic associations between lubricating oils and skin epitheliomas were noted in the 1920's among mule spinners* in the cotton industry in Great Britain. The relation was first established by Southam and Wilson,[715] who analyzed 141 cases of scrotal carcinoma observed at the Manchester Royal Infirmary over a 20-year period. Sixty-nine of the cases were in mule spinners in local cotton mills. Leitch[486] showed that 20% of the fatal cases of scrotal carcinoma in England occurred in mule spinners, an industry employing approximately 23,000 men. He found a yearly average of 11.8 cases of fatal cancer, or 50 fatalities per 100,000 spinners

*A *mule spinner* in the textile industry is an operator of a *mule,* a spinning machine that makes thread or yarn from fibers. Lubricant oils used in these machines were derived from oil shale and petroleum. The clothes of the spinner, which were changed infrequently, became saturated with oil from the oil mist, from the oily surface of the machine, and from spillage during maintenance.

employed per year. Southam[714] calculated the incidence of scrotal cancer at 250 cases per 100,000 spinners per year, and estimated that at that time 50 new cases of scrotal carcinoma were seen each year.

Henry[360] found that, of 3,753 cases of cutaneous carcinoma notified as occupational in Great Britain from 1920 to 1945, 1,389 occurred in the cotton industry. Of these, 1,296 (93.3%) were in persons who worked in the mule-spinning room. Another 48 persons who had been classified as having other occupations had been mule spinners at some time (however short) in their lives. Of the skin cancers in the cotton industry, 28.8% were on exposed surfaces, 10.7% on covered sites other than the scrotum, and 60.4% on the scrotum. In all cases, the carcinomas in the cotton industry were believed due to contact with mineral oils, presumably containing POM.

Skin cancers in mule spinners have rarely been reported outside the British Isles. Until Heller[357] studied the incidence of scrotal cancer among mule spinners in the United States in the 1920's, the disease had been reported only in immigrants. By studying hospital records and death certificates and questioning plant physicians, he was able to find records of only two cases of scrotal cancer in mule spinners who had not been employed in this occupation outside the country. He concluded that the incidence was insignificant in mule spinners in the United States and attributed this to the "refined character of the oil" used. Since 1953 in Great Britain, the use of noncarcinogenic oil, with a reduced polycyclic aromatic hydrocarbon content, has been obligatory under the Mule Spinning (Health) Special Regulation.[268] Only seven cases of skin cancer among mule spinners were reported since that time.[15] In 1966, one death from scrotal cancer was reported; the victim was a man who had been occupationally exposed before 1953.[16]

PETROLEUM REFINING

Heller[357] described 20 cases of cancer caused by industrial mineral oils in the United States. Eight were in employees of a refining company using crude oils from Illinois and Indiana. Eleven were gleaned from the records of the Memorial Hospital, New York, and the New York Skin & Cancer Hospital. Eight tumors were on the scrotum, and all except one of the others were on exposed parts. He was unable to document cancers in workers handling refined lubricating oils.

Hendricks *et al.*[358] and Lione and Denholm[497] studied the incidence of cancer among wax pressmen in one refinery from 1937 to 1957. Eleven cases of scrotal cancer were recorded, the incidence among those with 10 or more years of service being many times that among the general male population, although the incidences of other forms of cancer were not increased. These cases were directly associated with extensive, continuous, and prolonged exposure to slack or crude wax. Workers exposed to finished wax had no scrotal carcinoma. Only the pressmen were in contact with the aromatic oils. These data were consistent with the work of Smith *et al.*,[710] who demonstrated that only the aromatic portions of the crude wax were carcinogenic in animals.

LUBRICATING AND CUTTING OILS

Cruickshank and Squire[174] studied British workers exposed to mineral oils in the engineering industries. Of 138 workers using cutting oils, 60% of those exposed for more than 15 years had multiple hyperkeratoses on their hands and one had a scrotal cancer. They later investigated the records of scrotal carcinoma in the United Birmingham Hospitals between 1939 and 1948. Thirty-four cases had occurred—12 in those exposed to oil in the engineering industry, 13 in workers exposed to tar pitch, etc., and nine that could not be allocated to a definite etiologic association.

Cancer of the hands and forearms in those attending the United Birmingham Hospitals between 1941 and 1950 was studied by Cruickshank and Gourevitch.[173] Of 44 patients, 18 gave a history of occupational exposure to various oils and six of exposure to pitch, and three were in other occupations.

Mastromatteo[521] described six cases of squamous cell carcinoma in a single Canadian plant employing just over 1,000 workers. Five of the six cases occurred in machine-tool operators and were attributed to the carcinogenicity of cutting oils.

CREOSOTE

Creosote, a fractionation product of tar, has been associated with a number of reported cases of occupational cancer. The actual carcinogen has not been identified in the case reports, but this material contains aromatic hydrocarbons.

Reports of groups of cases include those of O'Donovan[569] (epitheliomas in four timber picklers) and of Bridge and Henry[89] (cancer in four timber picklers and six brick tile or pipe pressers). Henry[360] reported cases between 1920 and 1945 in 34 workmen, of whom 14 were employed in creosoting timber, eight in creosote storage, 11 at brick or pottery presses, and one in the manufacture of a creosote disinfectant.

ANTHRACENE

Anthracene is a tricyclic aromatic obtained from the crude oil of coal tar. A number of reports have described cases of cancer due to "anthracene." Leymann in 1917[495] noted that an annual report issued in 1902 by chemical plants in Oppeln, Silesia, recorded the occurrence of a variety of skin lesions in 22 of 30 workers in an anthracene plant. Three had undergone operations for scrotal cancers. Bridge and Henry[89] mention four cases of epithelioma in anthracene workers. Henry[360] notes the occurrence of five cases of epithelioma in workers at synthetic dyeworks engaged either in the purification or (in one case) the loading of boxes of anthracene. Repeated exposure of susceptible mice and rats to anthracene fails to produce skin cancer. Because it is generally held that man is less susceptible than rodents to polycyclic aromatic hydrocarbons, it is unlikely that anthracene itself is the carcinogen in anthracene plants; a product, more closely related to benzo[a]pyrene, is probably responsible.

SOOT

Percivall Pott described cancer of the scrotum in chimney sweeps in 1775. By 1788, laws regulating activities in the chimney sweep trade had been passed in England. Henry[359] cites an investigation of 1,631 cases of scrotal cancer, of which 121 were in chimney sweeps, 125 in metal workers, and 575 in textile workers. It is difficult to derive actual figures of incidence from these early reports, but Henry calculated the incidence of scrotal carcinoma from death certificates as 754.7 per million. Surprisingly, the Annual Report of H. M. Chief Inspector of Factories in 1964[15] noted that scrotal cancer in chimney sweeps was still occurring; the deaths of five former chimney sweeps due to scrotal cancer were reported between 1962 and 1964.

Factors That Influence POM Carcinogenesis in Man

Animal experiments have demonstrated that many factors can influence the susceptibility of the skin to specific identifiable carcinogens, such as benzo[a]pyrene and 7,12-dimethylbenz[a]anthracene. The prime factor of chemically induced cancer in man is the degree of exposure (how much and for how long), which depends, e.g., on the concentration of chemical carcinogen and the aggregate duration of exposure. Other factors can be regarded as cocarcinogenic.

Salaman and Roe[657] list the factors that may be considered as cocarcinogenic in various circumstances. When a carcinogen is applied to the skin, the vehicle may determine the rate of absorption and the amount absorbed. Factors that block detoxification or excretion of the carcinogen may also increase carcinogenesis. Altered physiologic states, such as hyperemia of the skin or subcutaneous tissue, may increase carcinogenesis, and agents that induce these changes may be classified as cocarcinogens. In experimental animals, carcinogenesis may be slowed or even inhibited by such conditions as dietary restrictions, stress, or altered hormonal balance.

CHEMICAL COCARCINOGENS

Suskind and Horton[739] discuss the influence of chemical accelerators like *n*-dodecane, a straight-chain hydrocarbon found in petroleum products. By applying a single dose of a strong carcinogen to mice and then applying dodecane, Horton *et al.*[396] were able both to decrease the latent time before onset of tumor formation and to increase tenfold the percentage of mice ultimately developing tumors.

The carcinogenic properties of combustion and distillation products may depend also on other aromatic substances, such as epoxides. An excellent review of animal experiments demonstrating the carcinogenicity of epoxides has been written by Van Duuren.[780]

The knowledge derived from experiments about the carcinogenic potency of petroleum and distillates on the skin has been summarized by Bingham and Horton.[57] A carcinogen, most probably a four- or five-ring polycyclic aromatic hydrocarbon, must be present. In occupational exposures, fractions not containing these compounds have not been shown to be carcinogenic. The potency of a fraction may be influenced to a great extent, however, by long-chain aliphatic and aromatic hydrocarbons (e.g., *n*-dodecane, dodecylbenzene, and diamylnaphthalene). In addition, some sulfur-containing compounds not precisely identified appear to influence and hasten the development of tumors. Some epidemiologic information is in close accord with these experimental observations. Thus, industrial exposures to coal tar and pitch high in polycyclic carcinogens result in tumors usually within 20–24 years of commencing that work. Spindle oils in which carcinogen concentration and accelerator concentration are low result in tumors that have a long latent period (50–54 years).[32] In the case of the paraffin wax processing using material that is low in carcinogen but high in accelerator, the latent period is as short (20 years) as in the case of exposures to coal tar, which is high in carcinogen and negligible in accelerator (Table 17-1).

ULTRAVIOLET RADIATION

It is known that ultraviolet radiation of wavelengths shorter than 320 nm is a significant carcinogen in itself and is responsible for most cases of skin cancer. Blum[64] summarizes the evidence on sunlight as an etiologic agent in human skin cancer. The limiting factor is genetic, i.e., the relative amount of melanin pigment in the skin.

The possible interactions between POM and ultraviolet radiation are particularly important. All the present evidence of such interactions is derived from animals. Epstein[231] has shown that ultraviolet radiation of 280–320 nm will accelerate 7,12-dimethylbenz[a]anthracene carcinogenesis in hairless mice. Using carcinogenic amounts of ultraviolet radiation (12.06×10^7 ergs/cm^2) in mice, a single application of 7,12-dimethylbenz[a]anthracene before initiation of ultraviolet exposures accelerated the rate of tumor appearance and their growth, as well as the incidence of tumors per mouse.

Santamaria and Giordano[660] shed further light on the relation be-

TABLE 17-1 Carcinogenesis in Man and Rodent: Relation of Latent Period to Carcinogen Content of Fossil-Fuel Fractions and Long-Chain Acceleration[a]

Material	Relative Carcinogen Content	Probable Percentage of Long-Chain Accelerators	Average Latent Period in Mice, weeks[b]	Average Latent Period in Man, years
Paraffin distillate (slack wax)	Low	High	20–35	20 or more
Spindle oil	Low	Low to moderate	60	40–50[c]
Coal tar	High	Negligible	12–27	23[c]
Pitch	Very high	Negligible	–	(23[c], bracketed with coal tar)
White mineral oil	None	Low	Noncarcinogenic	Noncarcinogenic

[a] Derived from Suskind and Horton.[739]
[b] Three applications per week.
[c] Based on data from Henry.[359,360]

tween ultraviolet radiation and polycyclic hydrocarbon carcinogenesis. They quote the conflicting results of previous studies on the action of ultraviolet radiation on polycyclic hydrocarbon carcinogenesis in relation to acceleration. They include the work of Findlay,[272] Maisin and De Jonghe,[508] Vlès *et al.*,[794] and Clark.[140] In relation to inhibition, they discuss the work of Doniach and Mottram,[211] Morton *et al.*,[545,546] Kohn-Speyer[442] (who reported no effect of ultraviolet radiation on cancer induction), Seelig and Cooper,[688] and Rusch *et al.*[646] In the papers cited, little attention is given to the total quantity of ultraviolet energy and wavelength, which are critical factors in carcinogenesis.

Santamaria and Giordano[660] found that the carcinogenic activity of polycyclic hydrocarbons was to a large degree associated with their photodynamic action, which can be demonstrated in *in vitro* experiments at molecular, subcellular, and cellular levels. They also studied changes in electrophoretic patterns of human serum proteins produced by various polycyclic hydrocarbons in the presence and absence of ultraviolet radiation. Again, there was a strong association between photodynamic activity and carcinogenic activity.

In vivo studies were then performed by painting benzo[a]pyrene on the skin of mice. Groups of mice were irradiated with various exposures of long-wavelength ($>$320 nm) ultraviolet radiation. Blum showed that, although these wavelengths are not carcinogenic in themselves, they will actively excite benzo[a]pyrene molecules. Small doses of ultraviolet radiation were found to increase tumor incidence, the most effective dosage being 8×10^{10} ergs/cm^2 twice a week. Greater doses were associated with inhibition of carcinogenesis and with more severe tissue damage. It was interpreted that this tissue damage, rather than alteration of the carcinogen, led to the inhibition of carcinogenesis.

Conclusions

The role of POM in the production of human skin cancer from occupational exposures seems well established. It appears that levels of exposure to carcinogens or cocarcinogens in industry are very different from those due to community air pollution. In most instances, the specific kind of POM has not been identified, nor has the importance of airborne transmission of the carcinogenic agent been considered in contrast with other routes of exposure.

The data do not allow the construction of any accurate dose–response relations, although carcinogenesis may occur at a high level of exposure to carcinogen or cocarcinogen. There is no information on the effect of clearly lower concentrations of chemicals, such as would be found in polluted community air.

The effect of other factors, such as cocarcinogens, is obviously important. This is illustrated by the effect of high concentrations of straight-chain hydrocarbons with low concentrations of polycyclic hydrocarbons in the production of shale-oil cancers in man.

Because the vast majority of human skin cancers are due to the effects of ultraviolet radiation, any factor that influences or is influenced by the biologic activity of ultraviolet radiation in carcinogenesis could be epidemiologically important. It is in this context that experiments showing synergism between ultraviolet radiation and POM have great significance.

Obviously, additional research is needed before the effects of ambient-air polycyclic compounds on the incidence of skin cancer can be assessed. Such research must be related to the actual conditions of human exposure and must take into account compounds other than POM that may be present in ambient air and the concomitant effects of ultraviolet radiation.

OCCUPATIONAL PULMONARY DISEASE

Studies of industrial populations have always been important in identifying etiologic factors in disease. Such studies are particularly valuable in establishing dose–response relations when the epidemiologic investigations of disease patterns are coupled with quantitative estimates of job-related exposures. There is ample industrial evidence that some polycyclic organic compounds are carcinogenic. Polycyclic organic matter is produced mainly by the combustion of fossil fuels, and the major industry-related cancer experience has been in industries that are heavily involved with the combustion or distillation products of coal. The first cancers related to POM were of the skin—e.g., scrotal cancer in chimney sweeps[379,600] and facial epithelioma in workers exposed to coke, coal, tar, and pitch.[104-106] However, a number of early investigators speculated that lung cancer was related to exposure to coal tar. In 1936, the Japanese described an unusual lung cancer experience in men engaged in coal carbonization for the production of gas: 12 of the 15 cases of cancer occurred in the lung.[429,430] At that time, lung cancer was relatively rare in Japan, accounting for only 3% of all malignant neoplasms. Kennaway and Kennaway[434] reported an approximately threefold excess lung cancer mortality for gas-production men, chimney sweeps, and several categories of gas workers from 1921 to 1938. The excess of lung cancer in gas workers was later confirmed by Doll *et al.*[206]

Lloyd[502] has found that the lung cancer death rate was 2.5 times higher than expected in coke-oven workers. Most of the lung cancers occurred in men who worked on top of the coke ovens: Those employed 5 or more years at full-time topside jobs had a tenfold excess risk of lung cancer.

Although it seems clear that there can be an excess risk of developing lung cancer in the coal-tar occupations, quantitative exposure data are very meager. There is only one study that provides some data on the incidence of lung cancer with respect to the magnitude of exposure to POM.[206,483] A selected population of 11,449 employees of the British Gas Works Industry was followed for an 8-year period. The study population included only employees who had at least 5 years of employment and were 40–65 years old at the beginning of the observation period. All but 0.4% of the men were successfully followed. The workers were classified into three broad categories of exposure—heavy, intermediate, and no exposure. Relative to the group without significant exposure, the heavily exposed workers had a 69% higher lung cancer incidence and, unexpectedly, a 126% higher death rate from bronchitis. There were no differences in smoking habits between the three groups of workers and the general population.

The concentrations of benzo[a]pyrene and other polycyclic aromatic hydrocarbons in gas-works retort houses of several types were measured.[483] The tarry fumes that escaped from retorts contained extremely high concentrations of polycyclic hydrocarbons, but, in general, men were exposed to these fumes only very briefly. The mean concentration of benzo[a]pyrene determined from long-period samples at sites representative of normal working conditions in three works was 3 $\mu g/m^3$, over 100 times the normal level in London. Above the retorts in an old horizontal retort house, the concentration was approximately 216 $\mu g/m^3$, about 10,000 times that in the city, and the "top-man" working there could be exposed to this in the normal course of his duty.

Figure 17-1 shows the above data as a crude dose–response curve relating the level of exposure to benzo[a]pyrene to the lung cancer mortality ratios of the urban nonsmokers, the average British gas-worker,[206,483] and the topside coke-oven worker.[502] Urban concentrations of benzo[a]pyrene vary markedly among cities, but it has

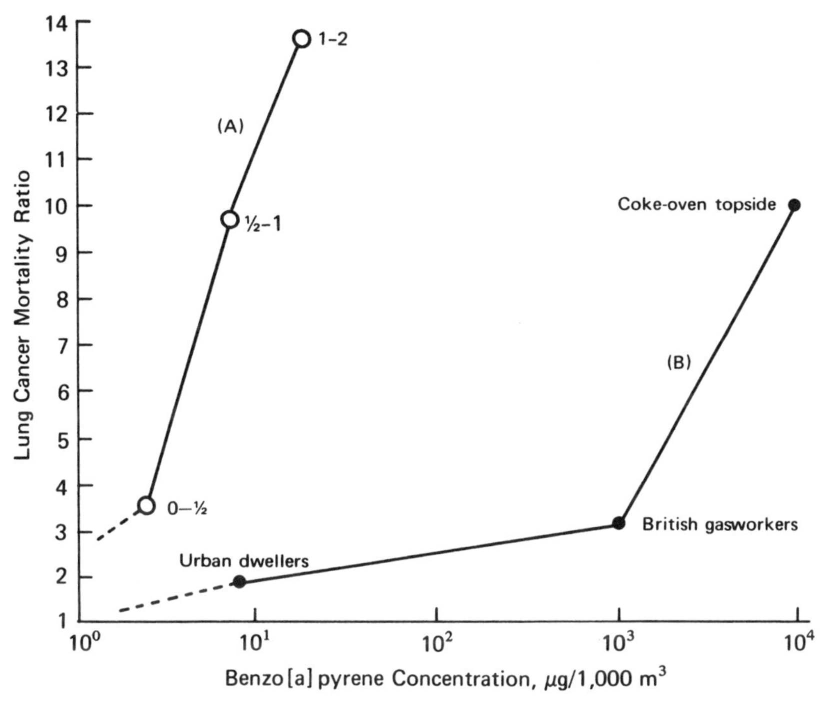

FIGURE 17-1 Dose–response curves: (A) for cigarette smokers (packs per day);[337] and (B) relating level of exposure to benzo[a]pyrene to lung cancer mortality ratios of urban nonsmokers, average gasworker, and topside coke-oven workers[206,483,502] (prepared by crude extrapolation of data).

been estimated roughly that the city dweller is exposed to a concentration of about 8.5 $\mu g/1{,}000\ m^3$. This estimate is plotted in Figure 17-1 with the corresponding mortality ratio of 2, which represents the risk relative to the rural non-cigarette-smoker. The British gasworkers experienced 69% more lung cancer than the general population and were exposed to an average concentration of 3,000 $\mu g/1{,}000\ m^3$ of benzo[a]pyrene.[206,483] For equivalence to urban atmospheric exposure, which takes place on a 24-hr/day basis, the occupational benzo[a]pyrene concentration is reduced by a factor of 3 to 1,000 $\mu g/1{,}000\ m^3$ and plotted in Figure 17-1 with the corresponding mortality ratio of 3.4 (i.e., 1.69×2.0). The increased mortality ratio, of course, represents the comparative response of the gasworker and general populations in which 68% and 61%, respectively, smoked cigarettes; in Figure 17–1, this excess mortality is being applied to urban nonsmokers. This is justified by the expectation that an added carcinogen exposure would constitute an even greater hazard to cigarette smokers than to nonsmokers, so that the degree of increased risk is probably overestimated for nonsmokers.

According to Lloyd,[502] the topside coke-oven workers have about a tenfold increase in lung cancer risk. There are no benzo[a]pyrene exposure data for this population, but the exposure is probably not greater than that found by Lawther *et al.*[483] above the gasworks retorts at 216,000 $\mu g/1{,}000\ m^3$. This value, corrected for a 24-hr/day exposure, is reduced by a factor of 3 to 72,000 $\mu g/1{,}000\ m^3$. This estimate is given in Figure 17-1 for coke-oven topside workers.

It is invariable in dose–response relations for both human and animal cancers that any increase in the dose of a carcinogen over that sufficient to cause a detectable response produces, at the very least, a proportional increase in cancer incidence. On this basis, the benzo[a]-pyrene lung cancer mortality ratio curve in Figure 17-1 lacks plausibility, because a dose increment of two orders of magnitude—from about 10 $\mu g/1{,}000\ m^3$ to 1,000 $\mu g/1{,}000\ m^3$—hardly increases the lung cancer mortality ratio of the average British gasworker relative to the urban dweller.

For comparison, a dose–effect curve for cigarette smokers is also shown in Figure 17-1. This curve is a plot of Hammond's data for lung cancer mortality ratios in American males 55–60 years old consuming 0–10, 11–20, and 21–40 cigarettes per day.[337] The maximal number of cigarettes smoked per day in each category (i.e., ½, 1, and 2 packs per day) is expressed in terms of urban-air benzo[a]pyrene concentrations that, according to the assumptions and calculations of Sawicki *et al.*,[667] produce equivalent lung exposures. The principal point to be noted is that, when the cigarette consumption increases by a factor of 4, from ½ to 2 packs of cigarettes per day, there is a rise in mortality ratio from 3.6 to 13.6.

Doubts can be raised about whether the twofold increase in lung cancer associated with urbanization and atmospheric pollution is due predominantly to benzo[a]pyrene. It may be important to distinguish between the type of pollution found in the retort houses and that in urban air. In a retort, coal is distilled in the absence of air, and the products include tar, carbon monoxide, other combustible gases, and some hydrogen sulfide, but very little sulfur dioxide or black smoke. When coal is burned in a domestic fire, distillation products, including tar, are emitted each time the fire is refueled, but combustion is accompanied by emission of smoke and sulfur dioxide. In more efficient heating appliances, little tar or smoke is produced, and the main products are carbon dioxide and sulfur dioxide. Urban exposures undoubtedly involve a greater duration than industrial benzo[a]pyrene exposures. The urban exposures begin at birth, and there is evidence that newborns are more sensitive to polycyclic aromatic carcinogens than adults.

Other considerations warrant caution in accepting benzo[a]pyrene as a major pulmonary carcinogen at current urban atmospheric concentrations. Lung cancer incidence has steadily increased since the 1940's, and yet, qualitatively, the carcinogen content of urban atmospheric pollution caused by combustion products of coal has been on the decline. Recent tissue-culture evidence indicates that the benzo[a]-pyrene content of particles recovered from city air accounts for less than 1% of its carcinogenic activity.[278]

NONOCCUPATIONAL NEOPLASTIC PULMONARY EFFECTS

Evidence of the incidence of environmentally related lung cancer in humans is derived largely from epidemiologic studies. Four groups of studies can be distinguished and will be discussed here: The first compares urban metropolitan populations with rural populations and examines the overall differences in lung cancer death rates, in most cases without examining specific etiologic factors; the second compares lung cancer death rates in migrants with those in their countries of origin and those in the countries to which they migrate and examines the changes in rates in the migrating population group, which change with changes in environment; the third compares demographic units—including countries, states in the United States, and cities or metropolitan areas—and studies the relation between the lung cancer death rates and various indices of pollution, using multiple-regression techniques in an attempt to separate the effects of environmental and other factors; and the fourth consists of sampling studies in which characteristics of lung cancer decedents are determined by family interviews and compared with the corresponding characteristics of the remainder of the population. Such studies offer the prospect of a relatively sharp discrimination between factors that are strongly related to lung cancer and factors that are only incidentally associated.

The general association between urbanization and increased lung cancer is not in question. Its relation to environmental rather than, say, genetic factors is almost equally certain. What characteristics of the urban environment are primarily responsible is a subject of controversy. The problem of identifying the causal factors is intensified by the difficulties of obtaining either accurate or extensive measures of exposure for most factors and the close association of urban factors with one another—e.g., an area high in benzo[a]pyrene will generally tend to be high in sulfur dioxide and hydrocarbons.

To enable us to summarize the results of different studies on a common scale, one of the common pollutants, benzo[a]pyrene, was chosen as an index of urban air pollution. Benzo[a]pyrene is taken as the primary index of air pollution, with the recognition that it is only one of the polycyclic organic materials in the air. The choice should not be taken to represent a conclusion that benzo[a]pyrene is the causal agent in urban lung cancer. Its selection as an index is plausible for several reasons: It appears in solid form in air, is usually adsorbed on particles, and therefore can be filtered and collected. It is relatively easy to measure and is well correlated with other POM. It has been found to be carcinogenic in animals, and it is suspected of being carcinogenic in man. The intent of the regression analyses in the present study is to establish a relation between urban air pollution levels (as indexed by benzo[a]pyrene) and lung cancer death rate.

The standard measure of benzo[a]pyrene concentration in air is the number of micrograms per 1,000 m^3 of air. A benzo[a]pyrene unit is defined as 1 $\mu g/1{,}000\ m^3$ of air (or 1 ng/m^3 of air).

Figure 17-2 shows the relation of smoke to cancer-producing substances (hydrocarbons). In an article published in 1963,[612] Pybus estimated that 2% of coal burned in London was discharged as smoke and

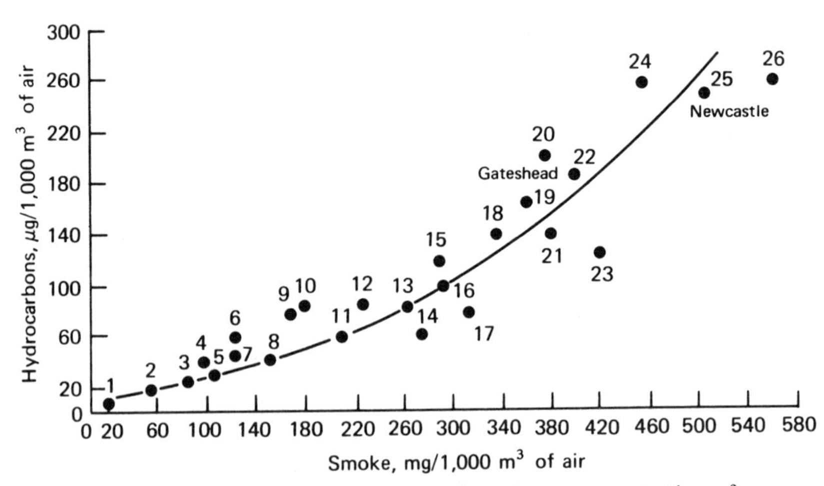

FIGURE 17-2 Relation of concentration of smoke to concentration of cancer-producing substances (hydrocarbons). (Based on data of Pybus.[612])

that 0.03% of the smoke was benzo[a]pyrene. Measurements by Sawicki *et al.*[667] and others show considerable seasonable variation, with winter concentrations some 10–20 times the summer concentrations. In over 100 U.S. sampling sites, Sawicki found an approximate log normal distribution of benzo[a]pyrene concentrations for urban sites, with a median winter–spring value of 6.6 μg/1,000 m^3. A similar distribution was found for nonurban sites, with a median value of 0.4 μg/1,000 m^3, as shown in Figure 17-3. In comparing monthly levels in areas with low and high benzo[a]pyrene, Sawicki found marked seasonal variations, with peak levels in the winter months. The relation of benzo[a]pyrene to season appears to be uniform in both high and low areas. The urban and rural distributions given by

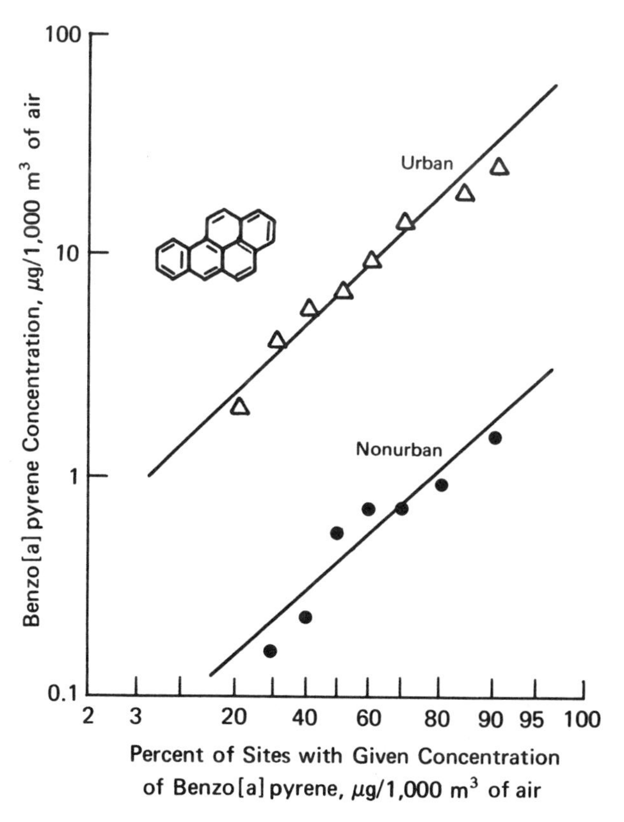

FIGURE 17-3 Frequency distribution of benzo[a]pyrene concentrations in the air in urban and nonurban sites, January–March 1959, based on composite samples. (Derived from Sawicki *et al.*[667])

Sawicki used January–March data. However, rough calculation from Figure 17-4 indicates that the January–March average is approximately equal to the annual average in both cases.

Ideally, we should have available the average concentrations of all suspect air pollutants for every environment in the study and for every year during the last several decades. Unfortunately, direct measurements of such pollutants have been rare in the past and are still sparse. Thus, in comparing countries, Stocks[729] uses such indices as amount of coal burned annually per capita and smoke or sulfur dioxide concentration. Where none of these is available, other indices, such as industrialization and gross national product, have been used. In this discussion, an attempt was made, wherever possible, to relate the measures given to benzo[a]pyrene—the primary index of air pollution.

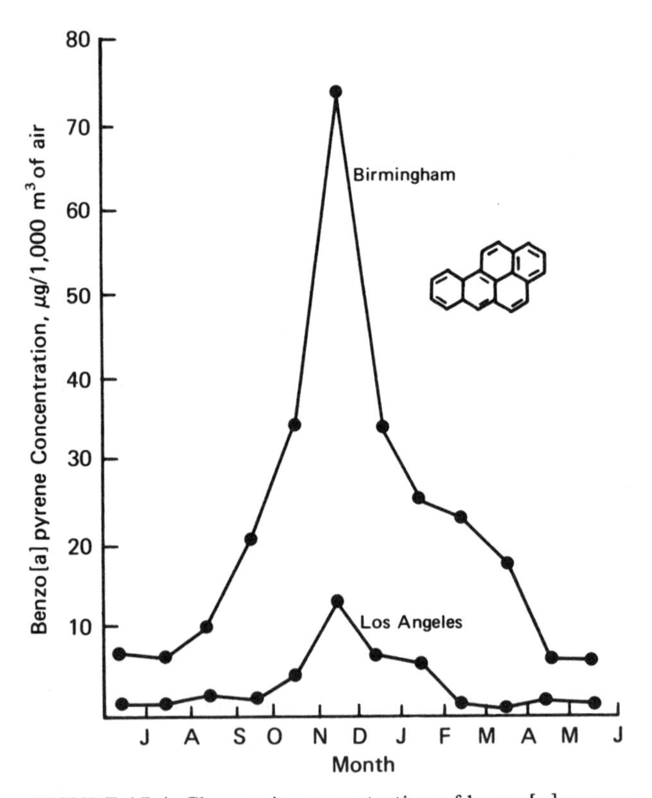

FIGURE 17-4 Changes in concentration of benzo[a]pyrene in the air monthly, July 1958–June 1959, Los Angeles and Birmingham. (Derived from Sawicki *et al.*[667])

Benzo[a]pyrene concentrations reported in the late 1950's, of course, may not be a good indicator of earlier or later pollution levels. The high benzo[a]pyrene concentrations in some areas have been reduced in recent years, owing to replacement of coal burning by other energy sources. Recent unpublished data from the National Air Surveillance Network show a distribution of benzo[a]pyrene concentrations in 1967 that is substantially reduced from that in 1959. The median January–March urban concentration in 1967 was approximately 2.5 units, in contrast with the 6.6 units in 1959; the comparable rural values are 0.4 and 0.2. The benzo[a]pyrene concentrations in different regions at a given time do give a reasonable index of long-term differences in pollution levels.

Types of Rates and Vital Statistics

Lung cancer is predominantly a disease of older men, and the magnitude of a lung cancer index depends heavily on the population base assumed. The studies discussed below vary considerably in the choice of index. The simplest index is, of course, an age–race–sex-specific lung cancer death rate, such as the rate for white males 45–54 years old. (Most studies and tabulations use ICD categories 160–164—cancer of the respiratory system—and, except where specifically noted, this is the group of deaths covered by the term "lung cancer.") Where a general index is called for, all races may be combined and the rate for, say, males over 35 may be used. The choice of 35 seems reasonable, in that few lung cancer deaths occur at lower ages. Because age distributions vary, it is preferable, if age-specific rates are available, to calculate the lung cancer death rate as age-adjusted to the age distribution of some specified standard population.

The direct method of standardization (which applies the age-specific rates to the distribution of ages in the standard population) is readily understood and generally appropriate when available. However, when the study population differs markedly from the standard population, the direct standardized rates may be subject to excessive uncertainty (e.g., a predominantly nonwhite study group standardized to the total U.S. population), and an indirect index may be preferred. The standardized mortality ratio (SMR) is simply the ratio of deaths observed in the particular group to the deaths to be expected if standard rates are applied. Often, the standard used is the mortality for the total group under study, with SMR's calculated for each subgroup.

For ready reference, Tables 17-2 and 17-3 show the age–race–sex

TABLE 17-2 U.S. Population, 1960, Based on U.S. Census

	No. Persons, 1,000's					
Sex and Race	Under 35 Years Old	35–44 Years Old	45–54 Years Old	55–64 Years Old	65–74 Years Old	75+ Years Old
White						
Male	44,931	10,564	9,114	6,850	4,702	2,206
Female	44,378	11,000	9,364	7,327	5,428	2,968
Nonwhite						
Male	6,513	1,192	979	686	414	181
Female	6,803	1,326	1,028	709	453	208
Total	102,625	24,082	20,485	15,572	10,997	5,563

distribution of the U.S. population (1960 census) and the age–race–sex-specific death rates for lung cancer (1959–1961 vital statistics). These data were used to calculate the indices of lung cancer mortality presented in Tables 17-4 through 17-6.

Characterization of Regions

The ascertainment of an urban or pollution factor requires the characterization of region of residence according to the expected de-

TABLE 17-3 Lung Cancer Death Rates (ICD 160–164), 1959–1961[a]

	No. Deaths per Million Persons				
Sex and Race	35–44 Years Old	45–54 Years Old	55–64 Years Old	65–74 Years Old	75+ Years Old
White					
Male	105	526	1,505	2,256	1,836
Female	34	98	173	271	399
Nonwhite					
Male	194	678	1,498	1,762	1,344
Female	40	128	209	253	284
Total	73	317	819	1,175	995

[a] Sources: U.S. Census of Population (U.S. Department of Commerce, Census Bureau) and Vital Statistics of United States (U.S. Department of Health, Education, and Welfare, National Office of Vital Statistics).

TABLE 17-4 U.S. Total Population, Lung Cancer Death Rate (ICD 160–164)[a] U.S. Race- and Sex-Specific Death Rates for Population of All Ages (per Million), 1959–1961[b]

Male		Female	
White	Nonwhite	White	Nonwhite
369	292	66	49

[a] Rates given are averages of three rates for total United States: 1959, 1960, and 1961; the 1959 rate did not include Hawaii deaths.
[b] Sources: Vital Statistics of United States (U.S. Department of Health, Education, and Welfare, National Office of Vital Statistics).

gree of its pollution problem. The most convenient characterization, although in many ways it is less than satisfactory, is provided by the designation of Standard Metropolitan Statistical Area (SMSA). An SMSA is a county or group of contiguous counties (except in New England) that contains at least one central city of 50,000 inhabitants or more or "twin cities" with a combined population of at least 50,000. In addition, other contiguous counties are included in an SMSA if, according to specific criteria, they are essentially metropolitan in character and are socially and economically integrated with the central city. In New England, towns and cities, rather than counties, are used in defining SMSA's. Central cities are those named in the titles of the areas. The entire territory of the United States has been classified as either metropolitan ("inside SMSA's") or nonmetropolitan ("outside SMSA's"). For a more detailed explanation and a listing of the component areas of each SMSA, see U.S. Department of Commerce.[765] Data on metropolitan–nonmetropolitan residence obtained from the Current Population Survey[776] are related to SMSA's as defined at the time of the 1960 U.S. Census, except as otherwise noted. SMSA's are in many cases heterogeneous with respect to population density and pollution. For example, the Chicago SMSA includes Kane and Porter

TABLE 17-5 U.S. Total Population, Lung Cancer Death Rate: Death Rates for Population Age 35 and Over (per Million)[a]

Male		Female		
White	Nonwhite	White	Nonwhite	Total
923	839	145	135	515

[a] Sources same as for Table 17-3.

TABLE 17-6 Lung Cancer Experience (ICD 160–164): Age-Specific Standardized Mortality Ratios for Total U.S. Population, 1959–1961[a]

	Standardized Mortality Ratio[b]				
	35–44 Years Old	45–54 Years Old	55–64 Years Old	65–74 Years Old	75+ Years Old
Males	1.56	1.71	1.84	1.89	1.81
Females	0.48	0.32	0.22	0.23	0.39

[a] Sources same as for Table 17-3.

[b] Standardized mortality ratio = $\frac{\text{expected deaths, 1959–1961}}{\text{observed deaths, 1959–1961}}$. Sex–age-specific expected deaths for 1959–1961 = age-specific death rate, standard population (U.S. total, 1960) × sex–age-specific population in 1960 × 3 (years).

Counties, both of which are predominantly rural. However, many important characteristics, such as cigarette sales, are available for SMSA's and not for more sharply defined urban areas.

A more relevant characterization is provided by the distinction between urban and rural areas. According to the 1960 Census definition, the urban population comprises all persons living in (a) places of at least 2,500 inhabitants incorporated as cities, boroughs, villages, or towns (except towns in New England, New York, and Wisconsin); (b) the densely settled urban fringe, whether incorporated or unincorporated, of urbanized areas; (c) towns in New England and townships in New Jersey and Pennsylvania that contain no incorporated municipalities as subdivisions and have either at least 25,000 inhabitants or 2,500–25,000 inhabitants and a density of at least 1,500 persons per square mile; (d) counties in states other than the New England States, New Jersey, and Pennsylvania that have no incorporated municipalities within their boundaries and have a density of at least 1,500 persons per square mile; and (e) other unincorporated places of at least 2,500 inhabitants.

Substantially the same definition was used for the 1950 Census, the major difference being the designation in 1960 of urban towns in New England and urban townships in New Jersey and Pennsylvania. In censuses before 1950, the urban population was defined to comprise all persons living in incorporated places of at least 2,500 inhabitants and areas (usually minor civil divisions) classified as urban by somewhat different rules related to population size and density. In all definitions, the population not classified as urban constitutes the rural population.

Changes in the size of the urban population from one census to another are affected by two components: growth in areas classified as urban at the beginning of the decade and reclassification of rural territory as urban. Between censuses, it is possible to obtain measures of only the first component from the Current Population Survey.[776] Regular publication of data on urban–rural residence from the Current Population Survey has been discontinued.

Urban–Rural Studies

A considerable number of studies have been carried out over the last 20 years to investigate lung cancer rates as they relate to urban living. In many of them, no measure of pollution was considered; and in most, no attempt was made to adjust for the cigarette factor. They provide, therefore, only a crude comparison between urban and rural lung cancer death rates.

Mancuso *et al.*[513] studied lung cancer death rates of urban and rural populations in Ohio from 1947 to 1951. He found the highest rates in the most urbanized areas and the lowest in rural counties. (Table 17-7).

Manos and Fisher[515] studied the age-adjusted mortality rates from 102 causes of death by degree of urbanization and by sex in the white population in the United States from 1951 to 1959 (Table 17-8). Among males, they found approximately twice the death rate in highly urbanized as in nonmetropolitan counties. The trend in females was much less marked.

Hoffman and Gilliam[383] examined lung cancer mortality distribution in the United States for 1948 and 1949 and reported urban–rural differences in white and nonwhite males and females, as shown in

TABLE 17-7 Comparison of Lung Cancer Death Rates in Ohio by Urbanization (Adjusted Rates, White Men, 25–64 Years Old, 1947–1951)[a]

Population Area	Standardized Mortality Ratio
Eight metropolitan counties (cities of 100,000+)	1.23
Seven other urbanized counties (50,000 or more persons in urban area)	0.82
Rural counties (remainder)	0.69

[a] Derived from Mancuso *et al.*[513]

TABLE 17-8 Age-Adjusted Lung Cancer Mortality Rates, 1951–1959 (of the Trachea, Bronchus, and Lung-Primary)[a]

	Mortality Rate (per 100,000 population)	
Population Area	White Males	White Females
Metropolitan counties with central city	11.5	1.6
Metropolitan counties without central city	9.0	1.3
All other counties	5.8	1.2

[a] Derived from Manos and Fisher.[515]

Table 17-9. In this study, urban mortality rates are approximately twice the rural rates in both white and nonwhite populations.

Curwen *et al.*[177] analyzed mortality due to cancer of the lung and larynx in the country boroughs of London and the urban districts and rural areas of different parts of England and Wales for the period 1946–1949.

As indicated in Table 17-10, standardized mortality ratios for cancer of the lung in both sexes and of the larynx in males increased with increasing population density and index of urbanization. In the same study, a comparison of districts designated as rural but varying in population density revealed a similar relation, as shown in Figure 17-5.

Levin *et al.*,[492] comparing lung cancer incidence in urban and rural areas of New York State exclusive of New York City, in 1949–1951, found the age-adjusted incidences per 100,000 of population to be as shown in Table 17-11. Comparison of the lung cancer incidences of metropolitan urban and metropolitan rural males reveals only a modest difference, but the difference in incidence between metropolitan urban and nonmetropolitan rural areas is striking.

TABLE 17-9 Lung Cancer Mortality Rates among U.S. Urban and Rural Populations, 1948–1949[a]

	Mortality Rate, Total Age-Adjusted (per 100,000 population)			
	White		Nonwhite	
Population Area	Male	Female	Male	Female
Urban	22.3	4.7	16.9	4.2
Rural	12.3	3.7	7.3	2.1

[a] Derived from Hoffman and Gilliam.[383]

TABLE 17-10 Standardized Mortality Ratios for Cancer of the Lung in England and Wales, 1946–1949[a]

Population Area	Standardized Mortality Ratio[b]	
	Males	Females
England and Wales	1.0	1.0
Greater London	1.37	1.32
Northern England	1.0	0.98
Midlands	0.93	0.93
Wales	0.79	0.69
Remainder	0.83	0.89

[a] Derived from Curwen *et al.*[177]
[b] Standardized mortality rates per 100,000 for males, 40.9; for females, 8.0.

Winkelstein *et al.*[825a] studied the relation of air pollution and economic status to various mortality indices, including lung cancer, in Buffalo and its environs. Although the overall lung cancer death rates for areas of differing pollution level were higher with higher pollution,

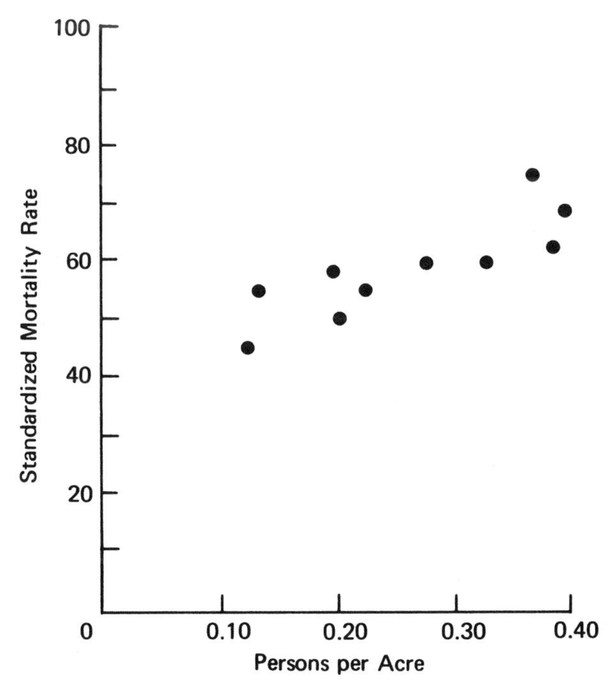

FIGURE 17-5 Lung cancer in rural populations of England and Wales. (Derived from Curwen *et al.*[177])

TABLE 17-11 Age-Adjusted Lung Cancer Rate in Urban and Rural New York State, 1949–1951 (Exclusive of New York City)[a]

Population Area	Age-Adjusted Lung Cancer Rate (per 100,000 persons)			
	Metropolitan		Nonmetropolitan	
	Males	Females	Males	Females
Urban	29.2	3.2	20.8	3.2
Rural	23.9	3.5	15.2	2.4

[a] Derived from Levin *et al.*[492]

the authors point out that pollution and low economic status are highly correlated and, in studying the rates specific for economic status, they failed to find any consistent relation between air pollution and lung cancer deaths.

Hagstrom *et al.*,[333] reporting in the Nashville air pollution study, found no association between pollution level and cancer of the lung. They point out that a real relation might be obscured in this study through the interaction with other factors, especially smoking, which was not ascertained. In this and the Buffalo study, the comparisons are of subareas within a fairly compact region. In consequence, the variation in exposure levels for men in the different subareas is rather less than is the case in comparisons between regions (especially in view of within-region mobility of workers). A real effect may therefore be correspondingly hard to distinguish.

Haenszel *et al.*[331] show a distribution in Iowa similar to that in New York when the age-adjusted cancer incidence per 100,000 of population is examined for urban and rural areas by primary site and sex, and then for urban and rural areas in metropolitan and nonmetropolitan counties with both sexes included (Tables 17-12 and 17-13).

Griswold *et al.*,[314] carrying out a similar type of analysis of data for lung and bronchus cancer in males from 1947 to 1951 in Connecticut, also showed an urban–rural difference of similar proportions, i.e., an urban incidence of 28.0 versus a rural incidence of 17.8.

Comparison of the New York, Iowa, and Connecticut studies reveals somewhat higher rates than in the United States as a whole, as shown in the study by Hoffman and Gilliam.[383] Particularly, the New York metropolitan–rural rates are much higher than the rates in the other two studies or the national rates and are comparable with the U.S. rates for urban white males. These data suggest that considerable care must be exercised in defining urban and rural populations. It would appear

TABLE 17-12 Age-Adjusted Lung Cancer Rate in Urban and Rural Areas of Iowa[a]

	Age-Adjusted Lung Cancer Rate (per 100,000 persons)			
	Male		Female	
Primary Site	Urban	Rural	Urban	Rural
Respiratory system	32.8	12.1	9.0	6.3
Larynx	2.9	1.4	0.9	0.0
Lung and bronchus	29.0	10.2	7.8	5.3
Other respiratory	0.9	0.5	0.3	1.0

[a] Derived from Haenszel *et al.*[331]

from these studies that the metropolitan–rural rate in New York State is not comparable with rates in other rural areas, such as Iowa.

Stocks,[727] correlating community size and lung cancer in the United Kingdom (1946–1949), found an increase in age-specific standardized mortality from lung cancer with increase in the number of dwelling units.

Prindle,[605] considering city size and regional differences in lung carcinoma and the percentage of habitual white male smokers, found results shown in Table 17-14.

The studies discussed above are subject to only limited interpretation, because they fail to consider the influence of smoking habits on the observed differences in death rates. Prindle's study suggests that urban and rural populations are comparable in smoking; thus, differences in cancer mortality do not appear to be explained by this factor. In any case, these studies consistently show, in males, urban lung cancer death rates approximately twice those found in the corresponding rural areas.

Disregarding for the present the possible differences in smoking habits, one may seek quantitative estimates of the relation between

TABLE 17-13 Age-Adjusted Lung and Bronchus Cancer Rates for Both Sexes[a]

	Age-Adjusted Rate (per 100,000 persons)	
Type of Population	Metropolitan Counties	Nonmetropolitan Counties
Urban	20.9	15.1
Rural	7.5	7.5

[a] Derived from Haenszel *et al.*[331]

TABLE 17-14 Death Rates for Malignant Neoplasm of the Trachea, Bronchus, and Lung among White Male Cigarette Smokers by Type of Area[a]

Population Area	Death rate (per 100,000 persons)	Percentage of Cigarette Smokers
Cities >1,000,000	29.4	48.5
Urban areas with 250,000–1,000,000	22.9	49.5
Urban areas <250,000	17.5	47.8
Rural areas	14.6	42.5

[a] Derived from Prindle.[605]

lung cancer death rates and measures of pollution. Taking the median January–March urban benzo[a]pyrene concentration of 6.6 μg/1,000 m^3 and rural concentration of 0.4 μg/1,000 m^3 of Sawicki *et al.*,[667] one may roughly associate a 100% increase in lung cancer death rate with a 6.2-unit increase in benzo[a]pyrene or an increase of approximately 15% in deaths per unit increase in benzo[a]pyrene. However, the 100% increase is associated with a contrast between the most heavily urban and the most rural environments, so that the pollution effect estimated from these studies should be somewhat less than 15%.

Migrant Studies

In the second group of studies, lung cancer death rates in migrants from one country to another were compared with those in their fellows in the home countries and with those in the people in the countries to which they had migrated. If such migrants can be considered as equivalent to random or representative samples of the populations of the home countries, then differences in death rates from those in the home countries can be ascribed to changes in environmental conditions.

Pollutants, including benzo[a]pyrene, vary considerably worldwide. Concentrations in Great Britain are much higher than those in the United States, and concentrations in Norway, Italy, New Zealand, South Africa, and Australia are considerably lower than those in the United States.[731]

Mancuso and Coulter[512] studied lung cancer death rates in male migrants 25–64 years old in Ohio; the smoking factor was not considered. Their results were as shown in Table 17-15. The findings suggest that the lung cancer death rates of migrants are between that of the mother country and that of the country to which they migrated. British immigrants had a lower death rate than that observed for the

TABLE 17-15 Lung Cancer Death Rates for Native and Immigrant Male Populations in Ohio, 25–64 Years Old[a]

	Lung Cancer Death Rate (per 100,000 persons)	
Population Group	Migrant Group	Nonmigrating Cohort Group
Native-born Americans	29	–
Immigrants from United Kingdom	32	55
Immigrants from Italy	19	16

[a] Derived from Mancuso and Coulter.[512]

English population, but greater than the rate for native Americans. Italian immigrants had a slightly greater lung cancer death rate than those remaining in Italy but considerably less than native white Americans and immigrants from the United Kingdom.

In a study by Haenszel[329] of migrants from specific countries, lung cancer incidences agreed in essence with those of Mancuso and Coulter in Ohio. The study also confirmed the observation that migrants coming from countries with lower lung cancer death rates showed a displacement in rates to a position intermediate between that observed for the home country and that for the country to which they migrated.

Eastcott's study[223] of migrants from the United Kingdom to New Zealand is important, in that it attempts to separate the effects of the urban and cigarette factors on cancer death rates. Without documenting individual smoking habits, it was determined that New Zealanders were heavier smokers than persons in the United Kingdom.

Migrants from the United Kingdom had a 35% higher risk of lung cancer than native New Zealanders if they came from the United Kingdom before the age of 30 and a 75% higher risk if they migrated after the age of 30. This was true, regardless of the fact that the migrants generally increased the number of cigarettes smoked after arriving in New Zealand.

Dean's studies[187,189] compared lung cancer rates in British subjects who migrated to South Africa and Australia with those in native-born South Africans and Australians. South Africans are among the heaviest cigarette smokers in the world, and British migrants to South Africa tended to increase their consumption of cigarettes markedly. In spite of this, they had a significantly lower lung cancer death rate than persons remaining in England but a higher rate than native South Africans, as shown in Table 17-16. The lung cancer mortality rates for

TABLE 17-16 Lung Cancer Death Rates for White Male Natives of England, Wales, and South Africa and United Kingdom Migrants to South Africa, 1947–1965[a]

	Annual Lung Cancer Death Rate (per 100,000 persons)	
Population Group	55–64 Years Old	65+ Years Old
Native white South Africans	50	112
United Kingdom migrants to South Africa	112	172
Native white United Kingdom	135	219

[a] Derived from Dean.[189]

men 40 years old or older, as indicated in the Australian study, are shown in Table 17-17. Variations in smoking practices (e.g., butt length, depth of inhaling, and number of puffs) are important in assessing the relation of cigarette smoking to lung cancer incidence. Dean, comparing United Kingdom and South African smokers, found that the average cigarette butt length was 25.3 mm for native white South Africans and 25.2 mm for immigrants from the United Kingdom. He also found that a greater percentage of South Africans inhaled deeply and that South Africans took more puffs per cigarette.

In Britain, Doll *et al.*[209] observed an average butt length of about 20 mm, whereas Hammond (quoted in Doll *et al.*), in the United States, noted a butt length of about 30 mm. Sampling methods for the two studies were very different. Collection was by mail in Great Britain, and butts discarded in public places (including restaurants) were collected in the United States; the results are therefore difficult to interpret.

TABLE 17-17 Age-Adjusted (40+ Years Old) Lung Cancer Death Rates, 1950–1958[a]

Population Group	Lung Cancer Death Rate (per 100,000 persons)
Native Australians	53
United Kingdom migrants to Australia	94
United Kingdom cohort group	154

[a] Derived from Dean.[189]

In another study, Reid *et al.*[628] considered lung cancer death rates in migrants from the United Kingdom and Norway and compared them with those in persons remaining in the home country and native-born Americans. Their results are summarized in Table 17-18. This large and carefully detailed study confirms the findings from other, previously cited studies. Lung cancer death rates of migrants are intermediate between those of native U.S. residents and those of persons in the home countries.

The results of these migrant studies are very similar to the results of the urban–rural studies reported, in that they show lung cancer death rates paralleling the general pollution levels in the areas in question. The data also suggest that exposure to pollution early in life may produce persistent effects. These studies lack data regarding the effect of the smoking factor. Consideration of heavy smoking habits in New Zealand, Australia, and South Africa, compared with the United Kingdom, however, provides some basis for concluding that the changes in lung cancer death rates are related to more than cigarette smoking, i.e., to an urban factor.

Regression Studies

Regression studies attempt to separate the effects of factors that differentiate urban and rural environments with the aim of identifying urban factors that might be held responsible for the difference in lung cancer death rates. The method usually adopted for attempting to separate the effects of different factors statistically is multiple regression. (See Appendix D for discussion of the regression analysis.)

TABLE 17-18 Age-Adjusted Death Rates from Lung Cancer in Great Britain, Norway, and the United States[a]

	Lung Cancer Death Rate (per 100,000 persons)	
Population Group	Males	Females
Great Britain residents	151.2	19.3
Great Britain-born U.S. residents	93.7	11.5
Norway residents	30.5	5.6
Norway-born U.S. residents	47.5	10.7
Native U.S. residents	72.2	9.8

[a] Derived from Reid *et al.*[628]

This method is subject to severe limitations, of which the following are most important:

1. Even if the relation has the assumed form, fluctuations in the data due to random effects of other variables make it difficult to determine the coefficients of the relation with reasonable precision. Thus, the failure to find a significant result, according to the usual statistical measures, is not necessarily evidence that there is no effect. A real effect may readily be hidden in the sea of fluctuations induced by extraneous variables.
2. Much more important from the point of view of confusion in interpretation is the disguising of effects actually due to one variable under the label of another. Particularly when variables are highly correlated (as is the case for most measures of urban crowding and pollution) and when there are substantial measurement errors (as is the case for measures of pollutants diffused through the air), the coefficient of a regression relation may be very difficult to interpret, and the addition or withdrawal of one variable in the equation may have profound effects on the coefficients for the others. In consequence, inappropriate analyses and interpretations are frequent, and no analysis can lay claim to positive assurance.

In spite of all these reservations, the regression method provides the best available technique for separating the effects of different variables, and the evidence was studied and interpreted by this method, subject to substantial qualification.

TABLE 17-19 Multiple Regression Analysis of Lung Cancer Death Rates for Males in 19 Countries and Cigarette and Solid-Fuel Consumption[a]

		Regression Coefficients		
Age Group, years	Average Death Rate (per million persons)	Constant (C_0)	Cigarettes, 1,000's per person per year (avg. = 1.76)	Solid Fuel, metric tons per person per year (avg. = 1.55)
Age-adjusted	749.3	330.0	110.0	144.0
25–34	10.0	2.8	2.0	2.0
35–44	73.2	9.7	23.0	15.0
45–54	427.6	164.0	78.0	80.0
55–64	1,377.2	704.6	138.0	276.0
65–74	1,939.3	810.0	321.0	361.0

[a] Derived from B. W. Carnow and P. Meier (unpublished data).

In 1966, Stocks[731] reported analyses of a number of variables related to lung cancer mortality in 19 countries. In view of the limited analyses provided by Stocks, additional regression calculations were performed for the same 19 countries. Cigarette consumption, known for the mid-1960's,[731] was used, as were age-specific death rates from Segi *et al.*[689] Stocks's 1955–1958 solid fuel consumption was used as a measure of pollution.

Earlier analyses of a number of variables reported by Stocks *et al.*[733] indicated that many were highly correlated with each other and some, such as liquid fuel, seemed not to be related to lung cancer or other indices of pollution. In accordance with these preliminary analyses, the two variables—amount smoked per capita (or amount produced) and solid-fuel consumption per capita—were chosen as the basic regression variables. Table 17-19 shows the results of the regressions of the form

$$Y = C_0 + C_1 X_1 + C_2 X_2,$$

in which Y = age-sex-specific lung cancer death rates from 1958 to 1959,
X_1 = cigarette consumption per capita, in thousands of cigarettes per year per person over 15 years old, and
X_2 = solid-fuel consumption per capita, in metric tons per person per year.

On the assumptions that the lung cancer death rate is related both to cigarette consumption and to solid-fuel consumption and that the effects are at least approximately additive, C_1 measures the increment in lung cancer death rate per unit increase in cigarette consumption, and C_2 measures the increment per unit increase in solid-fuel consumption. C_0 is the regression coefficient constant.

For cigarette smoking, the coefficient is approximately 15% of the average lung cancer death rate. For example, the coefficient for cigarettes in the male age-adjusted group is 110, which is 14.7% of the average rate of 749.3. Taken at face value, this suggests an increment in male lung cancer deaths of 15% per 1,000 cigarettes per year. To convert the rate per 1,000 cigarettes per year to a rate per cigarette per day, one multiplies it by 365/1,000, which gives a 5.4% increase in lung cancer death rate per cigarette per day. This corresponds to approximately a doubling in the lung cancer death rate corresponding to an increase in smoking of a pack (20 cigarettes) per day. This estimate is compatible with the variation in lung cancer death rates by smoking category reported by Hammond.[338]

For solid-fuel consumption, the regression coefficient is approximately 20% of the average lung cancer death rate (i.e., 144/749.3). This suggests an increment in male lung cancer deaths of 20% per metric ton of coal burned per year per capita. Although Pybus[612] estimated total benzo[a]pyrene released per ton of coal burned, there is no way to convert this to benzo[a]pyrene units.

The conclusion suggested by these results is that the products of solid-fuel combustion or of some variable highly correlated with solid fuel may be an important etiologic factor in lung cancer.

A study was undertaken by the Panel for the 48 contiguous states of the United States with the following independent variables:

Lung cancer death rate/million persons,

X_1 = cigarette sales per person over 15 years old (1963), and

X_2 = benzo[a]pyrene, in benzo[a]pyrene units (R. I. Larsen and J. B. Clements, unpublished data), weighting urban and rural values measured in each state according to the urban fraction of each state.

The results are shown in Table 17-20.

As noted in the analyses of regression coefficients for solid-fuel consumption in the 19-country study,[731] the regression coefficients for benzo[a]pyrene for white males are around 5% of the average rates (e.g., for age-adjusted white males, 47.5 is 5.5% of 867.5). Similar results are found for each age-specific group. This suggests that an increase in urban pollution corresponding to an average benzo[a]pyrene of 1 $\mu g/1,000\ m^3$ may result in an increase of 5% in the lung cancer death rate. The coefficients for nonwhite males are larger—about 15%. It should be kept in mind, in considering this group, that the nonwhite population in many states is distributed between urban and rural environments in a way quite different from the white majority. If, in fact, nonwhites are found in heavily polluted areas of industrial states but tend to be in rural areas of southern states, the proper benzo[a]-pyrene indices for nonwhite populations might be much more spread than those in Table 17-20, and an appropriate weighting might lead to a coefficient similar to that for whites.

The results for females, who generally exhibit much lower lung cancer death rates, are more variable and mostly nonsignificant, except for those 45–54 years old.

A final important qualification in comparing these results with those of the urban–rural studies is based on the different benzo[a]-pyrene levels used. The regression studies use 1969 levels that are

TABLE 17-20 Multiple Regression Analysis on Lung Cancer Death Rates (per 1,000,000 population) in the United States in the White and Nonwhite Population[a]

Age Group, years	Average Death Rate (per million persons)	Constant (C_0)	Tobacco Sales, $ (avg. = 28.8) (C_1)	Benzo[a]pyrene, μg/1,000 m³ of air (avg. = 1.38) (C_2)[b]
Male, white,				
age-adjusted	867.4	460.7	11.8	47.5
35-44	101.6	86.5	0.3	4.6
45-54	497.8	125.2	3.0	18.7
55-64	1405.8	894.4	14.4[c]	70.5
65-74	2064.7	835.4	35.3[c]	152.6[c]
Male, nonwhite,				
age-adjusted	844.6	184.2	16.5[c]	133.1[c]
35-44	132.3	22.8	1.6	46.4[c]
45-54	606.0	245.9	7.8	97.7
55-64	1367.6	141.6	34.0	179.0
65-74	1722.4	162.6	38.8[c]	318.8
Female, white,				
age-adjusted	129.5	92.1	1.0[c]	7.1[c]
35-44	32.9	12.7	0.8[c]	−1.1
45-54	86.5	36.4	1.1[c]	14.2[c]
55-64	163.1	113.5	2.0[c]	−6.2
65-74	255.2	184.7	1.7	16.3
Female, nonwhite,				
age-adjusted	156.3	151.5	0.1	1.2
35-44	34.3	66.0	−1.0	−1.1
45-54	85.8	32.4	−0.4	46.6[c]
55-64	184.9	−375.0	17.3[c]	44.5
65-74	391.7	1286.4	−19.2	−246.1

[a] Derived from B. W. Carnow and P. Meier (unpublished data).

[b] The estimate of benzo[a]pyrene concentration for each state was computed in the following manner: Benzo[a]pyrene concentrations were measured quarterly in each year from 1967 to 1969 in various urban and nonurban places in the United States. For each place, the four measurements in 1969 were averaged to obtain an estimate of the mean concentration for that year. For the few places with insufficient data in 1969, 1968 concentrations were used. All cities in a given state were categorized into (1) urban places whose population in 1960 was at least 1 million, (2) urban places whose population in 1960 was less than 1 million, and (3) nonurban places. The estimates of mean 1969 concentration were then averaged over all places with benzo[a]pyrene measurements in each of the three categories. Each of the resulting three averages was considered to be representative of all cities in the corresponding category, regardless of whether they had benzo[a]pyrene measurements. (In this case, all cities in the first category–i.e., with population of 1 million or greater–had measurements.) A weighted average of the three averages was obtained, using the corresponding populations as weights. This weighted average was considered to be the representative benzo[a]pyrene concentration for the entire state and was used as the second independent variable in the regression analyses of this table. In some states, no nonurban measurements were given; in those cases, the total population for the second category was taken to be the population of the entire state less the total population of all urban places with populations of at least 1 million each.

[c] Coefficient greater than 2 S.E.

judged to be, perhaps, no more than half the 1950 levels (which may actually be more relevant to current lung cancer deaths) quoted from Sawicki in the section on urban–rural studies. If we adjust our results correspondingly, we can conclude that, in place of 5% and 15%, coefficients of 2% or 3% for benzo[a]pyrene unit for whites and 7% or 8% for nonwhites may be more appropriate.

Sampling Studies

In sampling studies, values of the variables of concern, such as smoking habits and residence, are determined for individual subjects, and confidence in the interpretation of the results is correspondingly greater than in regression studies. The problem of disguising is still present, but to a far lesser degree.

Buell and Dunn[99] carried out a follow-up study of male American Legion veterans in California. They compared lung cancer death rates in subgroups classified by residence (urban, rural, inner-city) and smoking habits. A substantial difference was found between lung cancer death rates in the major cities and those in the smaller ones. Differences in smoking habits between residents of different cities were small, and, as shown in Table 17-21, differences in death rates between cities are only slightly modified by adjustment for smoking.

From unpublished data of R. I. Larsen and J. B. Clements, benzo[a]-pyrene concentrations for San Francisco and Los Angeles can be estimated at 1.1 and 1.8 units, respectively. Concentrations for other California cities are highly variable.

Dean,[188] in a study in Northern Ireland, determined smoking habits of lung cancer decedents and matched controls by interviewing their families. Lung cancer death rates for rural areas were consistently

TABLE 17-21 California Veterans Survey (American Legion): Lung Cancer Death Rates in Males, Age 25 and Older, Classified by Residence[a]

	Lung Cancer Death Rate (per 100,000 persons)		
Study Group	Los Angeles	San Francisco Bay Area	All Other California Cities
Adjusted for age only[b]	95.9	104.5	75.3
Adjusted for age and cigarette smoking[b]	95.4	102.0	75.5

[a] Derived from Buell and Dunn.[99]
[b] Adjusted to total American Legion population.

lower than those for urban areas in every smoking category, as shown in Table 17-22.

The levels of air pollution in Belfast have been measured, and the annual average benzo[a]pyrene concentration in England and Wales estimated from Stocks's data[728] is about 30 $\mu g/1{,}000\ m^3$. This high value is associated with a considerable difference in lung cancer death rates between inner Belfast and the country districts designated by Dean as truly rural. The differences in death rates as a percent of the Northern Ireland rates are 144%, 130%, 127%, and 87% for the four smoking categories, corresponding to 5%, 4%, 4%, and 3% increases, respectively, in lung cancer death rates per benzo[a]pyrene unit.

Hammond and Horn,[341] reporting on a 44-month follow-up of a sample population of U.S. white males, examined death rates from lung cancer. These were well-established cases exclusive of adenocarcinoma. The death rates by city size, adjusted for age and smoking history, are shown in Table 17-23.

From the Sawicki *et al.*[667] data using January–March data, which approximate the yearly average, benzo[a]pyrene concentrations for large cities and for rural areas can be estimated at 6.6 and 0.4 units, respectively. Thus, the relation between lung cancer death rate and estimated benzo[a]pyrene concentration is an increased rate of 52 – 39 = 13 per 100,000 of population, which is about 33% of the U.S. white male rate for 1969, corresponding to a difference in pollution level of 6.6 – 0.4 = 6.2 benzo[a]pyrene units, or a change of about 5% in the lung cancer death rate per unit.

Stocks and Campbell[732] found that increased lung cancer in rural areas was proportional to the number of cigarettes smoked. Studies in urban Liverpool revealed increases over the rural rates in every category of smokers; the differences were more pronounced in mild than in heavy smokers. The findings are shown in Table 17-24. The largest urban–rural differences in lung cancer death rates were in light smokers; there was only a modest effect in heavy smokers. The rural standardized death rates (SDR's) for the three categories of cigarette smokers were 70.8%, 36.3%, and 7.9% lower than the urban SDR's for light, moderate, and heavy smokers, respectively; the difference in benzo[a]pyrene concentration was 7.7 – 0.7 = 7 $\mu g/1{,}000\ m^3$.[732] This leads to estimates of increased lung cancer death rate per benzo[a]pyrene unit for light (10%), moderate (5.5%), and heavy (1.1%) smokers. These estimates cover a wide range and are considerably higher in the lighter smoking categories than is the case in most other studies. This discrepancy might reflect the unusually high benzo[a]pyrene concentrations in

TABLE 17-22 Age-Standardized Lung Cancer Death Rates for Persons Age 35 and Older[a]

No. Cigarettes Smoked per Day	Lung Cancer Death Rate (per 100,000 persons)								
	Inner Belfast	Outer Belfast	Urban Districts	Small Town	Environs of Belfast	Rural	Northern Ireland	Percent Difference[b]	Percent Difference per Benzo[a]pyrene Unit[c]
0	36	40	21	–	16	10	18	144	4.8
1–10	138	140	121	71	60	25	87	130	4.3
11–22	288	207	171	260	192	74	169	127	4.2
23+	509	430	515	772	716	173	383	88	2.9

[a] Derived from Dean.[188]

[b] Equals $100 \times \frac{\text{(inner Belfast)} - \text{(rural)}}{\text{total No. Ireland}}$.

[c] Equals $\frac{\text{percent difference}}{30\ (\mu\text{g benzo[a]pyrene per } 1{,}000\ \text{m}^3)}$.

TABLE 17-23 Lung Cancer Death Rates for White Males in the United States for Smokers and Nonsmokers by City Size[a]

Population Area	Lung Cancer Death Rate (per 100,000 persons): Adjusted for Age and Smoking History	No. Deaths Observed
City of 50,000+	52	83
City of 10,000–50,000	44	59
Suburb or town	43	67
Rural	39	52

[a] Derived from Hammond and Horn.[341]

the areas studied. It should, of course, be kept in mind that the number of deaths on which these rates are based is small. The apparent effect in nonsmokers is even greater (13%), but the numbers are clearly too small to draw any useful conclusions. Only one sampling site was used in each of the areas, and the urban–rural difference in benzo[a]pyrene for the communities studied by Stocks and Campbell is therefore uncertain.

Haenszel and associates[330,332] conducted a study in which a 10% sample of lung cancer deaths in white males and females was studied and interviews with family members were conducted. A sample of U.S. residents was also interviewed. Information collected included age, sex, smoking habits, location, and duration of residence. As shown in Table 17-25, the lung cancer death rate in males, adjusted for age and smoking history, is higher in urban areas than in rural areas, and the difference increases as the duration of residence increases. This finding gives added support to the view that the urban–rural difference in lung cancer mortality is related to direct environmental effects. Haenszel notes that, among males, the urban–rural difference is largest among heavy smokers. This finding was not duplicated in the data for females, nor was it the case in some other studies.[379,732] It may be noted that the lung cancer death rate for lifetime rural residents is half that for lifetime urban residents—in close agreement with the simple urban–rural studies discussed earlier.

An unusually detailed, well-documented study was reported in 1968 by Hitosugi.[379] The study was carried out in an area near Osaka including some regions of high industrial pollution and some with relatively low levels of pollution, as determined by measurements at sampling sites. The method used was to interview families of the 259

TABLE 17-24 Lung Cancer Death Rates from mid-1952 to mid-1954 According to Age, Smoking Category, and Population Area[a]

	Lung Cancer Death Rate (per 100,000 persons)													
	Age 45–54			Age 55–64			Age 65–74			SDR[b] Age 14–74				Percent Difference per Benzo[a]pyrene Unit[d]
Smoking Category	Rural	Mixed	Urban	Rural	Mixed	Urban	Rural	Mixed	Urban	Rural	Mixed	Urban	Percent Difference[c]	
Nonsmokers	0	0	31	0	0	147	70	0	336	14	0	131	89.3	12.8
Pipe smokers	0	0	104	30	59	143	145	26	232	41	25	143	71.3	10.2
Cigarette–light	69	57	112	70	224	376	154	259	592	87	153	297	70.7	10.1
Cigarette–moderate	90	83	138	205	285	386	362	435	473	183	132	287	36.2	5.2
Cigarette–heavy	117	214	205	626	362	543	506	412	588	363	303	394	7.9	1.1
Number of deaths	16	26	124	26	56	230	27	36	183	68	118	539		

[a] Derived from Stocks and Campbell.[732]
[b] Standardized death rate.

[c] Equals $100 \times \frac{\text{(urban SDR)} - \text{(rural SDR)}}{\text{urban SDR}}$.

[d] Equals $\frac{\text{percent difference}}{7.0\ (\mu g \text{ benzo[a]pyrene per } 1{,}000\ m^3)}$.

TABLE 17-25 Standard Lung Cancer Mortality Ratios in White Males in Urban and Rural Areas, Adjusted for Age and Smoking History[a]

Current Residence	Standard Mortality Ratio by Duration of Residence,[b] years					
	All Durations	<1	1-9	10-39	40+	Lifetime
Urban	113	166	107	117	177	100
Rural	79	154	88	83	75	50

[a] Derived from Haenszel *et al.*[330]
[b] Lung cancer base death rate = 78.2.

lung cancer cases reported over a 7-year period and also a random sample of 4,500 adults 35–74 years old. Factors recorded included age, sex, smoking habits, occupation, residence, and previous medical history. The results for males stratified by smoking habits are shown in Table 17-26, and the maximal benzo[a]pyrene, average sulfur dioxide, dust, and particulate matter for three degrees of pollution are listed in Table 17-27. Increases in lung cancer death rate are generally associated with the higher levels of pollution.

For the smoking categories, the change in lung cancer death rate (relative to the average for the total area) associated with the difference between high and low pollution is as follows: for 1–14 cigarettes daily, 10.6 to 23.5, an increase of about 120%; for 15–24 cigarettes daily, 14.7 to 27.0, an increase of about 90%; for 25+ cigarettes daily, 36.3 to 46.4, an increase of about 30%. The numbers of subjects are small and these percentage increases vary considerably, but a middle value may be taken as approximately 60%. The difference in maximal benzo[a]pyrene concentrations between the high- and low-pollution areas is 79.0 – 26.0 = 53.0 $\mu g/1{,}000\ m^3$ of air.

Although there are no data that reflect the seasonal variation in benzo[a]pyrene in the Osaka, Japan, area, data of Sawicki *et al.*[667] for two U.S. cities show great consistency and indicate that the average benzo[a]pyrene concentration over the year is approximately 30% of the yearly maximum (see Figure 17-3). Thus, an approximate conversion may be made from the maximum to an average of about 8.0 (= 0.3 × 26) for the lower-pollution areas and about 24.0 (= 0.3 × 79) for the high-pollution areas. The difference in average benzo[a]pyrene concentrations is thus estimated to be about 16 $\mu g/1{,}000\ m^3$. This corresponds to about a relationship between lung cancer death rate and benzo[a]pyrene of 4% (= 60%/16 μg).

TABLE 17-26 Lung Cancer Death Rates for Males and Females, Age 35–74, by Amount of Smoking and Level of Air Pollution[a]

Smoking Category, cigarettes daily	Lung Cancer Death Rate (per 100,000 persons)[b]							
	Males				Females			
	Low Pollution	Intermediate Pollution	High Pollution	Total	Low Pollution	Intermediate Pollution	High Pollution	Total
Nonsmoker	11.5 (5)	3.8 (1)	4.9 (1)	7.9 (7)	4.6 (15)	6.9 (12)	3.8 (6)	4.9 (33)
Ex-smoker	26.2 (11)	42.6 (7)	61.7 (7)	36.0 (25)	12.4 (2)	52.6 (2)	124.0 (3)	13.3 (6)
1–14	10.6 (9)	14.2 (10)	23.5 (14)	15.3 (33)	19.7 (13)	16.5 (6)	15.3 (5)	17.6 (24)
15–24	14.7 (18)	19.1 (17)	27.0 (17)	19.1 (52)	12.4 (1)	23.1 (2)	24.0 (1)	19.7 (4)
25+	36.3 (19)	15.8 (4)	46.4 (9)	44.0 (32)	– (0)	– (0)	– (0)	– (0)

[a] Derived from Hitosugi.[379]
[b] Numbers in parentheses are numbers of deaths.

TABLE 17-27 Levels of Specific Air Pollutants Corresponding to Low, Intermediate, and High Pollution Levels in Osaka, Japan[a]

Pollutant	Low	Intermediate	High
Dust fall, tons/km^2 per month	8.0	9.15	12.8
Sulfur dioxide, mg/day per 100 cm^2 of PbO_2	0.74	2.64	3.03
Suspended particulate matter, mg/m^3	0.19	0.22	0.39
Benzo[a] pyrene, μg/1,000 m^3 max.	26.0	31.0	79.0

[a] Derived from Hitosugi.[379]

Discussion

Although not all studies are completely consistent, it is clear that increased lung cancer mortality is generally associated with urban living. This relation does not appear to be accountable solely in terms of differences in smoking habits.

The attribution of the urban–rural difference in lung cancer to one or a few particular urban factors is far less certain, and, because the data are subject to considerable variability and extraneous influences, it is not possible to come to unequivocal conclusions. However, the high correlation of urban factors—such as solid-fuel consumption, smoke, and benzo[a] pyrene—suggests the reasonableness of using one of these factors as an overall index of urban pollution. Benzo[a] pyrene recommends itself as an index susceptible to direct measurement and known, in other contexts, as a potent carcinogen. Accordingly, the concentration of benzo[a] pyrene was used as an index of urban pollution. With fair consistency, most of the studies in the four categories discussed lead to an estimated increase of about 5% in the lung cancer death rates among males, corresponding to an increase in urban pollution represented by 1 μg of benzo[a] pyrene per 1,000 m^3.

It is essential to consider the evidence that does not conform to this pattern:

1. In most studies, relatively little effect of the urban environment on the lung cancer death rate for women is found. It is possible that a similar effect is present, but not clearly detectable, because of the generally much lower rates in women. An alternative explanation might be that the effects noted in men are due to occupational exposure or to travel through local regions of very high pollution.
2. In some regions—such as Finland (particularly Helsinki) and the

United States (New Orleans)—high lung cancer death rates are found, although the more familiar pollutants, including benzo[a]pyrene, are at a relatively low concentration. Special explanations have been suggested (e.g., wood smoke in Finnish saunas and coffee roasting in New Orleans), but it must be recognized that special explanations could probably be found for anomalies in almost any region.

3. The possibility that air pollution is in fact correlated with some entirely different variable (such as difference in smoking practices) that is the real cause cannot be excluded. None of these studies constitutes an experiment, and the more persuasive sampling studies are of limited extent. The measurements of air pollution, especially, are entirely inadequate to give reasonably good measures of individual exposure.

The Nashville and Buffalo studies, each carried out within a single region, found a strong correlation between cancer of the lung, bronchus, and trachea and economic status; but no consistent relation with air pollution was found.

Despite these reservations, a number of different types of studies do lead to fairly consistent estimates of a substantial effect of air pollution on lung cancer death rate, although more measurement of pollutants and far more extensive sampling studies should be undertaken. It appears both reasonable and prudent to take as a working hypothesis the existence of a causal relation between air pollution and lung cancer death rate at the rate of a 5% increase for each increment of pollution as indexed by 1 benzo[a]pyrene unit.

This hypothesis leads to the estimate that a substantial reduction in the pollution of highly urban environments would lead to a corresponding reduction in lung cancer death rate (e.g., a reduction of air pollution corresponding to a reduction of benzo[a]pyrene concentration from about 6 $\mu g/1{,}000\ m^3$ to around 2 $\mu g/1{,}000\ m^3$ might reduce the lung cancer death rate by about 20%). Similar benefits might be expected in all smoking categories.

Conclusions

Epidemiologic studies of lung cancer appear to show the following:

1. Lung cancer has emerged as the single greatest cause of cancer death in males and a significant cause of death in females in the United States, and its incidence has increased in the last 30 years.

2. A major etiologic factor appears to be cigarette smoking; however, smoking does not completely account for the increased incidence of the disease.

3. Urban dwellers have approximately twice as high an incidence of lung cancer as those living in rural areas. Within urban communities, the incidence is greater where more general industrial pollution is present.

4. Polycyclic organic matter, found in cigarette smoke in high concentrations, causes cancer of the lung and other organs in experimental animals, is present in the air in large quantities in industries whose workers have high lung cancer rates, and is present in the air of urban communities.

5. Generally, immigrants have an incidence of lung cancer between that noted for their countries of origin and that of the countries to which they migrate. The higher their ages when they migrate, the closer their rates are to those of their cohorts who remain in the countries of origin. In some of the studies, in which the home country had a much higher cancer rate, the rates in persons who left it decreased significantly, even though their cigarette smoking increased. These studies suggest a significant environmental effect operating early in life for lung cancer development.

6. A variety of types of epidemiologic studies lead to an estimate of the effect of pollution on lung cancer death rate of a 5% increase per unit increase in urban pollution as indexed by benzo[a]pyrene (one benzo[a]pyrene unit = 1 μg of benzo[a]pyrene per 1,000 m^3 of air).

NONOCCUPATIONAL NEOPLASTIC DISEASE OF OTHER ORGANS

The relation of POM to organs other than the lung is not well documented. In the Nashville Air Pollution Study,[333] carcinoma incidence was investigated in relation to sulfur trioxide, soiling dustfall, and 24-hr sulfur dioxide. In this study, socioeconomic factors in relation to carcinoma were also evaluated, but occupation and cigarette smoking were not considered. When the degree of exposure to air pollutants was kept constant, an inverse relation between socioeconomic class and mortality for cancer of the stomach, esophagus, and prostate was observed. For cancer of the bladder, there was a direct relation to socioeconomic class. The most consistent pattern was noted for suspended particulate matter, as measured by the soiling index. For the middle socioeconomic class, the death rate from all cancers combined was somewhat higher for the area of higher pollution, as measured by soiling. Direct relations were found between levels of pollution, as measured by soiling, and cancer of the esophagus, prostate, and bladder. Bladder death rates were highest for areas of high pollution for all four pollutants. For stomach cancer, significant differences were found to be in accord with degree of dustfall.

Levin *et al.*,[492] in a study in New York, showed an urban excess for cancer, not only of the respiratory system, but also of the esophagus, intestines, and rectum. Schiffman and Landau[677] demonstrated a significant correlation between some indices of air pollution and death rates from cancer of the stomach and esophagus, in addition to the lung, bronchus, and trachea. Dorn[214] revealed a lower mortality risk in nonmetropolitan, compared with metropolitan, counties for cancer of the bladder.

Winkelstein and Kantor,[824,825] in studies carried out in Buffalo, found a significant relation between prostatic and stomach cancer and particulate matter. There was no correlation, however, between lung cancer death rates and particulate matter. Cancer of the prostate appeared to be independent of economic status, at least in men under 70. In the middle economic grouping among white males 50–69 years old, the mortality rate for prostatic cancer was 2.7 times as high in the most polluted zone as in the least polluted. In regard to gastric cancer in white men and women 50–69 years old, mortality rates were almost twice as high in areas of high suspended particulate air pollution as in areas of low pollution. This appeared to be independent of the effect of economic status and was not apparently attributable to the ethnic distribution of the population in the study area.

Cohart[150] found a significant association between socioeconomic status of New Haven residents and the incidence of stomach cancer. The peak ratio of observed to expected stomach cancer cases occurred among males in the middle class, with a slight falling off among the poor. Among females, however, there was a statistically significant excess of observed to expected stomach cancer cases in the low socioeconomic class. Others have also shown a relation between the incidence of stomach cancer and social class.[145]

In general, the results for organ systems other than the lungs appear to be inconclusive. In most of the studies, smoking and dietary habits and pollution levels were not considered, making it impossible to evaluate the role of particulate POM in the induction of nonoccupational neoplastic diseases.

18

General Summary and Conclusions

SOURCES OF POLYCYCLIC ORGANIC MATTER

Polycyclic organic matter can be formed in any combustion process involving compounds of carbon and hydrogen. Naturally occurring POM emissions to the atmosphere do not appear to be significant. Major technologic sources of POM emissions include transportation, heat and power generation, refuse burning, and industrial processes.

The internal-combustion engine is a ubiquitous source of POM. Current efforts and projections of future control levels point toward a continuing decline in vehicular POM emissions.

The emissions from major stationary sources are poorly quantified. Available data suggest that coal-fired furnaces, coal-refuse bank burning, and coke production from the iron and steel industry account for the bulk of the nationwide POM emission inventory. Atmospheric POM concentrations are high in areas in which these processes are concentrated. Effective control procedures for these processes are lacking.

Current data suggest the following contributions of major source categories to the total national POM emission inventory (expressed in terms of annual estimated benzo[a]pyrene emissions): heat and power generation, 500 tons/year; refuse burning, 600 tons/year; coke production, 200 tons/year; and motor vehicles, 20 tons/year.

These data represent nationwide estimates based on extrapolations from individual source emissions. In specific areas, the relative contribution of any given source may differ significantly from that implied by the nationwide figures. For example, the vehicular source may be the major contributor in suburban areas where other major sources are absent.

ATMOSPHERIC PHYSICS OF PARTICULATE POLYCYCLIC ORGANIC MATTER

Polycyclic organic matter detected in the atmosphere has been identified with particulate matter, especially soot. It is uncertain whether POM condenses out as discrete particles after cooling or condenses on surfaces of existing particles after formation during combustion.

Particle size, surface area, and density are physical properties that have the greatest influence on the behavior of POM-containing aerosols. Generally, the particle-size spectrum of the atmospheric aerosol extends from less than 0.01 μm to greater than 10 μm. Very few measurements of mean surface area and density are available, but they are believed to range from 2 to 3 m^2/g and from 1 to 2 g/cm^3, respectively, for urban areas.

POM appears to be associated largely with particles less than 5 μm in diameter. Although large local variations are tabulated by the National Air Surveillance Network, suspended particulate-matter concentrations in U.S. urban areas are 100–200 $\mu g/m^3$, as measured by high-volume sampling. The benzene-soluble portion of this material is approximately 10% by weight, but can vary from 8 to 14% in urban areas. The POM component, as measured by benzo[a]pyrene, is less than the benzene-soluble fraction by a factor of 10 or more.

Because POM is carried by suspended particulate matter, its longevity in air depends on the lifetime of the carrier aerosol in air and on chemical alteration of POM itself. Initial estimates of atmospheric residence times of particles less than 5 μm in diameter exceed 100 hr under dry atmospheric conditions. Chemical reactivity in the presence of sunlight may lead to transition of POM to other material in several hours. Without sunlight, its lifetime may be much longer.

CHEMICAL REACTIVITY OF POLYCYCLIC AROMATIC HYDROCARBONS AND AZA-ARENES

Polycyclic aromatic compounds are highly reactive. Evidence suggests that they are degraded in the atmosphere by photooxidation, reaction with atmospheric oxidants, and sulfur oxides. Comparative data on reactions in solution, vapor, and adsorbed phases are very limited, and the great bulk of the available information pertains to solution reactions. In the few cases in which evidence is available, the reactions in other phases are similar. Reactions may be particularly facile when the compounds are adsorbed on such particulate material as soot. Chemical half-lives may be only hours or days under intense sunlight in polluted atmospheres.

Most of the likely atmospheric reactions produce oxygenated compounds from the hydrocarbons. Several mechanisms involving aromatic hydrocarbons and other pollutants may cause reactive oxidizing species to be delivered to genetic and other biologic material.

THEORETICAL ASPECTS OF CHEMICAL CARCINOGENESIS

Chemical carcinogens appear to transform normal cells directly into cancer cells. Chemical carcinogens may or may not "switch on" a latent oncogenic virus that is responsible for cancer induction. If the chemicals transform normal cells into cancer cells without the intervention of an oncogenic virus, they can do so either by a mutational or by a nonmutational mechanism. There is some evidence in favor of each possibility.

Tumors induced by carcinogenic hydrocarbons have individual antigens, and the response of a host to such stimuli is determined by its immunologic and hormonal status, its exposure to particular drugs, and its nutritional state. A variety of unknown host factors may influence the response to carcinogenic stimuli.

EXPERIMENTAL DESIGN IN CARCINOGENESIS TESTS

The proper design of a carcinogenicity testing program can depend on a wide variety of factors. Its purpose may be to determine safe concentrations of carcinogens, to identify agents with interesting properties for future investigation, to identify specifically potent carcinogens, or even to identify specifically weak carcinogens. Screening experiments with many dosages can be informative, even if the group sizes at individual dosages are limited. Data from a carcinogenicity experiment can be complex, but simplifying methods for analyzing the data may exist. Flexibility of analysis may be required to overcome unanticipated problems, for instance, a high rate of spontaneous tumors among controls. The existence of thresholds in carcinogenesis cannot be established solely by testing programs, and experimentation solely to such ends would be wasteful.

Neither epidemiologic nor experimental data are adequate to determine a safe dosage of any chemical carcinogen below which there will definitely be no tumorigenic response in humans. For these reasons, synthetic chemicals, such as food additives and pesticides, that are known to be carcinogenic must not be deliberately added to the environment. One must always insist on the lowest possible exposure to air pollutants, which contain a variety of defined and undefined carcinogens.

It is impossible to determine safe levels of human exposure to any known or unknown carcinogen on the basis of supposed no-effect levels in practical numbers of animals. Therefore, high dosages must be used to obtain statistically significant data. Intratracheal instillation of particles may serve as a model for human exposure to air pollutants. Human experience has provided valuable *post hoc* information from epidemiologic studies.

IN VIVO TESTS FOR CARCINOGENESIS AND COCARCINOGENESIS

The most reliable test systems for measuring carcinogenesis in mice and rats include application to the skin (mice), subcutaneous injection (also in hamsters), administration in feed, intraperitoneal injection, inhalation tests, and bladder implantation. Factors that influence hydrocarbon distribution in the host are particle size, retention and elution of particles, changes in ciliary movement, and mucous viscosity.

The polycyclic aromatic hydrocarbons and heterocyclic compounds constitute a group of known carcinogens that are present in the particulate phase of polluted air. However, the extent of the contribution of these agents to the incidence of human lung cancer is unknown. A variety of other compounds probably contribute to the human health hazard. These include tumor-promoting agents, cocarcinogens, noncarcinogenic initiating agents, and carcinogens other than particulate aromatic hydrocarbons. The last compounds, usually nonparticulate, include direct-acting alkylating agents, such as epoxides and lactones, and peroxides and hydroperoxides of olefins and of aromatic hydrocarbons. The role played by tumor-inhibitory or anticarcinogenic agents in the health effects of air pollutants is, at present, poorly understood.

Purified polycyclic compounds, such as benzo[a]pyrene, have produced tumors of the tracheobronchiolar tree or lung parenchyma only when adsorbed on particles and delivered below the larynx. In inhalation experiments, the addition of an irritant, such as sulfur dioxide, to an aerosol of benzo[a]pyrene has induced lung carcinomas in rats. Inasmuch as many potentially interacting influences are ubiquitous in polluted air–including solid particles, irritant chemicals, and gases–these may be cofactors as important for the induction of pulmonary cancer as the polycyclic hydrocarbons themselves.

MODIFICATION OF HOST FACTORS IN *IN VIVO* CARCINOGENESIS TESTS

An immunologic surveillance mechanism of a host against his own tumor exists, although some evidence suggests that it may be relatively weak. Many tumors have little or no immunogenicity, and this cannot be attributed entirely to immunoselection. Even when potentially immunogenic, a neoplasm may fail to immunize the host until late in its course. Furthermore, immunodepression does not regularly result in an increment in neoplasia. Other types of surveillance having little to do with specific acquired immunity may also constitute a major defense against incipient, chemically induced neoplasms.

It appears that newborn mice are often more susceptible than adult mice to chemical carcinogenesis (particularly in organs distal to the site of injection). This increased sensitivity, together with the requirement of smaller doses, suggests that the greater use of newborn mice for carcinogenicity tests of air pollution fractions would be advisable.

Subhuman primates are susceptible to experimental chemical carcinogenesis. Not all attempts to produce tumors with polycyclic hydrocarbons have succeeded, and the lack of success raises questions about resistance and the appropriate choice as to age and method of application. Primates of the suborder Prosimii appear to be more susceptible than other primates to carcinogenesis by these compounds, and the tumors have shorter latent periods than in other primates. Pulmonary carcinoma in simians has been produced by particles of beryllium salts, but not by polycyclic hydrocarbons. However, the latter compounds have produced lung cancer in prosimian primates by intratracheal instillation of particulate preparations.

Control of caloric intake to maintain normal body weight has been shown by human insurance statistics and experiments in mice to lower the hazard of developing cancer. Good reasons exist for an adequate intake of protein and vitamins to provide an optimal ability for tissues to detoxify the wide variety of environmental carcinogens to which man is exposed.

Because of a lack of sufficiently well-controlled experiments, no conclusive evidence is available to support, or refute, the hypothesis of a cocarcinogenic effect of respiratory infection. However, because respiratory infections are detrimental to a number of local and systemic defense systems, and because they have profound effects on cell proliferation and differentiation, the hypothesis of cocarcinogenicity of respiratory infections is attractive.

It has been found that ionizing radiation combined with polycyclic aromatic hydrocarbons produces an additive carcinogenic effect at various sites. The induction of pulmonary tumors by irradiation has been difficult to achieve because of difficulties in delivering the radiation to the lungs. Progress has been made through the use of polonium-210 adsorbed on hematite particles, and a combination of this and benzo[a]pyrene (also adsorbed on hematite) delivered by intratracheal instillation produced an additive carcinogenic effect.

DISTRIBUTION, EXCRETION, AND METABOLISM OF POLYCYCLIC HYDROCARBONS

No definitive study on the metabolism, tissue distribution, and excretion of carcinogenic hydrocarbons has yet been carried out. Radioactive labeling of carcinogenic hydrocarbons and fractionation of the radioactivity have been carried out. However, characterization of the compounds has not been achieved. Complete characterization of the metabolites and excretion products, tissue distribution, and binding to macromolecules has not yet been attempted.

A number of metabolites of compounds like benzo[a]pyrene have been identified. These include various dihydrodiols, phenols, quinones, and glutathione conjugates. It is likely that an epoxide is the metabolic precursor of these compounds. In the case of 7,12-dimethylbenz[a]-anthracene, there is metabolic hydroxylation of the methyl groups.

The metabolic products just referred to are produced primarily through aryl hydrocarbon hydroxylase, the drug-metabolizing enzyme system of the microsomes. This complex system is found in many tissues of many species and is inducible by polycyclic hydrocarbons and a variety of other compounds, such as pesticides and drugs. The enzyme system can also be inhibited by several compounds. The system can increase or decrease the toxicity of hydrocarbons, and it is probably responsible for their metabolic activation to a chemically reactive ultimate carcinogen.

IN VITRO APPROACHES TO CARCINOGENESIS

Two reliable and quantitative cell culture systems are now available for the study of hydrocarbon carcinogenesis. One uses hamster embryo cells in primary or secondary culture; the other uses cell lines derived from adult mouse prostates. These systems show an excellent correlation between the carcinogenic activity and the frequency of malignant transformed colonies produced by a series of hydrocarbons, making them potentially useful for screening carcinogenic activities of related compounds obtained from polluted air. Considerable fundamental information pertaining to the cellular and molecular mechanisms of chemical carcinogenesis has been obtained with these systems.

Organ cultures maintain, *in vitro,* differentiated tissue organization that resembles that in tissue in the intact animal. When organ cultures of human and mouse embryo tracheas are exposed to polycyclic hydrocarbons and fractions from polluted air, marked histologic alterations can be observed in the epithelial cells and their organization. Moreover, by this technique, selected human tissues in organ culture can be used to assess some of the biologic effects of air pollutants.

INDIRECT TESTS FOR DETERMINING THE POTENTIAL CARCINOGENICITY OF POLYCYCLIC AROMATIC HYDROCARBONS

The suppression of sebaceous glands in mouse skin after application of polycyclic hydrocarbons is not a reliable indicator of carcinogenicity. But it may have limited use in predicting the carcinogenicity of some groups of compounds, such as substituted benz[a]anthracenes.

The economy, rapidity, and simplicity of the photodynamic killing of paramecia, which can be conducted on less than 1-mg amounts of organic extracts, are attractive features. The data suggest that this bioassay provides a biologic index of potential carcinogenic hazard attributable to polycyclic compounds. However, evaluation of this concept demands correlated photodynamic, carcinogenic, and chemical studies on numerous samples and fractions of organic atmospheric pollutants collected from sources exemplifying a wide epidemiologic spectrum of incidence of respiratory tract cancer.

TERATOGENESIS AND MUTAGENESIS

Polycyclic hydrocarbons have not been shown to be teratogenic, although a number of other chemical carcinogens exhibit this biologic activity. The teratogenicity of community atmospheric pollutants and defined components thereof have not been tested as yet in mammalian species by inhalation or by parenteral administration.

A number of systems for determining the mutagenicity of atmospheric pollutants in animal species have been described. They include the dominant lethal assay, the host-mediated assay, and *in vivo* cytogenetics.

Although there is an association between mutagenic and carcinogenic activities in a number of compounds, there is as yet no proof that the two processes are closely related or that the mechanism of chemical carcinogenesis involves a somatic mutation.

Recent technical developments have made it possible to test the mutagenicity of chemical carcinogens in Chinese hamster cells in culture, by scoring for the production of drug-resistant or auxotrophic mutants. This leads to the possibility of studying mutagenesis and carcinogenesis simultaneously in the same cells.

VEGETATION AND POLYCYCLIC ORGANIC MATTER

No information was found to indicate that carcinogenic polycyclic hydrocarbons affected vegetation. However, absorption of polycyclic compounds by roots from contaminated solutions, by foliage from polluted atmospheres, and by aquatic plants from contaminated bodies of water increased the traces of these compounds already produced metabolically.

Burning of vegetation and some plant products may produce significant quantities of several carcinogenic hydrocarbons. Increased concentrations of these materials in organic soils and in sediments in large bodies of water suggest that many of the polycyclic compounds may be produced in decayed organic matter.

EFFECTS OF POLYCYCLIC ORGANIC MATTER ON MAN

There is clear evidence that airborne POM found in occupational settings—especially in relation to the products of burning, refining, and distilling of fossil fuels—is responsible for specific adverse biologic effects in man. The effects include cancer of the skin and lungs, nonallergic contact dermatitis, photosensitization reactions, hyperpigmentation of the skin, folliculitis, and acne. In concentrations found in urban or nonurban air, POM does not appear to cause any of those skin effects; similarly, there is no clear evidence that such materials as benzo[a]pyrene themselves in polluted air directly influence the pathogenesis of such nonneoplastic lung diseases as bronchitis and emphysema.

There is convincing statistical evidence of a dominant relation between cigarette smoking and lung cancer in man; one important factor in that relation is the polycyclic aromatic hydrocarbons, such as benzo[a]pyrene. Even in this lung cancer system, factors other than polycyclic aromatic hydrocarbons, such as phenols, may act as cocarcinogens or as accelerators.

Both animal experiments and epidemiologic data indicate that pulmonary cancer of environmental origin involves a complex series of factors and events in which polycyclic aromatic hydrocarbons constitute only one of the carcinogenic agents, that chemical cocarcinogens are also involved, and that the effects of particles, injurious gases, and coexistent viral and other pulmonary diseases must be considered.

The POM found in high concentration in cigarette smoke causes cancer of the lung and other organs in experimental animals; it is also present in the industrial environment, in which lung cancer rates are high, and is found generally in the air of urban communities. Examination of epidemiologic studies shows that, although a major factor in the causation of lung cancer in man is cigarette smoking, it does not account completely for the increased incidence of this disease. It appears that the incidence of lung cancer among urban dwellers is twice that of those living in rural areas; and within urban communities, the incidence is even greater where fossil-fuel products from industrial usage are highly concentrated in the air.

It appears, then, that there is an "urban factor" in the pathogenesis of lung cancer in man. The polycyclic organic molecule mentioned most prominently in this report has been benzo[a]pyrene. It was felt that benzo[a]pyrene could be used as an indicator molecule of urban pollution, implying the presence of a number of other polycyclic organic materials of similar structure that may also have some carcinogenic activity. The standard measure of benzo[a]pyrene concentration in the air is the number of micrograms per 1,000 m^3 of air. On the basis of epidemiologic data set against information regarding the benzo[a]pyrene content of the urban atmosphere, one can develop a working hypothesis that there is a causal relation between air pollution and the lung cancer death rate in which there is a 5% increase in death rate for each increment of urban air pollution. In this study, an increment of pollution corresponded to 1 μg of benzo[a]pyrene per 1,000 m^3 of air. On the basis of this assumed relation, a reduction in urban air pollution equivalent to 4 benzo[a]pyrene units (i.e., from 6 $\mu g/1{,}000\ m^3$ to 2 $\mu g/1{,}000\ m^3$) might be expected to reduce the lung cancer death rate by 20%. These data, however, are not to be interpreted as indicating that benzo[a]pyrene is the causative agent for lung tumors. There is much to support the idea of synergism or cocarcinogenesis, especially with respect to cigarette smoking. In addition, the carcniogenic significance of other polycyclic organic molecules in urban air pollution should be determined.

Prospective epidemiologic work correlated with analytic environmental surveillance has not been done that would provide insight into the true role of polycyclic aromatic hydrocarbons in atmospheric pollution as related to human disease in general and to lung cancer in particular.

19

Recommendations for Future Research

SOURCES OF POLYCYCLIC ORGANIC MATTER

1. Close scrutiny should be directed to deterioration effects of automobile control devices and the use of diesel-fueled vehicles under overloaded conditions.
2. Research into the effects of fuel compositions and of advanced emission control devices should be continued.
3. POM emissions from aircraft should be assessed.
4. Substitution of alternate fuels or more efficient combustion processes and discontinuance of coal-refuse storage practices seem to be appropriate methods for the restriction of coal-related POM emissions.
5. Emission associated with coke production requires additional research on control procedures and source analysis.

ATMOSPHERIC PHYSICS OF PARTICULATE POLYCYCLIC ORGANIC MATTER

1. Knowledge of the behavior of aerosols is essential to understanding the fate of POM in the atmosphere. Additional data relative to the

physical properties of atmospheric aerosols are needed. Simple and inexpensive instrumentation is required to obtain size–weight concentration data, particularly during short periods (minutes).

2. Further information should be obtained on the chemical and physical forms of POM in air, as well as details of its association with suspended particulate matter, especially with respect to particle size and chemical composition.

CHEMICAL REACTIVITY OF POLYCYCLIC AROMATIC HYDROCARBONS AND AZA-ARENES

1. More definite information on chemical half-lives under various conditions is essential.

2. Further research is needed on the products of chemical reaction under atmospheric conditions and on the possible biologic activity of these products.

3. Mechanisms that deserve further study are those involving aromatic hydrocarbons, which may cause reactive oxidizing species to be delivered to genetic and other biologic material.

STUDIES OF POLYCYCLIC ORGANIC MATTER IN ANIMALS, MAMMALIAN CELLS, AND VEGETATION

1. Improved methods for studying the genetics of mammalian cells should be developed.

2. Further biologic, biochemical, and molecular biologic research should be done in the fundamental mechanisms of chemical carcinogenesis, which could lead to the eradication of cancer by prophylaxis and possibly the reversion of cancers to normality.

3. Further exploration of the use of artificial atmospheres, such as benzo[a]pyrene and sulfur dioxide, for direct inhalation carcinogenesis tests is necessary, as well as the further use of POM adsorbed on particles to test a wide variety of air pollution fractions and subfractions.

4. The chemistry and biologic activities of such airborne cocarcinogens and tumor-promoting compounds as polyphenols and paraffinic hydrocarbons should be studied further, as should the activities of the oxidation products of airborne olefins and aromatic hydrocarbons—in particular, the chemical nature and carcinogenic

and other biologic properties of the epoxides, hydroperoxides, peroxides, and lactones.

5. An attempt should be made to demonstrate lung cancer in outdoor animals (birds) exposed to a highly polluted urban environment in which lung cancer in man is unduly high, comparing with a sheltered flock of the same origin and age distribution breathing purified air but otherwise under identical dietary, sanitary, and other conditions.

6. Greater use of newborn mice should be made in testing the carcinogenicity of fractions of polluted air.

7. Tests of domestically bred simian primates by the same methods that succeeded in prosimian primates and rodents should be conducted to provide data on their relative susceptibility to systemic, skin, and bronchial carcinogenesis.

8. A definitive study of the distribution, excretion, and metabolism of a polycyclic carcinogenic hydrocarbon should be conducted, with identification and characterization of all metabolites.

9. The microsomal aryl hydrocarbon hydroxylase system and the effects of various inhibitors should be assayed in a variety of human tissues.

10. A reliable source of standard preparations of polycyclic hydrocarbons and their metabolites and of specific inducers and inhibitors of the microsomal enzyme system should be provided.

11. Parallel tests for carcinogenicity of chemicals and fractions of polluted air should be conducted in animals and cell cultures to permit decisions on the usefulness of the *in vitro* systems.

12. Organ cultures should be used to study the histologic effects of POM on organized differentiated tissues, with particular attention to epithelial cells.

13. Increased emphasis on the testing of teratogenic activity of suspected carcinogenic fractions is needed.

14. Increased emphasis on testing for mutagenic activity of suspected carcinogenic fractions in animal systems is needed in order to gain further information on the relation between the processes of carcinogenesis and mutagenesis.

15. A screening committee should be appointed to establish the criteria for an environmental agent to be regarded as a carcinogen or tumor-initiator. A list of the environmental carcinogens and tumor-initiators should be compiled and made available.

16. The effects of traces of carcinogenic materials in vegetable foods on the incidence of cancer in man and animals should be

determined. The effects of long-term exposure to, and massive dosages of, polycyclic compounds on plant growth, development, and reproduction should also be investigated.

EPIDEMIOLOGIC STUDIES OF NONOCCUPATIONAL NEOPLASTIC PULMONARY EFFECTS

1. More precise quantitation of exposure and definition of ambient air concentrations of POM and other possible carcinogens, as well as all other major pollutants, is needed. This includes particulate matter, which may act as adsorbents and carriers of POM and irritant gases that, by interfering with and slowing pulmonary clearance, may increase the duration of contact between carcinogenic materials and bronchial mucous membranes.
2. Sampling and quantitation data are needed for every major city, so that a reasonable estimate of ambient air concentrations may be obtained.
3. Much greater documentation of cigarette smoking is needed. The exclusion from the 1970 Census of this major etiologic factor in disease is unfortunate. Valid estimates of cigarette consumption in major community areas, both urban and rural, in relation to lung cancer and other major disease entities are not easily available.
4. Additional sampling studies of cigarette smoking, occupation, and residence in well-defined populations are required.
5. Modern statistical methods should be used to examine the role of airborne carcinogens other than POM in disease and to investigate the relations of these and other environmental factors to nonpulmonary neoplasms.
6. The environment–cancer association is weakened by the finding of low benzo[a]pyrene concentration and high carcinoma incidence in a number of communities. Studies seeking other etiologic agents in these localities should be carried out.
7. More extensive investigation into the effects of airborne carcinogens should be carried out where they appear in industries in high concentrations. This would be particularly valuable if workers with heavy exposure to the same materials in different communities throughout the country were compared.
8. Migrant studies reveal strong epidemiologic evidence of a relation between airborne carcinogens and lung cancer. More detailed explora-

tion of the selection of the population subgroups that migrate is necessary to rule out selection as a possible factor in these differences.

9. The results of studies carried out for this report suggest the feasibility and desirability of further epidemiologic studies on airborne carcinogenesis.

Appendix A

Collection of Airborne Particles for Analysis of Polycyclic Organic Matter

Advances in methods of collection and separation of polycyclic aromatic hydrocarbons involving chromatography, absorption, and fluorescence spectroscopy have led to the development of a number of methods for the estimation of at least 25 polycyclic aromatic hydrocarbons. Of these, about eight are now estimated more or less routinely from ambient air samples in several laboratories in a number of countries. Data are available on the monthly and seasonal distribution of these compounds in an increasing number of cities, some of which show relatively high atmospheric concentrations. In the usual mode of separation, more or less quantitative information can be obtained on the concentrations of pyrene, fluoranthene, chrysene, benzo[a]-pyrene, benzo[e]pyrene, benzo[k]fluoranthene, benz[a]anthracene, perylene, benzo[ghi]perylene, coronene, and anthanthrene. Sawicki and co-workers[667] have examined the air of more than 130 urban and nonurban areas in all sections of the United States and found that the powerful carcinogen benzo[a]pyrene is universally present, in varying concentration.

The atmospheric concentrations of polycyclic aromatic hydrocarbons are so small, even in heavily populated cities, that the results are reported in micrograms per 1,000 m^3 of air sample. In general, benzo[a]pyrene increases markedly from a relatively low concentration in the summer months to a maximum during the coldest months of the heating season, the time of increased consumption of fuels for heat and power. Concentrations may vary from less than 1 $\mu g/1{,}000\ m^3$ in relatively clean air to over 100 $\mu g/1{,}000\ m^3$ in smoky, polluted air of a large city. Such trace quantities imply the sampling of extremely large quantities of air to collect enough smoke or particulate matter by filtration, or other means, for a sufficient sample for subsequent analysis. Usually a minimum of about 10,000 m^3 of air sample is required in order to collect by column chromatography a sufficient quantity of POM for determination of individual compounds by absorption spectroscopy.

SELECTION OF SAMPLING SITES

To assess the exposure of man to carcinogens in air, long-term average concentrations are considered to be of prime importance, whereas short-term fluctuations may have little meaning. Because airborne polycyclic aromatic hydrocarbons and other carcinogenic substances are derived from a variety of fluctuating sources, it is difficult to determine overall exposures of population groups except on the basis of mean levels of individual carcinogens and mixtures of known composition and potency over long periods.

In general, pollution levels in cities and suburban environments fluctuate widely, in accordance with prevailing meteorologic conditions. Other influences on pollutant concentrations include topography, nature and distribution of sources, distance from sources upwind and downwind, and seasonal and annual variations.

Selection of sites to ensure representative sampling must be undertaken with great care in order to incorporate the various influencing factors, at least on a sound statistical basis. An illustration of this approach is provided by the studies of smoke and polycyclic aromatic hydrocarbon content of the air in two pairs of European cities (Belfast and Dublin, and Oslo and Helsinki), as described by Waller and Commins.[800] They were undertaken as a pilot project under the supervision of the Cancer Study Group of the World Health Organization. Each of the four cities was divided into five areas having roughly equal populations, and the sampling sites were placed as close as possible to the centers of population density of the areas. Great care was taken in selecting sampling sites to avoid proximity either to individual sources of pollution or to open spaces. At each location, the samplers for the collection of suspended particulate matter (smoke and polycyclic aromatic hydrocarbons) were operated for a period of 12 months.

The general considerations involved in the selection and location of sampling sites have been discussed by Katz.[428] For the study of health effects, the sampling sites must ensure that the sample collections are truly representative of air that is actually breathed by the exposed population groups. Concentrations of pollutants vary with height of the sampling points above ground level, so results from sites on the roofs of buildings may differ materially from results from sites at breathing level. A representative number of sampling stations for a given area may be established after a preliminary study. This type of study should include information on the nature and magnitude of emissions from principal sources of pollution, a review of available climatologic and meteorologic data, and the gathering of some preliminary data on air pollutant concentrations in areas of severe and slight pollution.

SELECTION OF EQUIPMENT AND FILTERS

The general practice in sampling and collection of polycyclic aromatic hydrocarbons and related heterocyclic compounds of high molecular weight involves filtration of air to collect the suspended particulate matter. In the United States and Canada, the equipment used for this purpose is a high-volume sampler, which consists of a suction fan operated by a motor and equipped with a filter holder and calibrated air flow gauge or manometer. The filter is a rectangular fiber glass sheet of high collection efficiency and low resistance, 20 × 25 cm. Current samplers are exposed inside a louvered box that holds the filter surface, facing upward and horizontal, under a roof that protects the filter from rain and snow. The air sampling rate varies from about 1.2 to 1.7 m^3/min. Flash-fired fiber glass filters are used; they have an efficiency of about 99.97%, despite their low resistance to air flow. Particulate samples are normally collected over periods of 24–48 hr.

In West Germany, the Institut für Lufthygiene uses, in addition to a high-volume sampler, the BAT I,* the BAT II,* and the Draeger

*Air sampling instruments, similar to the high-volume sampler, that collect airborne dust or particulate matter; coarse particles larger than 10 μm are removed before passage of air sample through the system. Thus, only respirable particles are collected by the filters.

instrument. The air flow rate of the BAT I is 10 m^3/hr; of the BAT II, 100 m^3/hr; and of the Draeger instrument, 3 m^3/hr. The filters used in the studies described by Waller and Commins[800] were fiber glass, Whatman GF/A type, 12.5 cm in diameter, which collected particulate matter containing polycyclic aromatic hydrocarbons by means of low-volume samplers with an air flow rate of 5 m^3/day. Each sampler was operated continuously for a month to provide a sufficient sample of particulate material for gravimetric estimation. The determinations of polycyclic aromatic hydrocarbons were carried out on bulked samples for 6-month periods, mainly during the summer and winter seasons.

It has been shown that some polycyclic aromatic hydrocarbons are sufficiently volatile to evaporate during collection if long-period sampling procedures at low flow rates are used. Although this is true especially for the lower-molecular-weight hydrocarbons, it is impossible to collect anthracene, phenanthrene, pyrene, and fluoranthene efficiently. It is evident that accuracy requires collecting samples at comparatively high flow rates during short periods (about 24–48 hr) and analyzing them in the laboratory as soon as possible. In special cases of high air pollution levels, high-volume filters may become blocked in less than a day. It has been suggested that samples be collected at intermediate flow rates over a period of a week and that weekly samples be pooled to yield material adequate for the analysis of polycyclic aromatic hydrocarbons representing the four seasons of the year.

FILTRATION AND STORAGE

The analytic procedure recommended by the Intersociety Committee on Methods for Ambient Air Sampling and Analysis of the American Public Health Association involves the collection of particulate matter with high-volume air samplers on flash-fired fiber glass filters.[746,747] Before use, the filter should be washed thoroughly with pentane to make the blank as low as possible. The amount of particulate matter collected depends on many variables, such as the particulate loading in the atmosphere, sampler location, and volume and rate of air sampled. On the average, in an urban area, the sampler will collect approximately 250–350 mg of particulate matter while sampling 2,000–2,400 m^3 of air during a 24-hr period. Of this quantity of particulate matter, approximately 6–10% will be soluble in benzene.

Hence, this benzene extract or organic fraction will contain about 25 mg for an average high-volume sample. A quantity of benzene-soluble material amounting to 50–150 mg is needed for analysis of polycyclic aromatic hydrocarbons. To obtain this amount of material, it is necessary to pool the organic fractions of several individual fiber glass samples from a single site.

The procedure for weighing the filters before and after sample collection should be standardized to control these conditions, preferably at 25 C with a constant relative humidity below 50% in a conditioning chamber. The upper limit of particle size collected by the high-volume sampler is probably less than 50 μm in diameter in the standard shelter of the U.S. National Air Sampling Network (NASN). The size distribution of the material on the filter has not been measured directly. However, it can be deduced that about 90% of the weight of suspended particulate matter reported by the NASN can be associated with particles whose terminal settling velocities are less than that of a sphere approximately 8 μm in radius with a density of 1 g/cm^3.

With the BAT II instrument, the upper limit of size of the filtered particles is stated to be about 5–7 μm in diameter; and with the Draeger instrument, 10 μm. It is believed that sampling instruments that eliminate the coarser particles (i.e., larger than 10 μm), which cannot be inhaled, and collect the particles in the respirable range are best suited for evaluation of the hazard to human health of polycyclic aromatic hydrocarbons and related carcinogens. It is believed that no significant amounts of polycyclic aromatic hydrocarbons, especially those of high molecular weight, exist in the vapor phase in the ambient atmosphere. Consequently, they can all be collected by filtration of the particulate phase in appropriate sampling conditions.

The volatility of some polycyclic aromatic hydrocarbons is of concern in connection with filtration of particulate matter, especially at low flow rates over periods of several weeks or longer. Filtration media should consist of low-resistance fiber glass filters without organic binders. Membrane filters are generally unsuitable for air sampling for analysis of polycyclic aromatic hydrocarbons, because they have a high air flow resistance and undesirable solubility in organic solvents. Accurate calibration of flow meters or flow gauges is essential for the correct measurement of air volume during sampling.

Samples should be stored in the dark in cleaned glass vessels fitted with ground-glass stoppers, preferably in a refrigerator. The use of paper or polyethylene bags for storage is not recommended, because paper can absorb hydrocarbons from collected samples and polyethylene contains oils and antioxidants that may contaminate the samples. The organic fractions of samples remain stable for a period of 5 years if stored in a dark refrigerator. Exposure to light, especially ultraviolet, destroys some polycyclic aromatic hydrocarbons.[262]

STANDARDIZATION

Although much information is available on the atmospheric concentrations of polycyclic aromatic hydrocarbons of cities and industrial areas, the data derived by different methods of sampling and analysis cannot be directly compared or properly assessed. Samples collected by low-volume air flow filtration over periods of many weeks cannot be compared directly with those obtained by high-volume filtration over 1 or 2 days, owing to losses by volatilization of some polycyclic aromatic hydrocarbons in low-volume filtration sampling. For assessment of the potential hazards of airborne carcinogens to public health, only the respirable portion of the suspended particulate matter is useful. Samples in the required particle size range should be collected by similar methods of sampling and analysis.

A sampling system for the assessment of urban aerosol particle size–weight distribution has been described by O'Donnell *et al.*[568] The system includes a six-stage Andersen cascade impactor,[14] a backup filter, a vacuum pump, a critical-sized orifice, and a special cover that simulates entry conditions of the NASN high-volume sampler. After calibration of the Andersen sampler, particle size distributions can be determined directly from the particle weights on the six collection plates and the backup filter. To determine the total atmospheric particle concentration, the wall losses on the top impactor sieve of the Andersen sampler must be measured. Wall losses do not affect size distribution results with atmospheric suspended particles smaller than 9 μm in diameter, which is the size associated with deposition in human lungs. Particles larger than 9 μm can be determined by measuring these sieve losses. Alternatively, the fraction of particles larger than 9 μm may be estimated by taking the difference between the total atmospheric particle concentration, as measured by a high-volume sampler, and the total concentration of the particles smaller than 9 μm, measured by using five stages of the Andersen impactor and the backup filter.

Results for total concentration obtained using the Andersen sampler, corrected for sieve losses, were found to compare favorably with the total concentration measurements obtained simultaneously with the high-volume sampler. Only a slight error is caused by the differences in collection efficiencies between the high-volume filter and the backup membrane filter used with the Andersen sampler.

Table A-1 lists the effective cutoff diameters reported for particles collected on stages of the Andersen sampler during 24-hr sampling periods.[568] The average weight of particles collected in 24 hr was 137 μg/m³ of air sampled.

SUMMARY

High-volume filtration samplers are used routinely to collect atmospheric particulate matter on fiber glass mats for periods of 24 hr or more by the U.S. National Air Sampling Network. This type of sampling is adequate for determining the concentration of POM and of individual polycyclic compounds. However, such high-volume air samples do not provide information relative to aerosol particle size–weight distribution.

Emissions at various sources are sampled by filtration after passing through a cooling train. No information is available on sample losses in these sampling trains or on the errors associated with sample collection.

TABLE A-1 Effective Cutoff Diameters of Particles Collected on Stages of Andersen Sampler (24-hr period)

Stage	Effective Cutoff Diameter, μm	Stage Weight, mg/24 hr	Fraction of Sample Weight, %
Sieve 1	> 9.2	0.83	15.8
Stage 1	9.2	1.0	19.0
Sieve 2	–	0.13	2.5
Stage 2	5.74	0.51	9.7
Stage 3	3.3	0.40	7.6
Stage 4	1.76	0.32	6.1
Stage 5	0.98	0.25	4.7
Stage 6	0.50	0.33	6.3
Backup filter	< 0.50	1.49	28.3
Total	–	5.26	100.0

RECOMMENDATIONS

International standardization of sampling and analytic procedures in this field is urgently required, so that data collected in different countries can be evaluated. Only then can valid information be obtained in epidemiologic and biologic studies for the establishment of adequate criteria and standards of air quality on a worldwide basis.

Information on aerosol particle size–weight distribution is essential for the prediction of pulmonary responses in man after inhalation of aerosols.

Appendix B

Separation Methods for Polycyclic Organic Matter

In vivo bioassays have demonstrated that the greatest carcinogenic activity of organic pollutants is associated with the neutral fractions (nonacidic and nonbasic) that are enriched in polycyclic organic matter.[388,404] Present knowledge indicates that those fractions contain only two classes of compounds that include known animal carcinogens—the polycyclic aromatic hydrocarbons and their neutral nitrogen analogs, the aza-arenes (e.g., indoles and carbazoles). The basic aza-arenes constitute a third class of polycyclic organic pollutants that includes some known animal carcinogens.[346,699,700] Because the basic compounds have in general represented only a minor portion of organic pollutants, they have not been tested for carcinogenicity until now. Hence, it cannot be stated with certainty that the basic aza-arenes contribute to the overall carcinogenicity of the organic matter from polluted air. Nevertheless, until proved otherwise in animal experiments, the basic aza-arenes, polycyclic aromatic hydrocarbons, and neutral aza-arenes should be analyzed in evaluating the carcinogenic potential of organic pollutants.

STANDARD LABORATORY CONDITIONS

Precautions

All known carcinogenic polycyclic aromatic hydrocarbons absorb light at wavelengths of 350–450 nm and, in the presence of oxygen, may be oxidized.[464,587] It is recommended that the collecting device be protected from direct exposure to sunlight and from the radiation of fluorescent lamps. It is also advisable that analyses be performed in laboratories that are illuminated with yellow light (no radiation below 450 nm).

Hydrocarbon POM with four or more rings is quantitatively collected by filtering samples in normal atmospheric conditions with large-volume samplers.[154] Commins demonstrated that, at normal environmental temperatures (up to 30 C) and with proper storage, decomposition of even the most unstable aromatic hydrocarbon can be avoided.[154] During analysis, organic solvents have to be evaporated to concentrate the POM. The loss of POM during this procedure can be reduced or even avoided by working at pressures above 12 mmHg and at water-bath temperatures below 45 C. If one must concentrate the POM from large volumes, evaporation should take place in distillation columns operated at least at a 2:1 reflux ratio. Often, especially with gas or liquid chromatography, the POM solution has to be concentrated to a rather small volume—perhaps only a few milliliters—in which case the POM solution should be concentrated by freeze-drying.

Some basic aza-arenes, to be isolated, must be concentrated by extracting the POM solution with acids and then basifying with alkali, preferably at low temperature (e.g., by external cooling with ice water).

Internal Standards

Despite the greatest of precautions, there are some unavoidable losses during the analysis of POM. These losses may vary from 10 to 40%, depending on the sample and the experience of the chemist. One way to overcome these uncertainties is to standardize the analytic method. It has been demonstrated, however, that samples of polycyclic organic air pollutants can vary significantly in composition and concentration, even though the particles are collected at the same site but in different pollution conditions. Therefore, the best method to secure quantitative data for individual polycyclic organic compounds is to impose internal standards at the beginning of the analysis. This can be done by using the isotope-dilution method—a technique that yields quantitative data for individual hydrocarbons accurate to within 8%. Either the original sample is supplemented with one or several hydrocarbons not present in the original sample but concentrated with the test carcinogenic polycyclic organic compounds, or traces of a known carcinogenic hydrocarbon labeled with carbon-14 are added (tritium-labeled polycyclic aromatic hydrocarbons should not be used, because an exchange between tritium and hydrogen can occur during analysis).[387]

Reproducibility

In analyzing carcinogenic polycyclic hydrocarbons in the respiratory environment, some factors that can affect the reproducibility of analysis must be recognized, particularly the contamination of the sample with POM from solvents, adsorbents, and the environment and the poor quality of commercially available reference compounds. These sources of error in POM analysis cannot be overcome by the use of an internal standard and must therefore be avoided by purifying the solvents and agents to be used and by restricting the analytic work to laboratories that are ventilated with filtered air. To avoid possible cross-contamination, it is essential that reference compounds be purified in a separate laboratory under a properly working hood.

DISTRIBUTION

Distribution between Solvents

The organic matter, or "tar," in gasoline and diesel engine exhaust fumes is rich in aliphatic hydrocarbons. For example, an air pollution sample from a Detroit collection site with high traffic density consisted of more than 48% of nonpolar hydrocarbons (neutral fraction N-1).[388] The corresponding figure for the N-1 fraction from exhaust "tar" from a standard test-stand run of a conventional gasoline engine was more than 15%.[384] Because high concentrations of paraffins and olefins are known to reduce the efficiency of the column chromatographic separation of POM, it is advisable to separate the bulk of the aliphatic hydrocarbons from the POM. This can be done by distribution between cyclohexane and nitromethane,[389] cyclohexane and dimethylsulfoxide (D. Hoffmann and G. Rathkamp, unpublished data), *n*-hexane and dimethylsulfoxide, acetonitrile and *n*-hexane,[328] or nitromethane and carbon disulfide.[44] A good method involves distribution between cyclohexane and dimethylsulfoxide and back-extraction from the dimethylsulfoxide layer with cyclohexane after addition of water (D. Hoffmann and G. Rathkamp, unpublished data). With back-extraction, one has to evaporate only a relatively low-boiling solvent.

Countercurrent Distribution

Several reports are concerned with the enrichment of POM from environmental agents by countercurrent distribution. Demisch and Wright[195] suggest *n*-hexane and aqueous monoethanolammonium deoxycholate as a solvent pair, and Mold *et al.*[536] suggest cyclohexane and methanol–water (9 : 1) containing 0.83% of tetramethyluric acid. Selective separation systems like these may lead to the isolation of hitherto unknown carcinogenic hydrocarbons from the respiratory environment, especially if a 1,000-cell Craig countercurrent distribution apparatus is used. However, it is doubtful whether this method has practical value for routine analysis of polycyclic organic pollutants.

Extraction of Basic Aza-Arenes

If all the basic aza-arenes must be concentrated in one fraction, the concentrated solutions of the organic pollutants must be extracted with strong inorganic acids—perhaps as concentrated as 10–15% sulfuric acid. Lower acid concentrations do not lead to a quantitative separation of these compounds from the neutral and acidic organic pollutants.

COLUMN CHROMATOGRAPHY

Standard Conditions

The most widely used separation methods for the chemical analysis of polycyclic aromatic hydrocarbons and aza-arenes are column and ion-exchange chromatography. A large variety of adsorbents—such as alumina, silica gel, Florisil, and cellulose acetate—and various ion-exchange resins are selected for these separations. Adsorbents with uniform particle size, either 80–100 or 100–120 mesh, and column diameter : length ratios of at least 1 : 25 are suggested. Laboratory temperatures should be between 18 and 23 C and kept constant within a few degrees. (The flow rate of columns is known to change in laboratories whose temperatures change overnight or during weekends.)

The solvents to be used should be spectral-grade quality, or at least purified through a distillation column with a reflux ratio of 2 : 1. It is advisable to consult *Organic Solvents: Physical Properties and Methods of Purification* by Riddick and Bunger[631] for the purification of such solvents. Column separations generally begin with a low-boiling alkane as the elution solvent. It is advisable to dry even hydrocarbon solvents to avoid column deactivation (*n*-hexane can contain up to 0.01 vol.% water).

Several techniques are used for filling the column, mainly making a slurry of the adsorbent in a hydrocarbon solvent or filling the column partially with the solvent and pouring the adsorbent into it. In any case, air bubbles have to be avoided during packing and operation.

The flow rate of the column is best controlled with a Teflon stopcock. A fraction collector is essential for establishing reproducible separations. The separation can be based on either volume, timing, or change of fluorescence on exposure to ultraviolet light of 365-nm wavelength.

Maximal separation can be achieved by slowly and evenly increasing the polarity of the solvent (gradient elution, e.g., increasing the proportion of benzene in *n*-hexane with time). Further details on achieving reproducible column chromatographic enrichments or separations of POM are available in the literature.

Alumina

Alumina is one of the more strongly binding adsorbents. It is available in various grades, sizes, pH's, and activities. One of the better standardized preparations is Woelm alumina. In general, one should use alumina with relative activities between 26 and 32 for the separation of polycyclic aromatic hydrocarbons and neutral aza-arenes.[369] The ratio of POM concentrate to alumina will vary between 1 : 100 and 1 : 1,000, depending on the degree of enrichment desired. It should be possible to elute the polycyclic aromatic hydrocarbons in a reasonable time with *n*-hexane with an increasing concentration of benzene up to 25%. Some investigators have successfully separated polycyclic aromatic hydrocarbons on very long alumina columns.[143]

These conditions may well be best for a laboratory that needs to analyze only a few samples and is not specifically equipped for POM analysis. However, the investigator should be aware that the analysis may require a long time (5–10 days) and should therefore be performed isothermally. Although it is known that some decomposition may occur on neutral alumina, this is significant only if one is also interested in esters whose polarity is similar to that of polycyclic aromatic hydrocarbons. So far, alkaline alumina has not been applied to the analysis of aza-arenes.

Silica Gel

Except for the silica gel developed specifically for thin-layer chromatography, it is difficult to find commerically available silica gel that contains no fluorescent materials. Therefore, silica gel has to be purified, either by washing it on the column before the actual separation (which leads to partial deactivation) or by washing with a polar solvent and later reactivating it. The sample : adsorbent ratio varies between 1 : 50 and 1 : 500. Another disadvantage of silica gel is its relatively slow flow.

Florisil

Polycyclic aromatic hydrocarbons separate rather well on Florisil, despite the short elution time. Florisil is weakly cationotropic (weakly basic), which increases the possibility that traces of polycyclic aromatic hydrocarbons may be partially photooxidized. If one is interested in other neutral compounds with comparable polarity—such as indoles, some esters, and DDT and DDD—one should avoid using Florisil to prevent oxidation, hydrolysis, or dehydrohalogenation, respectively.

Chromatography on Cellulose and Modified Cellulose

Although the literature describes several powders of cellulose derivatives—such as xanthates, succinates, and acetates—as promising adsorbents, these chromatography materials so far have been tried only once for the separation of polycyclic aromatic hydrocarbons.[801]

GEL FILTRATION

Gel filtration is of great value when used to supplement column chromatography. During the last few years, gel filtration has led to the isolation and identification of several natural products, biologically active compounds, metabolites, and environmental agents.[599] Wilk *et al.*[819] first explored gel filtration for the separation of polycyclic aromatic hydrocarbons. Others have since found the system to be most valuable for the separation of neutral aza-arenes[385,386] and aromatic hydrocarbons.[624] The best results so far have been achieved by separating concentrates of aromatic hydrocarbons isothermally (at 32 C) on Sephadex LH-20 in long columns (ratio, about 1 : 1,000) with propanol-2 as solvent. One disadvantage of the gel filtration method is that it takes 1–3 days. Nevertheless, it is expected that gel filtration will become more widely used for the analysis of polycyclic organic pollutants.

THIN-LAYER CHROMATOGRAPHY

Since the development of thin-layer chromatography as a new analytic tool, E. Sawicki and associates[661,664,668] have skillfully applied this technique to air pollution research. Today, various polycyclic air pollutants are analyzed in accordance with the techniques developed by them. Included are the analyses of polycyclic aromatic hydrocarbons, aza-arenes, polycyclic carbonyl compounds, and phenols. Thin-layer chromatography is quick, reproducible, and inexpensive; in the hands of an experienced analyst, it leads to good separations of organic pollutants, as reviewed recently.[661]

One point for discussion is the great emphasis on thin-layer chromatography as a final step in isolating POM from urban air pollutants and then assaying the extracts by spectroscopic methods. Mass spectrometry and liquid chromatography, when applied as additional analytic tools for the identification of aromatic hydrocarbons and aza-arenes, have taught us that individual spots or bands from thin-layer chromatograms are sometimes, in fact, mixtures of two or more compounds. Furthermore, during thin-layer chromatography and paper chromatography, trace amounts of some polycyclic aromatic hydrocarbons and aza-arenes may be decomposed. Therefore, internal standards should be used to correct for the losses.

PAPER CHROMATOGRAPHY

Compared with thin-layer chromatography, paper chromatography, although requiring significantly more time, affords a better separation.

Of course, evidence of purity or homogeneity based solely on chromatographic methods should not be accepted as unequivocal. But either method can serve as a useful guideline.

The method of Tarbell *et al.*,[743] using Whatman No. 1 paper and dimethylformamide saturated with *n*-hexane, and the method of Spotswood,[718] using acetylated paper, result in good separation of polycyclic aromatic hydrocarbons. For the isolation of basic and neutral aza-arenes, Van Duuren *et al.*[786] have developed a new paper chromatography system. Some years ago, Sawicki[664] reviewed in detail the use of paper chromatography in air pollution research. A laboratory not equipped with a gas chromatograph with electron capture detection or a liquid chromatograph is well advised to use chromatography on acetylated paper for the final separation of the carcinogenic hydrocarbons benzo[a]pyrene, benzo[b]fluoranthene, benzo[j]fluoranthene, indeno[1,2,3-cd]pyrene, and chrysene.

PAPER AND THIN-LAYER ELECTROPHORESIS

E. Sawicki *et al.*[669] have developed methods for the separation of basic aza-arenes by paper and thin-layer electrophoresis. For some of the aza-arenes studied, encouraging degrees of separation were achieved, e.g., for the three benzacridines. These techniques are recommended as additional methods for the qualitative analysis of basic aza-arenes in polluted air.

Chemists, however, should be aware that paper, thin-layer, and column chromatography generally do not separate the individual alkylated polycyclic aromatic hydrocarbons from each other. This separation can be achieved only on gas chromatography columns with more than 25,000 theoretical plates and on liquid chromatography columns.

HIGH-SPEED LIQUID CHROMATOGRAPHY

Although liquid chromatography possesses many advantages for the separation and isolation of organic compounds, the method has remained unattractive owing to its long elution time and poor column efficiency. Recent work, however, has shown that the speed and efficiency of liquid chromatography can be greatly increased, to approach even that of gas chromatography. Compared with gas chromatography, high-speed liquid chromatography offers the advantage of operating at relatively low temperatures (<100 C), thereby preventing thermal rearrangements and decompositions. Depending on the chemical structure of the compounds being analyzed, a differential refractometer detector or an ultraviolet photometric detector (254 nm) is used. The detection limits for most compounds are 10^{-9} g for the refractometer detector and 10^{-10} g for the ultraviolet photometric detector for polycyclic aromatic hydrocarbons that have at least four condensed aromatic rings. At present, the use of high-speed liquid chromatography seems to be limited by the high cost of an efficient instrument with a high-pressure pulse-free liquid pump ($10,000+), the high cost of specific columns (which most laboratories are unequipped to prepare), and the rather cumbersome collection of reagents, solvents, and standards required for the determination of the identity and purity of unknowns. It appears for the time being that gas chromatography and liquid chromatography will be used to complement each other.

Separation of polycyclic aromatic hydrocarbons by liquid chromatography at 40 C and 80 atm on a 1-m column that is filled with a specific-surface-porosity support and coated with a hydrocarbon polymer has been reported.[436] A mixture of water and methanol (6:4) serves as the mobile phase. This system effectively separates the carcinogens chrysene, benzo[e]pyrene, and benzo[a]pyrene with retention times of 15, 22, and 27 min, respectively. Using 3-m columns with the same stationary phase described above for carcinogens, but under higher pressure and with a mixture of propanol-2 and water (1:4) as the solvent, one can separate all 10 major four- and five-ring aromatic hydrocarbons that are present in polluted air of New York City. Recently, Ledford *et al.*[484] separated arenes on a 1-m column filled with Durapak OPN, with 0.25% methylisobutylketone in heptane as the moving phase. On Corning CPG glass beads treated with octadecyltrichlorosilane as the stationary phase and acetonitrile as the moving phase, the same investigators completely separated benzo[e]pyrene (6 min), perylene (7 min), benzo[b]fluoranthene (8 min), and benzo[a]pyrene (9 min) and isolated a pure specimen of benzo[a]pyrene from a concentrate of tobacco smoke. These examples indicate the great potential of high-speed liquid chromatography for the analysis of carcinogenic polycyclic aromatic hydrocarbons; it most likely can also be applied to the analysis of carcinogenic aza-arenes. High-speed liquid chromatography is likely to become a major tool in the analysis of organic air pollutants.

GAS CHROMATOGRAPHY

The gas chromatograph is a versatile instrument for analyzing mixtures of organic air pollutants. Available at moderate cost ($2,000–8,000), gas chromatography (GC) combines simplicity of operation with high sensitivity. Useful application of GC to the analysis of POM dates from 1965, with noteworthy contributions from Wilmshurst,[821] Cantuti *et al.*,[115] DeMaio and Corn,[192] Carugno and Rossi,[120] Chakraborty and Long,[127] and Searl *et al.*[685]

Operating Principle

Gas chromatography vaporizes mixtures of organic compounds into an inert, mobile vapor phase (carrier gas) and then transports it through a long, narrow tube (column) containing an inert liquid (or soid) stationary phase in intimate contact with the mobile vapor phase. Organic molecules travel along the column according to their relative volatility and their physicochemical affinity for the liquid stationary phase at the temperature of operation. The temperature and liquid phase are selected for their ability to separate the sample mixtures into their pure constituent compounds; after a characteristic interval (retention time), the individual separated components emerge from the column and are monitored by a sensing device (detector) that permits the recording of their passage as a time-based analog signal rising above the detector background signal (base line).

Application to Polycyclic Organic Matter

Successful gas chromatographic analysis of POM is a highly desired goal because of its potential convenience, speed, and reproducibility.[192,821] By operating on a physical principle different from that of column or thin-layer chromatography, GC may separate mixtures of compounds that cannot be resolved by the other procedures. It should be recognized, however, that the complexity of the polycyclic organic fraction, consisting of many closely related compounds, nearly precludes reliance on a single separation technique. Thus, the use of GC in POM analysis must be considered as complementing the other techniques for qualitative and quantitative analysis.

Before the application of some refinements in 1965, analysis of POM by GC was not very satisfactory.[821] Indeed, the stringent requirements of resolution, stationary-phase thermal stability, and detector sensitivity demanded by the presence of trace levels of closely related, rather nonvolatile, polycyclic organic compounds in polluted air continue to challenge the analytic capabilities of GC.[685]

Detectors

Generally, the quantities of polycyclic organic compounds that can be sampled during a relatively short period are small. Therefore, only the most sensitive GC detectors can be considered for analysis of POM. For general hydrocarbon analysis, this has meant restriction to the use of the flame ionization detector (FID). The FID functions by burning the emerged (eluted), separated organic component in a hydrogen–oxygen flame. Intermediate combustion states of the organic molecule are ionized and appear as a current between two polarized electrodes. The current is then amplified into an electric signal suitable for display on an analog strip-chart recorder. The advantage of the FID is its wide dynamic range of linear sample response. The most recent versions of FID have sample sensing capability down to 10^{-10} g, almost as sensitive for POM as electron capture detection. Except for carbon atoms oxidized to the carbonyl state or beyond, the signal of the FID detector is directly proportional to the number of carbon atoms being burned, providing a quantitatively comparable detector response for most polycyclic organic compounds.

The second kind of detector of interest uses electron capture (EC), in which an eluted polycyclic organic component is irradiated by a beta emitter. Because of the high temperatures involved, EC detectors other than the conventional tritium version are required. When organic compounds capable of capturing some of the electrons emitted pass through the EC detector, their presence is detected by a diminution in intensity of the normal stream of electrons collected from the beta emitter. Whereas polycyclic organic compounds do indeed capture electrons, the EC cross section may be expected to vary with molecular structure; thus, the EC signal must be calibrated for each different molecule for quantitative analysis. Those who wish to avoid this procedure frequently use the EC detector in parallel (split stream) with the somewhat less sensitive but more general FID. With FID and EC detection, the variable response of EC according to molecular structure has been used to advantage by Cantuti *et al.*[115] to obtain tentative molecular identifications. A note of caution must be included for the use of EC on particulate samples collected from the atmosphere: EC detectors respond strongly to the polyhalogenated hydrocarbons used as pesticides. Hence, studies of comparative thin-layer chromatographic and gas chromatographic behavior between pesticides and POM may be necessary to avoid undesired interference from ambient pesticides in particulate matter.

When the organic compounds separated by GC must be collected for spectral identification, effluent stream splitters are generally satisfactory.[685] Occasionally, the less sensitive thermal conductivity detector is used[127] to avoid the possibility of complications caused by the stream-splitting inherent in the use of FID or EC. If this proves necessary, special minimal dead-volume (micro) thermal conductivity detector cells (e.g., those manufactured by Perkin-Elmer Corp., Norwalk, Conn.; or Carle Instruments, Inc., Fullerton, Calif.) are probably necessary.

Submicrogram samples of eluted GC fractions may be detected and collected with such detectors, whereas the conventional four-wire, semi-diffusion type of thermal conductivity cells cut off above the microgram level.

From time to time, it is to be expected that novel GC detectors will be introduced; however, unless they can match or exceed the performance of the detectors described here, their utility for GC analysis of POM will be inconsequential.

Columns

A discussion of the types of GC columns useful in the analysis of POM must include the nature of the liquid (or solid) partitioning phase and the supporting surface. The relative nonvolatility of POM is a major factor in its physicochemical association with airborne particulate substances. Therefore, it is not surprising that high temperatures are used in GC analysis of POM. The GC partitioning phases used for such purposes should have the highest thermal stability. Unfortunately, until recently, the available liquid phases that were thermally the most stable were barely adequate for the temperatures (>300 C) required. The silicones (e.g., SE-30, SE-52, and QF-1) commonly used for lipid analysis were often disposed in a dual-column configuration to compensate for the excessive liquid-phase bleed that occurs above 250 C. Obviously, columns operated in such a manner are going to be altered, if not depleted, after several days of use. Two conventional methods for reducing the overall bleed—lowering the column temperature and shortening the column—both work against effective POM analysis and represent, in part, the objections to work published before 1965 discussed by Wilmshurst.[821] What is required is a liquid partitioning phase that possesses the requisite thermal stability to permit higher operating temperatures and longer, more efficient (higher-resolving) columns.

Special precautions must also be taken in the selection of the appropriate materials for the liquid-phase support and for the column itself. Chromosorb W (Johns-Manville),[676,821] or its equivalent,[127,192] and Chromosorb G[685] appear to be satisfactory. There are preferences for minor variations of this flux-calcined diatomite, with one group favoring the acid-washed version,[82] another group a silanized version.[676] Inasmuch as no definitive study has been reported on the relative merits of the many diatomite versions available for analysis of POM at extremely low concentrations, it seems wise simply to use that which has received the most extensive deactivation—e.g., acid washing *and* silanization.[685] Wilmshurst[821] found the less reactive smooth spherical glass beads relatively poor in resolution.

If open tubular (capillary) columns are used, the column itself serves as the supporting agent for the liquid partitioning phase. The openness and linearity of carrier gas flow make capillary GC suitable for POM analysis, because low sample-carrying capacity of capillaries is not a problem. Wilmshurst[821] found Monel metal and copper unsatisfactory as column material, whereas type 304 stainless steel appeared more acceptable. Carugno and Rossi[120] used glass capillaries rather successfully. Because capillary columns used for POM analysis may have a short lifetime, and the preparation of a workable column is time-consuming, it would be unrealistic to claim that the preparation and use of these highly resolving columns is ideal for POM analysis. The subject must be studied further.

Future Prospects

There are several possibilities for improving the status of POM separation and analysis by GC. Two liquid partitioning phases were introduced recently: poly-*m*-phenoxylene (20 rings) and a polycarboarane-siloxane (Dexsil-300 GC), which are thermally stable up to 250 C and 350 C, respectively. Undoubtedly, analysis of POM will benefit from substitution of these phases in GC procedures. The use of inorganic phases (see Isbell and Sawyer[414]) for POM analysis is also an interesting possibility. Silanization of glass (or even stainless-steel) capillary walls can be used to minimize surface adsorptive or catalytic effects. Finally, if diatomite supports are disadvantageous for GC applications undertaken with packed columns, the recently introduced textured (silanized) glass beads may suffice.

Comments

1. Despite earlier limitations, gas chromatography shows increasing promise in qualitative and quantitative analysis of POM.
2. Mixtures of polynuclear organics may be separated and detected at nanogram levels if proper attention is given to deactivating the surfaces of columns and diatomite support material thoroughly.
3. Recently introduced partitioning phases now enable the relatively nonvolatile polycyclic organic compounds to be chromatographed at 300 C without damage to the partitioning phase.
4. Gas chromatography is best used in conjunction with other confirmatory spectral or chromatographic techniques.

COMBINATION OF SEPARATION METHODS

Compared with such environmental agents as tobacco smoke, petroleum, and gasoline, POM can be isolated from organic air pollutants and identified without great difficulty. In general, only two (at the most, three) steps are needed for the isolation of these agents from pollutants. In the past, combinations of column chromatography and paper or thin-layer chromatography were more widely used, with or without preliminary distributions of the organic matter between solvents. During the last 5 years, various combinations have been developed, including column or thin-layer chromatography with gas chromatography and column chromatography followed by combined gas chromatography and mass spectrometry.[388,573,661,664,668,676]

The choice of method for the analysis of POM depends on such factors as the degree of accuracy required by the scientist or the local air pollution control authority, the availability of professional and technical staff, the time allotted for analysis, and the availability of funds for equipment and research.

SUMMARY

For the separation of polycyclic aromatic hydrocarbons and aza-arenes, several precautions are suggested. These include the avoidance of photooxidation and losses during the evaporation of solvents. Quantitative data should be secured with the aid of internal standards under reproducible conditions.

Countercurrent distributions are cumbersome and of only limited value. The distribution of organic matter between solvents is most helpful for the separation of polycyclic aromatic hydrocarbons from paraffins. Higher concentrations of the latter negatively affect the chromatographic separation of POM.

Chromatography generally uses columns filled with alumina, silica gel, Florisil, or, occasionally, cellulose or modified cellulose. Separations should be completed under standard conditions, such as column ratios, ratio of material to adsorbent, solvents, and flow rates. Column chromatography results in high enrichment of polycyclic aromatic hydrocarbons, but rarely leads to complete separation. In recent years, gel filtration has been successfully explored.

Thin-layer chromatography is widely used; it works quickly, is inexpensive, and can be used in every laboratory. Its limitations lie primarily in the incomplete separation of mixtures of polycyclic compounds. Paper chromatography requires, in general, 12–18 hr, but gives better separations than thin-layer chromatography.

High-speed liquid chromatography offers several advantages, including speed, high resolution, very sensitive ultraviolet detectors (10^{-8} g for benzo[a]pyrene), and the use of lower temperature than gas chromatography. The technique needs further development. Nevertheless, it is predicted that in a few years it will be widely used, especially when a high-pressure pulse-free liquid pump becomes less expensive.

The most significant progress in the separation of POM has been achieved by gas chromatography. With various column lengths and diameters, liquid and solid phases and supports, and flame ionization and electron capture detectors, gas chromatography is a most versatile and sensitive technique. Polycyclic aromatic hydrocarbons may be chromatographed at nanogram levels if the column and support surfaces are properly deactivated. New partitioning phases now permit analysis at 300 C without interference from liquid-phase bleed.

POM is actually analyzed with a combination of separation and detection methods. The most widely used combination consists of column chromatography followed by thin-layer or paper chromatography. The most promising probably consists of column chromatography followed by gas chromatography or, even better, by gas chromatography plus high-speed liquid chromatography.

RECOMMENDATIONS

1. A national reference bank should be established to obtain and purify all the polycyclic aromatic hydrocarbons and aza-arenes that have been identified with certainty in the respiratory environment. Melting points, ultraviolet and infrared absorption spectra, emission and excitation spectra, relative gas–liquid chromatographic retention times, and mass spectra should be recorded for all the purified agents. If an agent has doubtful carcinogenic or tumor-initiating activity, the same laboratory should be able to supply enough of the purified hydrocarbon for biologic tests. The data so compiled should be made available to the scientific community. In some instances, small reference samples should be made available to air pollution control laboratories.
2. A screening committee should be appointed to establish the criteria for an environmental agent to be regarded as a carcinogen or tumor-initiator. A list of the environmental carcinogens and tumor-initiators should be compiled and made available.
3. A method should be developed for determining indirectly the major sources of carcinogens in polluted urban air and their contributions to the overall concentration.
4. A method should be developed for examining the concentration and composition profile of mixtures of polycyclic aromatic hydrocarbons in polluted air.

Appendix C

Detection, Identification, and Quantitation

ULTRAVIOLET ABSORPTION SPECTROSCOPY

Ultraviolet radiation is substantially absorbed by all polycyclic compounds. The degree of absorption of light energy (or wavelength) is a characteristic of a compound that may be used for identification and determination of its concentration in a medium. The measurement of such absorption has been widely used in air pollution analysis. The advantages of this technique include the commercial availability of high-quality spectrophotometers, the relative insensitivity of degree of absorption to trace impurities, the fact that the absorption spectrum of a mixture is usually the sum of the spectra of the components, the strength of ultraviolet absorption bands (which obviates the use of colorimetric reagents), and the high degree of sensitivity (which makes it possible to determine and quantitate microgram amounts of compounds). The disadvantages include the requirement of two or more separatory steps for reliable identification and particularly quantitation of air pollution samples and the low detection sensitivity (only one-tenth to one-thousandth that of fluorimetric analysis).

Two principal laws of light absorption are relevant here. The first, Bouguer's or Lambert's law, states that the proportion of light absorbed by a gas or liquid medium is independent of the intensity of the incident light and that each successive layer of the medium absorbs an equal fraction of the light that reaches it. The second, Beer's law, recognizes that light can be absorbed by a molecule in vapor or liquid phase only if the light collides with the molecule. The probability of such a collision is proportional to the number of absorbing molecules in the path of the light, and Beer's law states that the amount of light absorbed in each successive layer is proportional to the concentration of absorbing molecules and to the thickness of the layer. Combining these two laws gives

$$I = I_0 \; 10^{-\epsilon bc},$$

where I = intensity of light transmitted through the solution, international candles/cm^3
I_0 = original intensity of incident light, international candles/cm^3,
ϵ = molar extinction coefficient,
b = length of light path, cm, and
c = concentration of absorbing molecules, moles/liter.

The ratio $I : I_0$ is the transmittance of the medium; and the absorbance, the quantity usually measured by spectrophotometers, is

$$A = -\log(I/I_0) = \epsilon bc.$$

Note that the absorbance (A) is equal to the product of the concentration of the absorbing molecules, the path length of light in the medium, and the molar extinction coefficient, and thus is proportional to each. If more than one kind of molecule is present in the medium, the absorption is nearly always additive, i.e.,

$$A = b(\epsilon_1 c_1 + \epsilon_2 c_2 + \cdots).$$

Molar extinction coefficients of polycyclic organic molecules are tabulated in a number of sources.[51,139,175,280,378,388,473,671,761,762]

Spectral Characteristics of Polycyclic Organic Compounds

It is striking that all carcinogenic polycyclic aromatic hydrocarbons and aza-arenes identified in urban air[388] have highly structured ultraviolet absorption spectra.[139,761] Such spectra imply that the compounds are planar in the ground state.[51] If that planarity is required for their carcinogenic activity (a possibility that requires more study), then airborne carcinogens yet to be discovered will also have structured absorption spectra. Highly structured spectra will be helpful in detecting and identifying these compounds.

Clar's empirical classification of the bands in absorption spectra is simple and adequate for analytic spectroscopy.[139] The absorption spectrum of benzo[a]pyrene in Figure C-1 shows the three groups of bands as labeled by Clar: The α bands are weak ($\epsilon \gtrsim 10^2 - 10^3$) and usually at the low-energy (long-wavelength) side of the spectrum; they are sometimes partially covered by the neighboring ρ bands, which are more intense ($\epsilon \gtrsim 10^4$) and often have regular vibrational structure; the β bands occur at still higher transition energies (shorter wavelengths), are more intense ($\epsilon \gtrsim 10^5$), and have less vibrational structure than the other bands.

Although the total spectrum for each compound is unique, the individual features are not. For example, both chrysene and 4,5-methylene chrysene have α bands at 361 nm,[280] and both benzo[a]pyrene and benzo[ghi]perylene have ρ bands at 382 nm.[671] Therefore, identification and quantitation of a polycyclic compound on the basis of only one peak are unreliable.

Nonpolar inert solvents, such as pentane and cyclohexane, bring out the spectral structures of polycyclic compounds better than polar solvents, such as 95% ethanol, and are therefore the solvents of choice for spectroscopic identification and quantitation. Aza-arene spectra show a loss of structure and a shift to lower transition energies in acidic solvents, so analysis in both neutral and acidic solvents is useful.

Solvents appear to influence the positions of absorption bands slightly. Sawicki *et al.*[671] give λ_{max} (maximal wavelength) for the most intense α, ρ, and β bands of benzo[a]pyrene in pentane as 401, 382, and 296 nm, whereas the *Ultraviolet Atlas of Organic Compounds* places the maxima of benzo[a]pyrene in heptane at 404, 385, and 297 nm. These discrepancies in the positions of α and ρ bands are too large to be due to instrumental artifact and must be ascribed to solvent interaction.

Ultraviolet absorption spectra have been compiled and are available in various sources; the main ones are discussed briefly below.

1. *Polycyclic Hydrocarbons* by Clar[139] includes spectra of all known atmosphere-polluting carcinogenic polycyclic aromatic hydrocarbons. Spectra are plotted as log extinction coefficient versus wave-

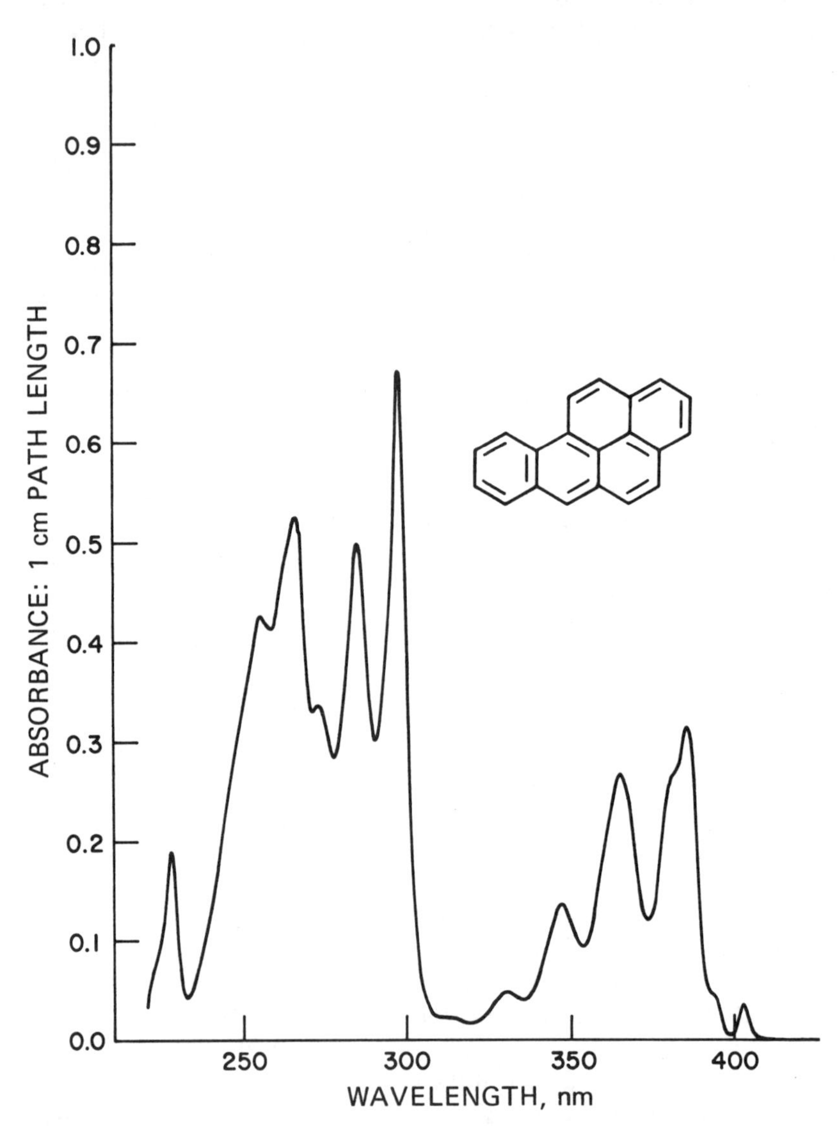

FIGURE C-1 The absorption spectrum of benzo[a]pyrene at a concentration of 10^{-5} M in cyclohexane (7.5 μg of benzo[a]pyrene). In Clar's notation, the peak at 403 nm is an α band, the peaks at 385, 364, 347, and 330 nm are ρ bands, and the peaks at shorter wavelengths are β bands.[139]

length. The extinction coefficient and wavelength values of major peaks are given in the figure legends and constitute a useful supplement to the spectra, which have small and varied scales. Clar does not include spectra of substituted aromatics or aza-arenes.

2. The *Ultraviolet Atlas of Organic Compounds*[761] is the most elegant compilation of spectra. Spectra are plotted as log extinction coefficient versus wave number (linear scale) and wavelength on transparent paper with a uniform scale throughout, which greatly facilitates comparison of spectra. Approximately half the known urban air carcinogens are included, and the format permits users to add spectra. Compounds are indexed by structural group, by formula, and alphabetically. Included with each spectrum are the structure of the compound, and values of the extinction coefficient, wavelength, and wave number of each peak.

3. *Ultraviolet Spectra of Aromatic Compounds* by Friedel and Orchin[280] contains spectra of nearly all the known urban air carcinogenic polycyclic aromatic hydrocarbons, several aza-arenes, and many substituted compounds. The spectra are plotted as log extinction coefficient versus wavelength, and the scales are uniform and convenient. The introduction contains clear, concise discussions of nomenclature, different methods of plotting spectra, effects of substituents and solvents on spectra, and multicomponent analysis. Unfortunately, this excellent book is out of print.

4. *Organic Electronic Spectral Data,*[762] presently a six-volume set, provides references to spectra published from 1946 to 1961 (additional volumes covering more recent publications are in preparation). These volumes list compounds by formula and name as designated by Chemical Abstracts Index System, giving solvent λ_{max} (log e) and bibliographic reference. This series, which already lists over 100,000 spectra, will be increasingly valuable as a guide to spectra in the literature.

Three other compilations, although less useful, should be mentioned. *Absorption Spectra in the Ultraviolet and Visible Region,* edited by Láng,[473] is a continuing series of spectra that are not arranged in any systematic way and are of unknown reliability. The series does not include even such well-known pollutants as benzo[a]pyrene, benzo[e]-pyrene, and chrysene. Hirayama's *Handbook of Ultraviolet and Visible Absorption Spectra of Organic Compounds*[378] contains two tables, one giving absorption maxima from chemical structure, and the other, absorbing chromophore from absorption maxima. The nomenclature and structure classification in this book are unfamiliar, and the spectra are pre-1960 and of uncertain quality. Because the data given are unreliable, one would need to consult the original literature. Hirayama's book is both less comprehensive and less current than *Organic Electronic Spectral Data.* Finally, there are references in the literature[175] to the Sadtler Research Laboratories* collection of ultraviolet spectra, reputed to include 28,000 spectra, but this collection has proved inaccessible.

Spectrophotometric Techniques

Many good dual-beam, ratio-recording spectrophotometers are available commercially. With only occasional calibration and maintenance, these instruments can be used continually to yield accurate and reliable data.

Wave number or wavelength calibration is generally straightforward; holmium oxide filters and mercury vapor discharge lamps provide suitable spectra with little effort.[621] A generally accepted method for calibration of photometric accuracy is not yet available. Several chemical standards have been recommended;[621] potassium chromate in potassium hydroxide has proved useful for testing photometric accuracy and estimating stray light.[531] The ease and persistence of spectrophotometric intensity calibrations are clear advantages of spectrophotometry relative to fluorimetry or spectrophotofluorimetry.

Because many polycyclic aromatic hydrocarbons and aza-arenes have absorbance peaks as narrow as 4 nm, their spectra must be measured with instruments able to resolve 0.5 nm or better. Otherwise, the spectra will be distorted in that true peak height will not be obtained. Correct spectra, free from artifacts due to the spectrophotometer, are essential for comparison with published spectra or tables. The use of single-beam nonrecording instruments to obtain correct spectra is tedious and can be recommended only if time and manpower are abundant.

Many polycyclic compounds are fluorescent, so care must be taken during absorption measurements to minimize the ratio of fluorescence to transmitted light reaching the detector. Sample absorbance should be kept below 1.0, and the sample should be placed well away from the light detector. These precautions are often ignored in reports of polycyclic compound absorption spectra.

Extracts of collected particulate matter prepared by any solvent

*3316 Spring Garden St., Philadelphia, Pennsylvania.

system contain a plethora of compounds that must be separated before quantitative spectrophotometry is possible. Column chromatography yields fractions in solvents that may or may not be suitable for spectral analysis. (Column chromatography is discussed in Appendix B.) After chromatography, it is usually good practice to evaporate the separating solvent at low temperature and redissolve the residue in a solvent, like pentane or cyclohexane, that is inactive and transparent in ultraviolet radiation up to 40,000 cm^{-1} (250 nm).[3] The solvent should be chosen to reveal spectral structure and to permit comparison with standard spectra.

When separations are performed by thin-layer chromatography, extraction from the layer is common practice; the choice of extracting solvents is governed as above. A blank portion of the thin layer, at the same distance relative to the solvent front (R_f) as the selected test spot, should be extracted to provide the reference for dual-beam spectrophotometry.

About 10 times greater sensitivity than usually reported for spectrophotometry may be obtained by taking advantage of microcells and limiting final extraction volume to about 50 μl. Adapters for such microcells are commercially available.

In situ spectra from compounds on paper or thin-layer chromatograms are sometimes used to detect and identify POM. Diffuse-reflectance accessories to spectrophotometers may be used to obtain reflectance spectra of spots on various adsorbents.[666] Also, repeated scanning of chromatograms with scanners set at incremental wavelengths permits plotting of reflectance spectra. Such spectra have several major drawbacks: The relation between reflectance and absorbance is unclear and is influenced by the choice of adsorbent; the spectra are uncorrected; there is very little information on reflectance spectra in the literature; and reproducibility is poor. For these reasons, reflectance spectra are much less useful than absorption spectra of extracts.

Interpretation of Spectra

Spectra are often distorted by the presence of impurities or interfering compounds; present practice in the interpretation of "unknown" spectra appears to rely on the interpreter's experience more heavily than is desirable. Incomplete separations are the major problem in interpreting ultraviolet absorption spectra of air pollution extracts. Interpretation of spectra of pure compounds is not a problem.

A satisfactory separation should yield an unknown fraction whose absorption spectrum can be matched closely to the spectrum of a pure reference compound. A close match implies that between the reference and test compounds the absorbance ratio of any two peaks in one spectrum does not differ by more than 10% from the absorbance ratio of the two corresponding peaks in the other spectrum. Such close agreement is not usually found, even by procedures recommended very recently.[666] For example, thin-layer chromatography of an air sample extract of benzo[a]pyrene shows a fraction whose A_{385}/A_{297} ratio is 0.28; the corresponding ratio of pure benzo[a]pyrene is 0.43. From the data of Sawicki *et al.*,[671] this ratio should be 0.50; the *Ultraviolet Atlas*[761] data give the ratio as 0.47. These latter two values are in satisfactory agreement with the pure benzo[a]pyrene ratio, but differ significantly (∿35% difference) from the value for the benzo[a]pyrene fraction of the particulate air sample extract. Thus, with the presently available separation methods, the accuracy of spectrophotometric analysis of air sample extracts is questionable. The calculated weights of the benzo[a]pyrene fraction, based on absorbance values at 382, 362, and 295 nm, are 1.4, 1.7, and 2.2 μg. The reproducibility claimed for determinations at 382 nm is ± 7%.[666]

Partial separation into fractions containing several compounds each may be adequate for the detection and identification of a particular compound if the absorption spectrum has one or more peaks characteristic of that compound. A rough estimate of the amount of the compound may be made by the base-line method.[665] Such estimates, however, should be labeled as only qualitative (± 50%), particularly if only one absorption peak is used. There are two compelling reasons for not trusting base-line estimates derived from only one peak: Dozens of organic compounds have absorption peaks at any given wavelength,[378] and drawing a base line under a peak presupposes that the background is smooth, whereas most polycyclic compounds have highly structured spectra. For reliable detection and identification, at least two absorption peaks must be used, preferably from different spectral bands.

Once a compound has been identified in an extract and its molar extinction coefficient gleaned from the literature (or from a spectrum of a pure sample), the calculation of the compound concentration is straightforward, provided that all the components of the extract have been identified[280] or, if not all components are known, that the fraction of the absorbance at any given wavelength due to the specific compound of interest is known.

Ideally, quantitation is based on single-component extracts. But in practice, it may be necessary or desirable (to save time) to work with multicomponent systems, in which case the experience and judgment of the analyst carry a heavy burden. If a separation yields a close match between spectra of standard and separated compounds, spectrophotometric quantitation should be accurate to within 10%. It would be well if compilations of *Air Quality Data*[773] indicated the method of analysis with the expected accuracy of separation and analysis. For this compilation, the method included separation by thin-layer chromatography followed by fluorescence analysis. The overall accuracy, limited by the characteristics of collection, was determined by controls to be ± 20% (E. Sawicki, personal communication). Publication of benzo[a]pyrene concentrations calculated to three or four significant figures constitutes an unwarranted faith in the analytic accuracy.

Knowing the concentration of a given compound in the final extract should enable the analyst to determine the amount initially present in the air sample. The entire extraction and separation must be quantitative if the spectrophotometric data are to yield accurate concentrations.

LUMINESCENCE SPECTROPHOTOMETRY

Luminescence analysis has been reviewed in several publications recently,[38,60,82,363,577,835] and the reviews should be consulted for detailed presentation of the matters raised here.

The books by Parker[577] and Hercules[363] give special attention to techniques of luminescence analysis; Parker presents a great deal of experimental detail and a list of earlier textbooks and monographs. Becker[38] discusses the theory of luminescence processes and, like Parker, offers a convenient source of data. Sawicki published a review of fluorescence analysis as it is related to the identification of air pollutants.[663] Pringsheim's classic text[606] on luminescence analysis remains a useful reference source of qualitative observations of visible fluorescence. Luminescence analysis is a valuable technique when combined with thin-layer chromatography.

The early literature is replete with inaccurate and false data resulting from experimental error, so great care must be exercised in evaluating published information on luminescence spectra, lifetimes, and quantum yields. Furthermore, reports of analytic sensitivity and presentations of spectra are not readily compared. Complete knowledge of the methods and criteria used in purifying standard compounds is required to evaluate and compare reported data. Valuable reference sources of luminescence data have been published by Schmillen and

Legler[678] and Zander.[835] Convenient, uncritical guides to the literature have been compiled by Passwater[583] and Lipsett.[498]

"Fluorescence" is defined as the luminescence emitted in a radiative transition from the lowest excited singlet state to the ground state. Luminescence arising from a radiative transition from the lowest excited triplet state to the ground state is termed "phosphorescence." Measurements of luminescence involve the recording of two types of spectra—emission spectra and excitation spectra. An emission spectrum is obtained by irradiating, or exciting, a compound at the wavelength of maximal absorption while scanning for emission at wavelengths longer than that used for excitation. Excitation spectra are obtained by analyzing the emission at the wavelength of maximal emission while irradiating the compound with wavelengths shorter than that of maximal emission. Excitation spectra are very similar to absorption spectra; in fact, excitation spectra can be converted to absorption spectra, and, in the case of a pure compound, the two should be identical. Fluorescence and phosphorescence can be experimentally differentiated: Fluorescence is a relatively fast process and has a half-life of less than about 10^{-7} sec; phosphorescence is a relatively slow process and has a half-life of over 10^{-4} sec. Depending on solvent and temperature, the fluorescence band more or less overlaps the electronic absorption band in polycyclic aromatic hydrocarbons and aza-arenes, in many cases appearing as the mirror image of the absorption band. Phosphorescence bands appear at lower energy than the fluorescence bands. The energy difference between the para-band (^{1}La in the Platt nomenclature and U in the Moffitt nomenclature) absorption maximum and the onset of phosphorescence (^{3}La band) is about $10\text{–}12 \times 10^3$ cm^{-1} for the polycyclic aromatic hydrocarbons and most aza-arenes of interest. These differences permit selective observation of one or the other process by proper choice of wavelength or by the use of shutter systems. It is often possible to observe both the singlet–singlet and the very weak singlet–triplet absorption processes by phosphorescence excitation techniques. Experimentally, mirror-image symmetry is not always observed between the singlet–triplet excitation band and the phosphorescence emission band.[520]

Instrumentation

Conventional instrumentation in emission spectroscopy consists of an excitation train (comprising a high-intensity light source, a monochromator or filter, and a quartz sample cuvette or sample tube that may be placed in a quartz Dewar vessel for low-temperature work) and an analyzing train (typically, a monochromator and a photomultiplier). In monitoring phosphorescence or delayed fluorescence selectively, a shutter system is used for alternately exciting and viewing the sample. The sample may be viewed at the face on which it is being irradiated, at right angles to the direction of irradiation, or from behind the cell in a line with the direction of irradiation. The type of geometric arrangement used depends largely on the kind of specimen to be examined—e.g., dilute solutions and gases, concentrated solutions, opaque solids, and frozen solutions. Inner-filter effects, which are instrumental artifacts due to excessive absorption of the exciting light or absorption of the luminescence emitted, vary with the type of geometric arrangement chosen and the specimen being examined. A thorough discussion of the advantages and disadvantages of each type of geometry in relation to each kind of specimen can be found in Parker.[577 (pp. 220–233)] It should be noted that self-absorption influences the apparent position of emission bands, as well as their intensity,[554] and the geometry chosen is an important factor in this regard.

Many methods are available for alternately chopping the excitation and luminescence beams. Mechanical chopping, using rotating slotted disks or slotted cans, is most common. Electrooptical devices, such as Kerr cells, have not been extensively used, owing to the unacceptable attenuation of luminescence intensity when the cell is supposedly "open," or transparent. Electronic gating of both lamp and photomultiplier has been accomplished with some success. Although gating of the excitation source leads to difficulties, gating of the photomultiplier alone is acceptable in combination with a mechanical excitation shutter, and this arrangement allows flexibility of time resolution in viewing luminescence spectra. Winefordner has published a discussion of time-resolved phosphorimetry and its major advantage, increased selectivity of analysis.[823] Time resolution is obtainable with a slotted-disk spectrophotometer, but most commercial devices use a slotted rotating can, and analytic applications of time-resolved instrumentation are rare.

Digital data acquisition has been applied to luminescence analysis,[217,499] although no attempt has been made to automate a spectrophotometer. Aside from the obvious value of being able to automate the lengthy process of correcting and evaluating spectra, digital techniques permit the mathematical extraction of data from complex spectra of mixtures. Methods and applications of the mathematical techniques have been published.[499,809]

An increase in excitation intensity can lead to an increase in sensitivity of luminescence analysis; monochromaticity of the excitation source also determines, in part, the selectivity and sensitivity of analysis. These factors, taken together with the sensitivity of fluorescence analysis to scattered light and the limitations imposed by the bandwidth of the excitation source, would lead one to believe that lasers would be especially useful in luminescence studies. Some work has been done using lasers.[575,698] A convenient tunable source, the dye laser, has only recently become commercially available, and laser sources may be used more widely in the future.

Most commercially available instruments are suitable for qualitative luminescence analysis, although not all are capable of the high resolution required in some low-temperature techniques. These instruments are not all obtainable with attachments for low-temperature phosphorescence and delayed-fluorescence studies; however, little thought appears to have gone into this aspect of luminescence analysis, and such attachments are generally inadequate. Another deficiency in commercial instrumentation is the stray light characteristics of the monochromators used in the less expensive devices. Because the scattered and otherwise unwanted light output might depend heavily on the monochromator wavelength settings, spurious luminescence bands are often observed. These problems can be overcome through the use of filters, as long as the resulting attenuation in luminescence intensity is acceptable. An instrument by Farrand Optical can record both corrected excitation and emission luminescence spectra at room temperature, as well as at liquid nitrogen temperature. Instruments manufactured by American Instrument and G. K. Turner are available with attachments for automatic sensitivity correction of luminescence spectra. These instruments are especially suitable for quantitative luminescence analysis.

Sample Preparation

Luminescence processes are highly sensitive to impurity effects. Care is required in separating and purifying samples to obtain the necessary sample purity and to prevent the introduction of interfering substances. The solvent should be chosen with regard to ease of purification and inertness, in addition to physical and spectroscopic properties. Samples should be prepared in a clean laboratory; interfering substances like cigarette smoke are readily picked up in concentrations sufficient to impair sensitivity or produce spurious spectra. Most samples containing polycyclic aromatic hydrocarbons require efficient degassing to remove oxygen before luminescence analysis, inasmuch as oxygen severely quenches emission. Oxygen quenching affects room-temperature analysis to a greater extent than low-temperature luminescence spectroscopy, because it is a diffusion-controlled process.

One approach toward minimizing impurity effects involves the use of plastic hosts. Temperature quenching of fluorescence and phosphorescence appears to be determined mainly by the viscosity-dependent migration of contaminants in samples,[757] and high-viscosity plastics promise to make ambient-temperature luminescence spectroscopy more broadly useful. For instance, the intensity of phosphorescence emission of dibenz[a,h]anthracene is found to be higher in a plastic matrix at room temperature than in low-temperature glass.[290] In general, cross-linked matrices, like polycarbonate, polystyrene, and polymethylmethacrylate, are most useful. There are some drawbacks in the use of plastic samples: Emission spectra of compounds in plastic matrices taken at room temperature are usually broad and diffuse; and changes in relative intensities of bands and in spectral band positions are noted. The possibility of increased luminescence intensities and relative insensitivity to temperature effects, however, make the use of cross-linked plastic matrices attractive. An added advantage in the use of plastic samples is that phosphorescence samples and sample holders may be made so as to obtain the same reproducibility in sample positioning as that obtained in ambient-temperature fluorescence or absorption measurements. This seemingly trivial problem has often been a source of error.

Any separation procedure may be used before analysis of the components of a mixture by emission spectroscopy. Of particular importance is the analysis of mixtures after separation by thin-layer chromatography and paper chromatography. Sawicki and Sawicki have published a review of this method.[661] Fluorescence measurements have been made in this way on polycyclic aromatic hydrocarbons, aza-arenes, phenols, and ketones. In general, a compound or suitable ionic species or reaction product of the compound may be identified *in situ* by fluorescence color and position of the fluorescent spot. Quantitative analysis has usually been carried out after further elution of the components from the chromatogram.

Tentative methods based on column or thin-layer chromatography combined with fluorescence analysis have been proposed for quantitative analysis of polycyclic aromatic hydrocarbons—in particular benzo[a]-pyrene and benzo[b]fluoranthene—in atmospheric aerosols.[748-750] The thin-layer chromatographic method has been used to elute the organic fraction from particulate matter.[749] Benzo[a]pyrene is identified on the thin-layer chromatography plate by fluorescence methods and is further eluted and analyzed as the cationic salt in sulfuric acid with a fluorimeter. Lower limits of determination are reported as 3 ng of benzo[a]pyrene with a spectrophotofluorimeter and 10 ng with a filter fluorimeter. Accuracy to within 5–10% is reported.

Thin-layer chromatographic fluorescence analysis has unique promise in speed and convenience if complete qualitative *and* quantitative *in situ* analyses can be performed. Several schemes have been published.[661,663] In a recent analysis,[756] anthracene, phenanthrene, pyrene, fluoranthene, chrysene, benz[a]anthracene, benzo[a]pyrene, benzo[ghi]perylene, dibenz[a,h]anthracene, and coronene were separated, identified, and quantitatively analyzed by thin-layer chromatography combined with *in situ* fluorescence excitation and emission measurements. Limits of detection for these compounds were reported to be within 0.1–1 μg. Several complications remain in the direct analysis of luminescence from thin-layer chromatography plates. Light scattering and luminescence from the plate itself are particularly troublesome in such techniques. Luminescence from a crystalline or adsorbed compound often differs drastically from that of the same compound in liquid solutions or in low-temperature glass. Finally, luminescence spectra of separated components on thin-layer chromatography plates are often found to differ significantly from luminescence spectra of standard compounds obtained under the same conditions. At present, unambiguous identifications from *in situ* luminescence spectra appear to require considerable experience.

Luminescence Analysis

Enough information is contained in low-temperature luminescence spectra so that comparisons with spectra obtained from chemical standards may aid in identifying unknown compounds or chromatographic bands or spots. Information can also be obtained concerning functional groupings attached to the molecule by noting whether the compound is predominantly fluorescent or phosphorescent.

Vibrational analysis of luminescence spectra of polycyclic aromatic hydrocarbons has been reviewed.[567] Although there is no specific method of achieving optimal results, several factors that affect the resolution of the band structure can be identified. The observation of characteristic vibrational progression lines is favored by low temperature, high viscosity, low solute concentration, high component purity, and the use of nonpolar solvents, such as *n*-alkanes.

Variation in pH certainly changes the nature of the solute, especially if ionic species can be produced, and spectral changes are expected, which may aid in analysis. A review of pH effects in fluorescence and phosphorescence analysis has been published.[682] Some polycyclic aromatic hydrocarbons and aza-arenes are readily analyzed, at appropriate pH, as the anionic or cationic salts. It is of special interest that many compounds become more acidic or basic upon photoexcitation. At appropriate pH, optical absorption or photoexcitation spectra may be characteristic of neutral molecules, whereas the photoluminescence is characteristic of the cationic or anionic species. Spots on thin-layer chromatography plates can be rapidly identified by exploiting this property. For instance, 7H-benz[d,e] anthracen-7-one undergoes fluorescence characteristic of its cation on a thin-layer chromatography plate treated with trifluoroacetic acid. Selective agents, such as trifluoroacetic acid and tetraethylammonium hydroxide, facilitate similar identification of acridines and carbazoles.

In sufficiently dilute solutions and in the absence of interfering substances (compounds that absorb light in the same region as the compound of interest and compounds that quench or sensitize luminescence of the compound of interest), emission intensity is a linear function of concentration. Excitation intensity and detector response are complicated functions of instrument characteristics; for analytic work, it is common to use calibration curves determined in conditions exactly the same as those to be encountered in the analysis. For polycyclic aromatic hydrocarbons, linear calibration curves are generally obtained for concentrations below 10^{-5} mole/liter when using the hydrocarbon para-band (^{1}La, Platt; U, Moffitt) or β band (^{1}Bb, Platt; X, Moffitt) for excitation and for concentrations below 10^{-3} mole/liter when using the α band (^{1}Lb, Platt; V, Moffitt). Phosphorescence curves are found to be linear over a greater range of concentration than are fluorescence curves.

The spectroscopic procedures themselves are highly sensitive and accurate. The limiting factors in the analysis of polycyclic matter appear to be the reliability and efficiency of the extraction and separation procedures used before luminescence analysis. The overall accuracy of analytic procedures involving spectroscopic measurements is complicated by the difficulty in distinguishing between many of the polycyclic compounds. For instance, benzo[a] pyrene, benzo[ghi]-perylene, and benzo[k] fluoranthene exhibit very similar electronic spectra. If the sample preparation procedures cannot be relied on to separate the components of a mixture thoroughly, then the use of no one spectrophotometric technique can be considered completely reliable. Common practice involves the measurement of one band in one type of spectrum (i.e., absorption, fluorescence, or phosphorescence), and this cannot be considered acceptable.

Luminescence Sensitization and Quenching

Absorption of energy by one compound, called the "donor," and energy transfer from the excited donor to another compound, called the "acceptor," may lead to luminescence of the acceptor. This type of luminescence is known as "sensitized luminescence." It requires an overlap of the emission spectrum of the donor with the absorption spectrum of the acceptor and an excited-state energy of the acceptor that is lower than that of the donor. The donor is known as the "sensitizer."

Sensitization methods have unique potential for monitoring small quantities of compounds. Parker demonstrated the determination of impurities in polycyclic aromatic hydrocarbon standards by sensitized delayed fluorescence.[577] If several impurities are present, it is possible to excite the one with the lowest triplet energy first, and it should be possible to excite a limited number of impurities progressively and resolve the spectra with mathematical techniques. As an example, anthracene and benz[a]anthracene may be determined in pyrene by energy transfer to both impurities, using pyrene as the sensitizer. The presence of anthracene can be independently determined, using acridine orange as a sensitizer. High concentrations of components are required and solutions must be deoxygenated, which makes this method tedious. Zander[835] has proposed doing this in crystallized aromatic hydrocarbons at low temperature—a technique that is much more sensitive and does not require deoxygenation. No applications to air pollution analysis have been reported.

Organic electroluminescence[364,836] can be used both quantitatively and qualitatively for the analysis of such polycyclic compounds as anthracene, 9-phenylanthracene, 9,10-diphenylanthracene, pyrene, and coronene. Electroluminescence techniques have many disadvantages, compared with photoluminescence and other analytic methods. Luminescence may arise by annihilation of radical ions with other oxidizing or reducing agents. High concentrations (about 10^{-3} M) are usually required, although spectra have been observed at very low concentration (about 10^{-7} M). A potential for handling complex mixtures without separation of components seems apparent. Electroluminescence has been observed for polycyclic aromatic hydrocarbons, substituted derivatives, and heterocyclics.

Organic electrochemistry presents unique opportunities for the analysis of complex mixtures[49,514,589] but does not appear to have been exploited in air pollution analysis. Polarographic oxidation and reduction potentials are available for many polycyclic hydrocarbons,[49,514,595] and, although several compounds may have the same reduction potential, organic polarography would be a uniquely rapid method of quantitative analysis if coupled with other techniques for unambiguous identification of components.

Quenching techniques have also been demonstrated to be of value in pollution analysis.[633,675] By using compounds like acidic or alkaline nitromethane, acetophenone, and carbon disulfide, the luminescence of polycyclic compounds can be selectively quenched, leading to the analysis of azaheterocyclic compounds in the presence of polycyclic hydrocarbons, aromatic amines, and imino heterocyclic compounds.

Evaluation of Sample Purity

Excitation spectrophotometry is an extremely sensitive form of absorption spectroscopy. A change in the spectral distribution of luminescence when the wavelength of excitation is changed or the inability to match the absorption and excitation spectrum is an indication of impurity. Because the problem of purity in both the luminescence standards and samples can be a source of error in luminescence studies, more widespread use of excitation spectrophotometry for analysis of sample purity would be desirable.

Luminescence decay, simple to measure in phosphorimetry, is also an excellent index of purity. A logarithmic plot of detector response is linear if the substance is pure (in the absence of heavy-atom solvents or solvent inhomogeneity). A good discussion of the technique of making lifetime measurements has been published recently by Demas and Crosby.[194]

Correction of Spectra

No comparisons may be made between spectra obtained and analyses performed in different laboratories, unless data related to the calibration curve are known and reported. Many procedures have been advanced for constructing calibration curves.[363,577] The most convenient calibration procedures rely on the use of luminescence standards. There is little agreement on the choice of an appropriate compound. Quinine bisulfate in 1 N sulfuric acid is most widely used.

Luminescence spectra must be corrected for lamp intensity, which decreases with wavelength; for optical elements, such as mirrors, monochromator gratings, and lenses; and for response of photomultiplier tubes in the detection system. If analysis is carried out with fixed molecular orientation, which is common when the solute is dissolved in high-viscosity solvent and adsorbed on a surface or joined to a polymer by chemical reaction, corrections must be made for the polarization of luminescence, because the horizontal and vertical components of light are not transmitted equally through monochromators. Numerical analysis indicates that large errors can result if this correction is not applied.[576]

Several methods of plotting corrected spectra are used.[230,577] The most common is to plot relative quanta per unit frequency interval against either wavelength or wave number.

Complications and Limitations

Fluorescence analysis is particularly sensitive to light scatter and Raman emission. Low-temperature luminescence techniques are influenced by bubbling and "snow" in the coolant, cracking of the sample glass, and repeatability of sample placement. All emission techniques are influenced by luminescence of solvents and sample cells, instrument noise, accuracy and repeatability of wavelength settings, scatter of light in monochromators, and random fluctuation in source intensity.

The occurrence of photochemical dimerizaton and oxidation reactions places limitations on sensitivity and applicability of luminescence analysis. Photochemical reactivity of the luminescent compound limits the intensity of excitation that can be used. Photochemical products may also contribute positive or negative error in quantitative analysis, because some products are themselves luminescent. Photochemical reactivity, then, may be a special problem in luminescence sensitization, in that a relatively large amount of energy may be transferred to the compound under study with this technique. Photochemical reactions involving other components of the sample may occur. Parker and Hatchard have reported such a reaction of benzo[a]pyrene and pyrene with a polymer host matrix, polyvinylpyridine butylbromide.[578]

Proper choice of solvent, pH, viscosity, and addition of some metal ions in particular cases increase both the sensitivity and the selectivity of analysis. Luminescence intensity and band position are functions of solvent and temperature. Fluorescence can be quenched through complex formation and van der Waals interactions with polar solvents.[224] The fluorescence bands of 9,10-diazaphenanthrene, phenanthrene, naphthalene, and pyrene are found to blue-shift (and sharpen) below 120 K owing to solvent orientation effects.[20,418,570] Failure to compensate for temperature fluctuations or to permit temperature equilibrium is a serious source of error. A study of the effects of temperature fluctuations in low-temperature glasses has been published by Leubner.[490] Van Duuren has published reviews of temperature and solvent effects on luminescence spectra of aromatic compounds.[781,782] A review of environmental effects in fluorescence analysis was recently published.[810]

Excimer (excited dimer) formation is a special problem in solid luminescence. In liquid solutions, it is sensitive to both temperature and solvent at low hydrocarbon concentrations,[395] and significant errors can occur in sensitive analytic work.

Quenching by oxygen is a serious problem and may be very selective. Berlman[50] and Parmenter and Rau[580] have published useful data on quenching of luminescence by oxygen. Pyrene fluorescence is especially susceptible to oxygen quenching. Although luminescence is not influenced by some metal ions,[311] the presence of heavy metal anions, such as Br^- and I^-, and especially of paramagnetic ions, such as Co^{2+} and Mn^{2+}, does alter luminescence characteristics in specific cases.[39] Metal ions favor intersystem crossing from the excited singlet state to the excited triplet state and therefore often enhance phosphorescence intensity at the expense of fluorescence intensity. Thus, addition of some metal ions or compounds, such as ethyliodide and ethylbromide, to solutions containing the compound under investigation often enhances the sensitivity of analysis.

Luminescence properties of crystalline compounds on dry thin-layer chromatography plates depend as much on the crystal structure as on the compound. Efficient energy-transfer processes in crystals ensure that fluorescence bands will be diffuse, and excimer luminescence, seen as a structureless and broad band, is often noted. There is reason to believe that crystal phosphorescence should be too weak to be observed, if it occurs at all. Crystal luminescence is very sensitive to lattice distortions and dislocations. Band position and intensity, fine structure, and photochemical reactivity are drastically altered when a crystal is "stressed."[37,590] Intensity of fluorescence depends on the crystal size and on the presence of oxygen.[724] Some complications can be handled by annealing or by carefully choosing solvents to control crystal size, but these factors present difficult problems for *in situ* analyses of crystalline or solid-state systems. The luminescence properties of crystals are beyond the scope of this discussion and have been reviewed elsewhere.[61,822] Birks and Cameron have published a useful paper, which discusses crystal fluorescence of many carcinogenic hydrocarbons.[62]

Comparison of Absorption and Emission Methods

In terms of utility in analytic applications, luminescence and absorption spectrophotometry are compared in three major respects: ease, selectivity, and sensitivity. Although progress has brought a high degree of sophistication to luminescence instrumentation and procedure, the analyst is required to control directly many instrumental variables that are often controlled automatically in absorption spectrophotometers. In general, there are no standard methods of obtaining and reporting luminescence spectra, although many proposals regarding method, instrumentation, and presentation have been put forward. Therefore, more care and effort are required of the analyst in interpreting, reporting, and especially comparing luminescence data. However, the analyst has control over more experimental conditions in measuring luminescence. The wavelength of excitation and the wavelength of viewing luminescence can be independently selected. These, as well as conditions of solvent, temperature, and concentration of added sensitizers or quenchers, can be chosen so as to favor the observation of light emission of a particular type or from a particular compound. This has been used to advantage in the analysis of complex mixtures of polycyclic aromatic compounds without prior separation of the components. For instance, benzo[a]pyrene may be detected and quantitatively measured in the presence of as many as 40 polycyclic compounds under appropriate conditions.[671,672] Indeed, measurements of polycyclic compounds in the air have been made in this manner.[579]

The sensitivity of luminescence analysis may be reported in a variety of ways, leading to some confusion in attempting to make comparisons without considering details of instrumentation and procedure. Theoretical and experimental comparisons have been made of the limits of detection of various polycyclic aromatic hydrocarbons by several methods.[126,523,835]

It is clear that emission techniques are more sensitive than absorption spectrophotometry and that sensitivity depends in part on the instrumentation used.[126,835] Concentrations in the nanogram-per-milliliter range appear detectable by luminescence techniques. Absorption techniques appear to be less sensitive by a factor of 10^2–10^3. This may be compared with the subnanogram sensitivities claimed for gas chromatographic and mass spectrometric techniques. Whereas the data indicate that phosphorescence analysis and fluorescence analysis are competitive procedures, in general, many factors combine to make them complementary.

It is well to note that a combination of photoluminescence with other techniques may lead to even more selective and sensitive analyses. It appears that several techniques offer promise of quick routine analyses of complex systems, but they have yet to be exploited.

Summary and Conclusions

Emission and absorption spectral analysis of polycyclic organic compounds offers nanogram sensitivity and accuracy within 10%.

Clean separations are critical. Failure to achieve clean separation results in substantial analytic uncertainty. Some techniques of emission analysis have potential for circumventing extensive separation and should be pursued. There is not yet a good combination of separation and spectral analysis that permits analytic accuracy to within 25%. Compilations of air quality data should include methods of separation and analysis and estimates of the accuracy of both.

INFRARED AND RAMAN SPECTROSCOPY

Infrared and Raman spectra reveal features characteristic of individual bands or functional groups. Strong infrared bands are related to vibrations that cause changes in dipole moment; strong Raman bands are related to vibrations that cause changes in molecular polarizability. Therefore, the two spectra are complementary and are often used together. Although single-ring compounds, which exhibit considerable symmetry, have been extensively investigated, polycyclic compounds have received only scant attention from vibrational spectroscopists.[622]

The uniqueness of individual bands is unsettled. It is not known whether there are spectroscopically active vibrations that are characteristic of such ring structures as chrysene or benzo[a]pyrene as a whole. Vibrational bands that are related to individual functional groups are not unique, but the total spectrum of a molecule is unique. Thus, the chief use of vibrational spectroscopy will be in identification of unknowns by revealing the presence of carbonyl or other functional groups and by matching known and reference spectra.

Band strengths in vibrational spectroscopy are one hundredth to one tenth as great as those in electronic (ultraviolet or visible) spectroscopy. Infrared band strengths are often not proportional to concentration. Infrared quantitative analysis requires large samples and careful calibration. Raman band strengths are proportional to concentration, but are even lower in general than infrared band strengths and require hours of exposure time to obtain spectra.

No special techniques have been developed for infrared or Raman analysis of polycyclic compounds. In the few cases where vibrational spectra are mentioned, common techniques appear to have been used.

Internal reflection spectroscopy is a technique whereby the sample, layered on an interface between two transparent media, is tested by analyzing the infrared light as it is totally internally reflected at the interface. This technique yields spectra that are essentially the same as conventional infrared spectra. The advantages lie in the small (microgram) amount of material required and the ability to use liquid or solid samples free of any suspending "matrix." This technique has been applied to pesticide[365] and asbestos[564] analysis and appears to be useful for trace analysis of any nonvolatile compound.

The disadvantages of vibrational spectroscopy are the relatively weak bands, the fact that infrared band strengths are not proportional to concentration, the requirement for a vibrationally transparent medium, and the lack of unique polycyclic spectral features. The disadvantages far outweigh the advantages of these techniques. Internal reflection spectroscopy may become useful and deserves more attention. At present, either ultraviolet absorption or fluorescence is preferable to vibrational spectroscopy for analysis of POM.

MASS SPECTROMETRY

Organic mass spectrometry is practical over a range of sensitivities somewhat greater than that available with the more convenient and less expensive ultraviolet spectrophotometers and spectrofluorimeters. Resolution with the less expensive equipment is to the nearest integral mass number, so the molecular weights of the parent molecule and its characteristic fragmentation products are expressed in nominal atomic mass units (amu). Because sensitivity and resolution are inversely related, this type of resolution affords the highest sensitivity, extending to 10^{-11} g. As resolution is increased to the range where millimass (10^{-3} amu) differences in the various elemental masses may be discerned, the ultimate sample sensitivity may fall to 10^{-7} g. Briefly, the difference between low mass resolution and high resolution is the difference between determining the nominal molecular weights of the parent molecule and daughter fragments and determining the atomic composition of these entities. In many cases, it is desirable that the latter be determined directly and unambiguously for the identification to be unequivocal.

Operating Principle

Organic mass spectrometry is based on high-vacuum ionization of organic molecules, generally by bombardment with 70-eV electrons. The ions thus generated under vacuum may be deflected and focused variously by electric or magnetic fields such that inertial (mass) differences allow the ions to be dispersed and displayed according to their own masses. Under electron bombardment at 70 eV, most organic molecules receive more than enough energy to dislodge an orbital electron, thereby creating positively charged molecular ions. The excess kinetic energy thus imparted to the molecular ion may then be released by various mechanisms that involve rupture of the parent ion into smaller neutral and ionized fragments. Those fragments and the parent molecular ion constitute the principal features of the mass spectrum, which is simply a statistical record of the abundance of ions at particular masses throughout the mass range scanned. Because of the tendency of an ionized molecule or fragment to decompose according to its intrinsic structural features, the statistically consistent fragmentation pattern at 70 eV reveals structural detail that in many instances can be used to reconstruct the original molecule.

If fragmentation predominates over the stability of the molecular ion to the extent that the molecular ion cannot be discerned (e.g., in the case of substituted carbazoles),[98] alternative approaches to ionization may be desired. Presently, they include low-ionizing-voltage (7–15 eV) electron bombardment, field ionization, and chemical ionization procedures, all of which enhance the stability of the molecular ion by minimizing its surplus kinetic energy.

Application to Analysis of Polycyclic Organic Matter

The use of mass spectrometry itself in the analysis of POM is analogous to the use of other spectroscopic methods, such as ultraviolet fluorescence, in the examination of chromatographic fractions.[685] In the analysis of polycyclic organic chromatographic fractions, mass spectrometry possesses the advantage over ultraviolet of being able to detect all organic species present, not just those with the appropriate configuration of conjugated unsaturation. Hence, mass spectrometry may be used to determine the purity of separated polycyclic organic compounds.[146,685] However, except for some aza-arenes,[98] the molecular stability of polycyclic organic compounds on ionization is so great that the excess kinetic energy imparted by the 70-eV electrons can be accommodated within the ionized molecule. This results in a minimum of molecular fragmentation in the mass spectrometer and, except for molecular ions with multiple charges, a minimum of distinguishing qualitative features among spectra of isomeric polycyclic organic compounds. Thus, mass spectrometry is useful chiefly in determining the molecular weights and purity of polycyclic organic chromatographic fractions. In this regard, the low-ionizing-voltage technique has been used to simplify the spectra taken on mixtures by producing mainly the unfragmented, singly charged molecular ion.[417,695,701] If high-resolution mass spectrometry is available, exact unambiguous atomic compositions of the molecular ions present in mixtures may be determined.[417,695]

In the handling of samples for mass spectrometric analysis, polycyclic organic compounds separated by various forms of chromatography present little difficulty. Because they are generally so involatile, they may be analyzed by the most sensitive of sampling techniques—the direct-insertion probe. In the case of paper or thin-layer chromatography, if absolutely necessary, the polycyclic organic sample may be analyzed without prior elution from the chromatographic matrix—i.e., the matrix plus sample may be inserted directly into the evacuated mass-spectrometry ionization chamber, where the sample may be slowly vaporized by moderate heating. If polycyclic organic compounds are collected from the gas chromatograph in melting-point capillaries, the tubes themselves may be introduced with the probe.[525] Because the sample is at least partially consumed, mass spectrometry should be performed after other nondestructive spectral analyses have been satisfied.

Coupled Gas Chromatography and Mass Spectrometry

As indicated above, mass spectrometry and other spectroscopic procedures generally are applied to polycyclic organic compounds in combination with chromatographic separations. Naturally, the more manipulations required in the various stages, the more time-consuming the overall procedure, with attendant chances for losses and contamination. Efficient direct coupling of chromatographic separation with spectral identification would circumvent such mechanical problems and accelerate the analysis. With the recently introduced improvements in gas chromatographic analysis of polycyclic organic compounds and with an inert, efficient sample-transferring interface, the use of coupled gas chromatography and mass spectrometry (GC–MS)[501,525,526,806] for the analysis of POM is now logical and timely.

Coupled GC–MS, combining the high separational capability of gas chromatography with the molecular definitiveness of mass spectrometry, affords unparalleled speed, sensitivity, and completeness of identification of the constituents in trace amounts of complex mixtures. In fact, the ability of coupled GC–MS to produce analytic data "on the fly" far exceeds the ability of technicians to process the data. Generally, moderate to heavy GC–MS sample processing requires the assistance of automated data acquisition and reduction to utilize the capacity of GC–MS systems fully. Sampling sensitivity depends on whether low or high resolution is used. As to the choice between low- and high-resolution GC–MS, it is almost always preferable to run the low-resolution GC–MS profile before high-resolution analysis, inasmuch as it is not possible to obtain statistically reliable ion abundances to the same degree of sensitivity with high-resolution mass spectrometry as with low-resolution mass spectrometry. High-resolution mass spectrometry represents a greater investment in apparatus; nevertheless, whenever the unambiguous elemental composition of any compound and its fragmentation products is required[417,695] and there is sufficient sample (more than 10^{-7} g), it is the method of choice.

Summary

Mass spectrometry affords a sensitive means of determining the probable identity and relative purity of polycyclic organic chromatographic fractions. Generally, this has been accomplished by the low-voltage ionization method, which produces spectra containing mainly singly charged, intact molecular ions. The masses determined for these ions indicate the nominal molecular weights of the species present when low-resolution spectra are taken. If high-resolution spectrometry is used, the empirical formulas of the species are also determined. Recent success in coupling gas chromatography with mass spectrometry suggests benefits in applying this sensitive and selective dual technique to the analysis of polycyclic organic material.

NEW AND PROMISING SPECTROSCOPIC TECHNIQUES

Fourier transform spectroscopy combines an old optical instrument—the older mathematical apparatus of Fourier analysis—with modern computer technology. The interferometer, in contrast with the spectrometer, is a nondispersive device in which light is first split into two beams and then recombined after the beams have traveled paths of somewhat different lengths. By measuring the recombined-beam energy as a function of the split-beam path-length difference, one obtains an interferogram, which may be transformed into the more familiar spectrum plot of energy versus frequency (or wavelength) or may be used as is. Fourier transform spectroscopy may be applied throughout the electromagnetic spectrum, but the infrared region has received much of the attention. The most striking advantage of the technique using new commercial instruments is its high speed. It can be coupled to gas chromatograph output to obtain spectra of fractions as they pass by in about 11 min, in contrast with the hours required with dispersing instruments.

Correlation spectroscopy is a dispersive technique whereby several spectral features are scanned for at once; spectrometers are modified by the addition of multiple slits, screens, and rotating refractive plates. The attractive feature of this technique is the possibility of selective detection of individual compounds in mixtures,[184] which could reduce the time and effort now required for separation.

Electron spectroscopy measures energy of electrons ejected from the sample under bombardment with x rays[362] or with photons of energy greater than about 10 eV.[55] The energy of the ejected electrons is governed by the identity of the atoms and their environment (e.g., oxidation state) in the sample molecule. For this reason, this technique is best for structural determinations. The sample must be stable in a vacuum of 10^{-5} torr; microgram quantities are adequate for analysis. Hercules[362] states that the technique may be used to obtain element ratios accurate to within about 5%.

CONCLUSIONS

It is not possible to point to either absorption or emission spectroscopy as clearly superior for use as a standard procedure for spectral analysis of air samples. There are too many interrelating uncertainties of cost, complexities of instrument operation and interpretation, interference of trace compounds, and uncertainties of separation methods. New spectroscopic methods—such as chemiluminescence, electron spectroscopy, Fourier transform spectroscopy, and correlation spectroscopy—promise procedures that will circumvent separation difficulties.

Appendix D

Regression Analysis

The method of multiple regression, widely used to estimate the influence of one or more variables on some quantity of interest, is subject to serious limitations of interpretation when the observed variables cannot be controlled by the observer and can be associated in unknown ways with other hidden but influential variables. The method does, nonetheless, enable one to apply a standard numerical expression to trends in the data that would otherwise have to be described more informally and with greater difficulty.

The procedure is most easily understood in the case of a single independent variable, such as amount of smoking (cigarettes per annum per person over age 15), whose relation to the quantity of interest, such as age-standardized lung cancer death rate, is sought. Figure D-1 shows the graph of the 48 lung cancer death rates versus the 48 tobacco sale values for the 48 contiguous states discussed in Table 17-20. Although there is considerable scatter, there is also an appreciable tendency for states with high tobacco sales to have high lung cancer death rates. A variety of methods might be used to express the average relation. If, however, it is decided to express the relation as a linear function, one plausible choice of best-fit line is the one that minimizes the overall discrepancy in the sense of least squares, i.e., if

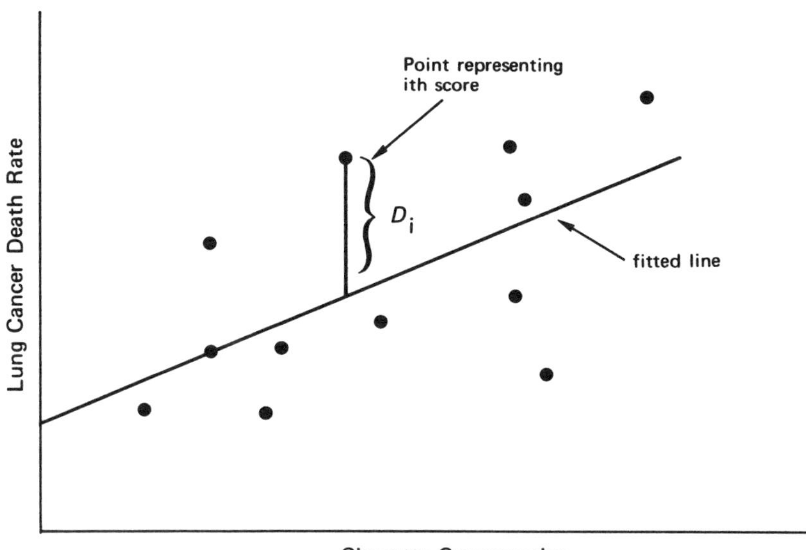

FIGURE D-1 Average relation between lung cancer death rate and cigarette consumption for 48 contiguous states.

D_i = [lung cancer death rate (LCDR) for state i] –(value predicted from fitted line). There is a unique line that minimizes the quantity $S = \Sigma D_i^2$. This line is called the fitted regression of Y (lung cancer death rate) on X (cigarette consumption), and its slope is called the regression coefficient (of Y on X).

One reasonable approach to the question of whether the slope found in this way is "significant" is to ask how likely it is–given the values of $Y_1, \ldots, Y_{48}$, and those of $X_1 \ldots, X_{48}$, which are at hand–that a completely random pairing of Y's with X's would give as steep a slope as that provided by the actual pairing. To a good approximation, that question can be answered by comparing the regression coefficient with its standard error. If the slope is greater than twice the calculated standard error, it is likely that something beyond random pairing of Y's with X's is going on (in the sense that fewer than 5 times in 100 would such a random pairing give as steep a slope).

Two generalizations of the above procedure are, in fact, used in the text. First, it is hoped to fit a linear relation between Y and two or more X's, e.g., X_1 = cigarette consumption, and X_2 = average benzo[a]pyrene concentration. Now there is a three-dimensional plot, with coordinate axes representing X_1, X_2, and Y. The trend of the points is now fitted by passing a plane (rather than a line) through the scatter, choosing the unique plane for which $S = \Sigma D_i^2$ is minimal, where D_i = (LCDR) – (value predicted from fitted plane).

Second, because for some states some of the population groups (e.g., female, nonwhite) are very small, the relation between Y and the X's might be expected to be less pronounced than in states where the population is larger. Appropriate weights, w_i, can be assigned in such a case, and one then seeks to minimize $S_w = \Sigma w_i D_i^2$.

The general principles of the analysis remain unchanged.

It is appropriate to emphasize that the regression method constitutes merely one of a number of possible ways to fit a linear function to the data. It has an advantage over some other methods, in that it is easy to calculate a standard error to use in judging whether the regression coefficient is larger than a random pairing of Y's with X's might plausibly provide. The method in no way precludes effects of hidden variables, nor does it, in itself, imply any causal connection, even when the existence of a relation is beyond doubt.

References

1. Aalbersberg, W. Ij., G. J. Hoijtink, E. L. Mackor, and W. P. Weijland. Complexes of aromatic hydrocarbons with strong Lewis acids. J. Chem. Soc. 1959:3055–3060.
2. Aalbersberg, W. Ij., G. J. Hoijtink, E. L. Mackor, and W. P. Weijland. The formation of hydrocarbon positive ions in strong proton donors. J. Chem. Soc. 1959:3049–3054.
3. Abelev, G. I. Antigenic structure of chemically-induced hepatomas. Progr. Exp. Tumor Res. 7:104–157, 1965.
4. Abell, C. W., and C. Heidelberger. Interaction of carcinogenic hydrocarbons with tissues. VIII. Binding of tritium-labeled hydrocarbons to the soluble proteins of mouse skin. Cancer Res. 22:931–946, 1962.
5. Adams, R. N. Anodic oxidation pathways of aromatic hydrocarbons and amines. Accounts Chem. Res. 2:175–180, 1969.
6. Adams, R. N. Electrochemistry at Solid Electrodes. (Monographs in Electroanalytical Chemistry and Electrochemistry.) New York: Marcel Dekker, Inc., 1969. 402 pp.
7. Adamson, R. H., R. W. Cooper, and R. W. O'Gara. Carcinogen induced tumors in primitive primates. J. Nat. Cancer Inst. 45:555–559, 1970.
8. Agricola, G. De Re Metallica. Translated by H. C. Hoover and L. H. Hoover. New York: Dover Publications, Inc., 1950. 672 pp.
9. Alekseeva, T. A. The use of luminescence spectral methods for study of the fractional composition of aromatic dispersed bitumens. Tr. Vses. Nauch. Issled. Geologorazved. Neft. Inst. 33:271–277, 1962. (in Russian)

10. Allen, M. J., E. Boyland, and G. Watson. Experimental bladder cancer, pp. 34–35. In British Empire Cancer Campaign. Thirty-fourth Annual Report, 1956. London: British Empire Cancer Campaign, 1956.
11. Allison, A. C., M. E. Peover, and T. A. Gough. Polarographic measurements of electron donation and acceptance by carcinogenic compounds. Nature 197: 764–765, 1963.
12. Altshuller, A. P., and J. J. Bufalini. Photochemical aspects of air pollution: A review. Environ. Sci. Technol. 5:39–64, 1971.
13. Alwens, W., E. E. Bauke, and W. Jonas. Auffallende Häufung von Bronchialkrebs bei Arbeitern der chemischen Industrie. Arch. f. Gewerbepath. u. Gewerbehyg. 7:69–84, 1936.
14. Andersen, A. A. A sampler for respiratory health hazard assessment. Amer. Ind. Hyg. Assoc. J. 27:160–165, 1966.
15. Annual Report of H.M. Chief Inspector of Factories on Industrial Health, 1964. London: H.M. Stationery Office, 1969. 31 pp.
16. Annual Report of H.M. Chief Inspector of Factories on Industrial Health, 1966. London: H.M. Stationery Office, 1967. 45 pp.
17. Arcos, J. C., M. F. Argus, and G. Wolf. Chemical Induction of Cancer. Structural Bases and Biological Mechanisms. 4 vols. (2nd ed.) New York: Academic Press Inc., 1968.
18. Auerbach, O., E. C. Hammond, D. Kirman, and L. Garfinkel. Effects of cigarette smoking on dogs. II. Pulmonary neoplasms. Arch. Environ. Health 21:754–768, 1970.
19. Ayres, S. M., and M. E. Buehler. The effects of urban air pollution on health. Clin. Pharmacol. Ther. 11:337–371, 1970.
20. Baba, H., and C. Mugiya. Temperature dependence of fluorescence spectra and fluorescence polarization of some organic compounds in polar solvents. The Franck–Condon excited states. Bull. Chem. Soc. Jap. 43:13–19, 1970.
21. Badger, G. M. Mode of formation of carcinogens in human environment. Nat. Cancer Inst. Monogr. 9:1–16, 1962.
22. Badger, G. M., and T. M. Spotswood. The formation of aromatic hydrocarbons at high temperatures. Part IX. The pyrolysis of toluene, ethylbenzene, propylbenzene and butylbenzene. J. Chem. Soc. 1960:4420–4427.
23. Bailey, E. J., and N. Dungal. Polycyclic hydrocarbons in Icelandic smoked food. Brit. J. Cancer 12:348–350, 1958.
24. Bailey, P. S. The reactions of ozone with organic compounds. Chem. Rev. 58:925–1010, 1958.
25. Bailey, P. S., J. E. Batterbee, and A. G. Lane. Ozonation of benz[a]anthracene. J. Amer. Chem. Soc. 90:1027–1033, 1968.
26. Bailey, P. S., J. E. Keller, D. A. Mitchard, and H. M. White. Ozonation of amines. Adv. Chem. Ser. 77:58–64, 1968.
27. Bair, W. J. Inhalation of radionuclides and carcinogenesis, pp. 77–101. In M. G. Hanna, Jr., P. Nettesheim, and J. R. Gilbert, Eds. Inhalation Carcinogenesis. AEC Symposium Series, No. 18. Washington, D.C.: U.S. Atomic Energy Commission, 1970.
28. Baldwin, R. W., and C. R. Barker. Tumour-specific antigenicity of aminoazodye-induced rat hepatomas. Int. J. Cancer 2:355–364, 1967.
29. Baldwin, R. W., C. R. Barker, and M. J. Embleton. Immunology of carcinogen-

induced and spontaneous rat tumours, pp. 503–505. In J. Dausset, J. Hamburger, and G. Mathé, Eds. Advance in Transplantation. Proceedings of the First International Congress of the Transplantation Society, Paris, 27–30 June 1967. Copenhagen: Munksgaard, 1968.

30. Baldwin, R. W., and M. J. Embleton. Immunology of 2-acetylaminofluorene-induced rat mammary adenocarcinomas. Int. J. Cancer 4:47–53, 1969.
31. Bang, F. B. Effects of invading organisms on cells and tissues in culture, pp. 151–261. In E. N. Willmer, Ed. Cells and Tissues in Culture. Methods, Biology, and Physiology. Vol. 3. London: Academic Press Inc., 1966.
32. Barger, W. R., and W. D. Garrett. Surface active organic material in the marine atmosphere. J. Geophys. Res. 75:4561–4566, 1970.
33. Barnett, E. deB., J. W. Cook, and H. H. Grainger. Studies in the anthracene series. Part III. J. Chem. Soc. 121:2059–2069, 1922.
34. Bateman, A. J. Mutagenic sensitivity of maturing germ cells in the male mouse. Heredity 12:213–232, 1958.
35. Bateman, A. J. Testing chemicals for mutagenicity in a mammal. Nature 210: 205–206, 1966.
36. Batsakis, J. G., and H. A. Johnson. Generalized scleroderma involving lungs and liver with pulmonary adenocarcinoma. Arch. Path. 69:633–638, 1960.
37. Baum, E. J. Photochemistry of organic crystals—anthracene. In J. N. Pitts, Jr., Ed. Excited State Chemistry. New York: Gordon & Breach, Science Publishers, Inc. (in press)
38. Becker, R. S. Theory and Interpretation of Fluorescence and Phosphorescence. New York: John Wiley & Sons, Inc., 1969. 283 pp.
39. Beer, R., K. M. C. Davis, and R. Hodgson. Formation of excited charge transfer complexes in the quenching of anthracene fluorescence by anions. J. Chem. Soc. 1970D: 840–841.
40. Begeman, C. R. Carcinogenic aromatic hydrocarbons in automobile effluents, pp. 163–174. In Vehicle Emissions (Selected SAE Papers). (Paper 440C presented January 1962 at the SAE Automotive Engineering Congress.) New York: Society of Automotive Engineers, Inc., 1964.
41. Begeman, C. R., and J. M. Colucci. Benzo(a)pyrene in gasoline partially persists in automobile exhaust. Science 161:271, 1968.
42. Begeman, C. R., and J. M. Colucci. Polynuclear Aromatic Hydrocarbon Emissions from Automotive Engines. SAE Paper 700469. Detroit: Society of Automotive Engineers, 1970. 13 pp.
43. Bell, J. Paraffin epithelioma of the scrotum. Edinburgh Med. J. 22:135–137, 1876.
44. Bell, J. H., S. Ireland, and A. W. Spears. Identification of aromatic ketones in cigarette smoke condensate. Anal. Chem. 41:310–313, 1969.
45. Bentley, H. R., and J. G. Burgan. Polynuclear hydrocarbons in tobacco and tobacco smoke. Part I. 3:4-Benzopyrene. Analyst 83:442–447, 1958.
46. Bentley, H. R., and J. G. Burgan. Polynuclear hydrocarbons in tobacco and tobacco smoke. Part II. The origin of 3:4-benzopyrene found in tobacco and tobacco smoke. Analyst 85:723–727, 1960.
47. Berenblum, I., N. Haran-Ghera, and N. Trainin. An experimental analysis of the "hair cycle effect" in mouse skin carcinogenesis. Brit. J. Cancer 12:402–413, 1958.

48. Berenblum, I., and P. Shubik. The persistence of latent tumour cells induced in the mouse's skin by a single application of 9:10-dimethyl-1:2-benzanthracene. Brit. J. Cancer 3:384–386, 1949.
49. Bergman, I. The polarography of polycyclic aromatic hydrocarbons and the relationship between their half-wave potentials and absorption spectra. Trans. Faraday Soc. 50:829–838, 1954.
50. Berlman, I. B. Handbook of Fluorescence Spectra of Aromatic Molecules. New York: Academic Press Inc., 1965. 258 pp.
51. Berlman, I. B. On an empirical correlation between nuclear conformation and certain fluorescence and absorption characteristics of aromatic compounds. J. Phys. Chem. 74:3085–3093, 1970.
52. Berwald, Y., and L. Sachs. *In vitro* cell transformation with chemical carcinogens. Nature 200:1182–1184, 1963.
53. Berwald, Y., and L. Sachs. *In vitro* transformation of normal cells into tumor cells by carcinogenic hydrocarbons. J. Nat. Cancer Inst. 35:641–661, 1965.
54. Best, E. W. R. A Canadian Study of Smoking and Health. Ottawa: Department of National Health and Welfare, 1966. 137 pp.
55. Betteridge, D., and A. D. Baker. Analytical potential of photoelectron spectroscopy. Anal. Chem. 42:43A–44A, 46A, 48A, 50A, 52A, 54A, 56A, Jan. 1970.
56. Biesele, J. J. Mitotic Poisons and the Cancer Problem. Amsterdam: Elsevier Publishing Co., 1958. 214 pp.
57. Bingham, E., and A. W. Horton. Environmental carcinogenesis: Experimental observations related to occupational cancer, pp. 183–193. In W. Montagna and R. L. Dobson, Eds. Advances in Biology of Skin. Vol. VII. Carcinogenesis. Oxford: Pergamon Press, 1966.
58. Bingham, E., A. W. Horton, and R. Tye. The carcinogenic potency of certain oils. Arch. Environ. Health 10:449–451, 1965.
59. Biological effects of asbestos. Ann. N.Y. Acad. Sci. 132(Art. 1):1–766, 1965.
60. Birks, J. B. Photophysics of Aromatic Molecules. New York: John Wiley & Sons, Inc., 1970. 704 pp.
61. Birks, J. B. Scintillations in organic solids, pp. 433–508. In D. Fox, M. M. Labes, and A. Weissberger, Eds. Physics and Chemistry of the Organic Solid State. Vol. 2. New York: John Wiley & Sons, Inc., 1965.
62. Birks, J. B., and A. J. W. Cameron. Crystal fluorescence of carcinogens and related organic compounds. Proc. Roy. Soc. London 249A:297–317, 1959.
63. Blifford, I. H., Jr. Tropospheric aerosols. J. Geophys. Res. 75:3099–3103, 1970.
64. Blum, H. F. Sunlight as a causal factor in cancer of the skin of man. J. Nat. Cancer Inst. 9:247–258, 1948.
65. Blumenthal, H. T., and J. B. Rogers. Spontaneous and induced tumors in the guinea pig, with special reference to the factor of age. Progr. Exp. Tumor Res. 9:261–285, 1967.
66. Blumer, M. Benzpyrenes in soil. Science 134:474–475, 1961.
67. Bock, F. G. Early effects of hydrocarbons on mammalian skin. Progr. Exp. Tumor Res. 4:126–168, 1964.
68. Bock, F. G., and G. E. Moore. Carcinogenic activity of cigarette-smoke con-

densate. I. Effect of trauma and remote X irradiation. J. Nat. Cancer Inst. 22:401–411, 1959.

69. Bock, F. G., G. E. Moore, and S. K. Crouch. Tumor-promoting activity of extracts of unburned tobacco. Science 145:831–833, 1964.

70. Bock, F. G., and R. Mund. A survey of compounds for activity in the suppression of mouse sebaceous glands. Cancer Res. 18:887–892, 1958.

71. Bock, F. G., and R. Mund. Evaluation of substances causing loss of sebaceous glands from mouse skin. J. Invest. Derm. 26:479–487, 1956.

72. Bogacz, J., and I. Koprowska. A cyto-pathologic study of potentially carcinogenic properties of air pollutants. Acta Cytol. (Baltimore) 5:311–319, 1961.

73. Bolling, H. Sostanze cancerogene nei cereali sottoposti ad essiccazione con gas di combustione. Tec. Molitoria 15:137–142, 1964.

74. Bonser, G. M., E. Boyland, E. R. Busby, D. B. Clayson, P. L. Grover, and J. W. Jull. A further study of bladder implantation in the mouse as a means of detecting carcinogenic activity: Use of crushed paraffin wax or stearic acid as the vehicle. Brit. J. Cancer 17:127–136, 1963.

75. Boren, H. G. Carbon as a carrier mechanism for irritant gases. Arch. Environ. Health 8:119–124, 1964.

76. Borneff, J., and R. Fischer. Cancerogene Substanzen in Wasser und Boden. VIII. Untersuchungen an Filter-Aktivkohle nach Verwendung im Wasserwerk. Arch. Hyg. Bakt. 146:1–16, 1962. (summary in English)

77. Borneff, J., F. Selenka, H. Kunte, and A. Maximos. Experimental studies on the formation of polycyclic aromatic hydrocarbons in plants. Environ. Res. 2:22–29, 1968.

78. Bornstein, R. D. Observations of the urban heat island effect in New York City. J. Appl. Meteorol. 7:575–582, 1968.

79. Boubel, R. W., and L. A. Ripperton. Benzo(a)pyrene production during controlled combustion. J. Air Pollut. Control Assoc. 13:553–557, 1963.

80. Boutwell, R. K. Some biological aspects of skin carcinogenesis. Progr. Exp. Tumor Res. 4:207–250, 1964.

81. Boveri, T. The Origin of Malignant Tumors. Translated by M. Boveri. Baltimore: The Williams & Wilkins Co., 1929. 119 pp.

82. Bowen, E. J. Luminescence in Chemistry. London: D. Van Nostrand Co., Ltd., 1968. 254 pp.

83. Boyland, E., and P. Sims. Metabolism of polycyclic compounds. 24. The metabolism of benz[a]anthracene. Biochem. J. 91:493–506, 1964.

84. Boyland, E., and P. Sims. The metabolism of benz[a]anthracene and dibenz[a,h]anthracene and their 5,6-epoxy-5,6-dihydro derivatives by rat-liver homogenates. Biochem. J. 97:7–16, 1965.

85. Boyland, E., P. Sims, and C. Huggins. Induction of adrenal damage and cancer with metabolites of 7,12-dimethylbenz(a)anthracene. Nature 207:816–817, 1965.

86. Boyland, E., and G. Watson. 3-Hydroxyanthranilic acid, a carcinogen produced by endogenous metabolism. Nature 177:837–838, 1956.

87. Braun, A. C. The Cancer Problem. A Critical Analysis and Modern Synthesis. New York: Columbia University Press, 1969. 209 pp.

88. Braun, A. C., and H. N. Wood. The plant tumor problems. Adv. Cancer Res. 6:81–109, 1961.

89. Bridge, J. C., and S. A. Henry. Industrial cancers, pp. 258–268. In Report of the International Conference on Cancer, London, 17th–20th July, 1928. New York: William Wood and Company, 1928.
90. Bridges, B. A., and J. Huckle. Mutagenesis of cultured mammalian cells by x-radiation and ultraviolet light. Mutat. Res. 10:141–151, 1970.
91. Bridges, B. A., J. Huckle, and M. J. Ashwood-Smith. X-ray mutagenesis of cultured Chinese hamster cells. Nature 226:184–185, 1970.
92. Brookes, P., and C. Heidelberger. Isolation and degradation of DNA from cells treated with tritium-labeled 7,12-dimethylbenz(a)anthracene: Studies on the nature of the binding of this carcinogen to DNA. Cancer Res. 19: 157–165, 1969.
93. Brookes, P., and P. D. Lawley. Evidence for the binding of polynuclear aromatic hydrocarbons to the nucleic acids of mouse skin: Relation between carcinogenic power of hydrocarbons and their binding to deoxyribonucleic acid. Nature 202:781–784, 1964.
94. Brown, R. D. A theoretical treatment of the Diels–Alder reaction. Part I. Polycyclic aromatic hydrocarbons. J. Chem. Soc. 1950:691–697.
95. Brues, A. M. Critique of the linear theory of carcinogenesis. Science 128: 693–699, 1958.
96. Bryan, G. T., and P. D. Springberg. Role of the vehicle in the genesis of bladder carcinoma in mice by the pellet implantation technic. Cancer Res. 26:105–109, 1966.
97. Bryan, W. R., and M. B. Shimkin. Quantitative analysis of dose-response data obtained with carcinogenic hydrocarbons. J. Nat. Cancer Inst. 1:807–833, 1941.
98. Budzikiewicz, H., C. Djerassi, and D. H. Williams. Structure Elucidation of Natural Products by Mass Spectrometry. Vol. I. Alkaloids. San Francisco: Holden-Day, Inc., 1964. 233 pp.
99. Buell, P., and J. E. Dunn, Jr. Relative impact of smoking and air pollution on lung cancer. Arch. Environ. Health 15:291–297, 1967.
100. Buell, P., J. E. Dunn, Jr., and L. Breslow. Cancer of the lung and Los Angeles-type air pollution. Prospective study. Cancer 20:2139–2147, 1967.
101. Burdette, W. J. The significance of mutation in relation to the origin of tumors. A review. Cancer Res. 15:201–226, 1955.
102. Burrows, I. E., and A. J. Lindsey. Formation of polycyclic aromatic hydrocarbons by pyrolysis of simple aliphatic hydrocarbons. Chem. Ind. 1961:1395.
103. Burstein, N. A., K. R. McIntire, and A. C. Allison. Pulmonary tumors in germfree mice: Induction with urethan. J. Nat. Cancer Inst. 44:211–214, 1970.
104. Butlin, H. T. Three lectures on cancer of the scrotum in chimney-sweeps and others. Lecture 1. Secondary cancer without primary cancer. Brit. Med. J. 1:1341–1346, 1892.
105. Butlin, H. T. Three lectures on cancer of the scrotum in chimney-sweeps and others. Lecture 2. Why foreign sweeps do not suffer from scrotal cancer. Brit. Med. J. 2:1–6, 1892.
106. Butlin, H. T. Three lectures on cancer of the scrotum in chimney-sweeps and others. Lecture 3. Tar and paraffin cancer. Brit. Med. J. 2:66–71, 1892.
107. Cahnmann, H. J., and M. Kuratsune. Determination of polycyclic aromatic

hydrocarbons in oysters collected in polluted water. Anal. Chem. 29:1312–1317, 1957.
108. Campbell, J. A. Cancer of skin and increase in incidence of primary tumours of lung in mice exposed to dust obtained from tarred roads. Brit. J. Exp. Path. 15:287–294, 1934.
109. Campbell, J. A. Influenza virus and incidence of primary lung tumours in mice. Lancet 2:487, 1940.
110. Campbell, J. A. Lung tumours in mice. Incidence as affected by inhalation of certain carcinogenic agents and some dusts. Brit. Med. J. 1:217–221, 1942.
111. Campbell, J. M., and R. L. Cooper. The presence of 3,4-benzpyrene in snuff associated with a high incidence of cancer. Chem. Ind. 1955:64–65.
112. Campbell, J. M., and L. Kreyberg. The degree of air pollution in Norwegian towns. Brit. J. Cancer 10:481–484, 1956.
113. Campbell, J. M., and A. J. Lindsey. Polycyclic aromatic hydrocarbons in snuff. Chem. Ind. 1957:951.
114. Campbell, J. M., and A. J. Lindsey. Polycyclic hydrocarbons extracted from tobacco: The effect upon total quantities found in smoking. Brit. J. Cancer 10:649–652, 1956.
115. Cantuti, V., G. P. Cartoni, A. Liberti, and A. G. Torri. Improved evaluation of polynuclear hydrocarbons in atmospheric dust by gas chromatography. J. Chromatogr. 17:60–65, 1965.
116. Cardiovascular Diseases in the U.S., Facts and Figures. New York: American Heart Association, 1965.
117. Carnow, B. W. Pulmonary disease and air pollution. A Chicago problem. Chicago Med. 71:581–586, 1968.
118. Carozzi, L. Le cancer professionnel. Arch. Elect. Med. 42:85–93, 118–142, 155–160, 1934.
119. Cartwright, J. Particle shape factors. Ann. Occup. Hyg. 5:163–171, 1962.
120. Carugno, N., and S. Rossi. Evaluation of polynuclear hydrocarbons in cigaret smoke by glass capillary columns. J. Gas Chromatogr. 5:103–106, 1967.
121. Casarett, G. W. Experimental radiation carcinogenesis. Progr. Exp. Tumor Res. 7:49–82, 1965.
122. Cattanach, B. M. A genetical approach to the effects of radiomimetic chemicals on fertility in mice, pp. 415–426. In W. D. Carlson and F. X. Fassner, Eds. Effects of Ionizing Radiation on the Reproductive System. Proceedings of an International Symposium held at Colorado State University, Fort Collins, Colorado. New York: The Macmillan Co., 1964.
123. Cattanach, B. M., C. E. Pollard, and J. H. Isaacson. Ethyl methanesulfonate-induced chromosome breakage in the mouse. Mutat. Res. 6:297–307, 1968.
124. Cavalieri, E., and M. Calvin. Molecular characteristics of some carcinogenic hydrocarbons (benzpyrene/dimethylbenzanthracene/3-methylcholanthrene/nucleic acids). Proc. Nat. Acad. Sci. U.S.A. 68:1251–1253, 1971.
125. Cavill, G. W. K., A. Robertson, and W. B. Whalley. The chemistry of fungi. Part VIII. The oxidation of methylene groups in compounds analogous to O-dimethylcitromycin. J. Chem. Soc. 1949:1567–1570.
126. Cetorelli, J. J., W. J. McCarthy, and J. P. Winefordner. The selection of

optimum conditions for spectrochemical methods. IV. Sensitivity of absorption, fluorescence, and phosphorescence spectrometry in the condensed phase. J. Chem. Educ. 45:98–102, 1968.

127. Chakraborty, B. B., and R. Long. Gas chromatographic analysis of polycyclic aromatic hydrocarbons in soot samples. Environ. Sci. Technol. 1:828–834, 1967.

128. Chamberlain, A. C. Interception and retention of radioactive aerosols by vegetation. Atmos. Environ. 4:57–78, 1970.

129. Charlson, R. J., N. C. Ahlquist, and H. Horvath. On the generality of correlation of atmospheric aerosol mass concentration and light scatter. Atmos. Environ. 2:455–464, 1968.

130. Chassevant, F., and M. Héros. Recherche du benzo3,4,pyrène dans le café vert et torréfié et dans les sous-produits de torréfaction. Café, Cacao, Thé 7:349–358, 1963.

131. Chen, T. T., and C. Heidelberger. Cultivation *in vitro* of cells derived from adult C3H mouse ventral prostate. J. Nat. Cancer Inst. 42:903–914, 1969.

132. Chen, T. T., and C. Heidelberger. *In vitro* malignant transformation of cells derived from mouse prostate in the presence of 3-methylcholanthrene. J. Nat. Cancer Inst. 42:915–925, 1969.

133. Chen, T. T., and C. Heidelberger. Quantitative studies on the malignant transformation of mouse prostate cells by carcinogenic hydrocarbons *in vitro*. Int. J. Cancer 4:166–178, 1969.

134. Chouroulinkov, I., P. Lazar, C. Izard, C. Libermann, and M. Guérin. "Sebaceous glands" and "hyperplasia" tests as screening methods for tobacco tar carcinogenesis. J. Nat. Cancer Inst. 42:981–985, 1969.

135. Chu, E. H. Y. Mammalian cell genetics. III. Characterization of X-ray-induced forward mutations in Chinese hamster cell cultures. Mutat. Res. 11: 23–34, 1971.

136. Chu, E. H. Y., E. G. Bailiff, and H. V. Malling. Mutagenicity of chemical carcinogenesis in mammalian cells, pp. 62–63. In Tenth International Cancer Congress, Houston, 1970. Abstracts. Houston, Texas: Medical Arts Publishing Co., 1970.

137. Chu, E. H. Y., and H. V. Malling. Mammalian cell genetics. II. Chemical induction of specific locus mutations in Chinese hamster cells *in vitro*. Proc. Nat. Acad. Sci. U.S.A. 61:1306–1312, 1968.

138. Chu, E. W., and R. A. Malmgren. An inhibitory effect of vitamin A on the induction of tumors of forestomach and cervix in the Syrian hamster by carcinogenic polycyclic hydrocarbons. Cancer Res. 25:884–895, 1965.

139. Clar, E. J. Polycyclic Hydrocarbons. 2 vols. New York: Academic Press Inc., 1964. 974 pp.

140. Clark, J. H. The effect of long ultraviolet radiation on the development of tumors induced by 20-methylcholanthrene. Cancer Res. 24:207–211, 1964.

141. Clark, W. E., and K. T. Whitby. Concentration and size distribution measurements of atmospheric aerosols and a test of the theory of self-preserving size distributions. J. Atmos. Sci. 24:677–687, 1967.

142. Clayson, D. B. Chemical Carcinogenesis, pp. 164–171. Boston: Little, Brown and Co., 1962.

143. Cleary, G. J. Discrete separation of polycyclic hydrocarbons in air-borne particulates using very long alumina columns. J. Chromatogr. 9:204–215, 1962.
144. Clemmesen, J. Bronchial carcinoma—a pandemic. Dan. Med. Bull. 1:37–46, 1954.
145. Clemmesen, J., and A. Nielsen. The social distribution of cancer in Copenhagen, 1943–1947. Brit. J. Cancer 5:159–171, 1951.
146. Clemo, G. R. Some constituents of city smoke. Tetrahedron 23:2389–2393, 1967.
147. Clemo, G. R., and E. W. Miller. The carcinogenic action of city smoke. Chem. Ind. 1955:38.
148. Coffin, D. L. Health Effects of Airborne Polycyclic Hydrocarbons. Presented at the Symposium on Human Health and Vehicle Emissions. Detroit: Society of Automotive Engineers, 1971. 23 pp.
149. Cohart, E. M. Socioeconomic distribution of cancer of the lung in New Haven. Cancer 8:1126–1129, 1955.
150. Cohart, E. M. Socioeconomic distribution of stomach cancer in New Haven. Cancer 7:455–461, 1954.
151. Colucci, J. M., and C. R. Begeman. Carcinogenic air pollutants in relation to automotive traffic in New York. Environ. Sci. Tech. 5:145–150, 1971.
152. Colucci, J. M., and C. R. Begeman. Polynuclear Aromatic Hydrocarbons and Other Pollutants in Los Angeles. Presented at the 2nd International Clear Air Congress, Washington, D.C., Dec. 6–11, 1970.
153. Colucci, J. M., and C. R. Begeman. The automotive contribution to airborne polynuclear aromatic hydrocarbons in Detroit. J. Air Pollut. Control Assoc. 15:113–122, 1965.
154. Commins, B. T. Interim report on the study of techniques for determination of polycyclic aromatic hydrocarbons in air. Nat. Cancer Inst. Monogr. 9: 225–233, 1962.
155. Commins, B. T. Polycyclic hydrocarbons in rural and urban air. Int. J. Air Pollut. 1:14–17, 1958.
156. Commins, B. T., R. L. Cooper, and A. J. Lindsey. Polycyclic hydrocarbons in cigarette smoke. Brit. J. Cancer 8:296–302, 1954.
157. Conney, A. H. Pharmacological implications of microsomal enzyme induction. Pharmacol. Rev. 19:317–366, 1967.
158. Cook, J. W., and R. H. Martin. Polycyclic aromatic hydrocarbons. Part XXIV. J. Chem. Soc. 1940:1125–1127.
159. Coomber, J. W., and J. N. Pitts, Jr. Singlet oxygen in the environmental sciences. VIII. Production of O_2 ($'\Delta g$) by energy transfer from excited benzaldehyde under simulated atmospheric conditions. Environ. Sci. Technol. 4:506–510, 1970.
160. Cooper, R. L., and A. J. Lindsey. Atmospheric pollution by polycyclic hydrocarbons. Chem. Ind. 1953:1177–1178.
161. Cooper, R. L., and A. J. Lindsey. 3:4-Benzpyrene and other polycyclic hydrocarbons in cigarette smoke. Brit. J. Cancer 9:304–309, 1955.
162. Cooper, R. L., A. J. Lindsey, and R. E. Waller. The presence of 3,4-benzpyrene in cigarette smoke. Chem. Ind. 1954:1418.

163. Corbett, T. H., C. Heidelberger, and W. F. Dove. Determination of the mutagenic activity to bacteriophage T4 of carcinogenic and noncarcinogenic compounds. Molec. Pharmacol. 6:667–679, 1970.
164. Corn, M. Nonviable particles in the air, pp. 47–94. In A. C. Stern, Ed. Air Pollution. Vol. 1. Air Pollution and Its Effects. (2nd ed.) New York: Academic Press Inc., 1968.
165. Corn, M. Urban aerosols: Problems associated with evaluation of inhalation risk. In Assessment of Airborne Particles. Proceedings of the 3rd University of Rochester Conference on Environmental Toxicology, June 1970. Springfield, Ill.: Charles C Thomas, 1971.
166. Corn, M., T. L. Montgomery, and N. A. Esmen. Suspended particulate matter: Seasonal variation in specific surface areas and densities. Environ. Sci. Tech. 5:155–158, 1971.
167. Crocker, T. T. Effect of benzo[a]pyrene on hamster, rat, dog and monkey respiratory epithelia in organ culture, pp. 433–443. In M. G. Hanna, Jr., P. Nettesheim, and J. R. Gilbert, Eds. Inhalation Carcinogenesis. Proceedings of a Biology Division, Oak Ridge National Laboratory, conference held in Gatlinburg, Tenn., Oct. 8–11, 1969. AEC Symposium Series 18. Oak Ridge, Tenn.: U.S. Atomic Energy Commission, Division of Technical Information, 1970.
168. Crocker, T. T., J. E. Chase, S. A. Wells, and L. L. Nunes. Preliminary report on experimental squamous carcinoma of the lung in hamsters and in a primate (*Galago crassicaudatus*), pp. 317–328. In P. Nettesheim, M. G. Hanna, Jr., and J. W. Deatherage, Jr., Eds. Morphology of Experimental Respiratory Carcinogenesis. Proceedings of a Biology Division, Oak Ridge National Laboratory, conference held in Gatlinburg, Tenn., May 13–16, 1970. AEC Symposium Series 21. Oak Ridge, Tenn.: U.S. Atomic Energy Commission, Division of Technical Information, 1970.
169. Crocker, T. T., and B. I. Nielsen. Effect of carcinogenic hydrocarbons on suckling rat trachea in living animals and in organ cultures, pp. 765–787. In L. Severi, Ed. Lung Tumours in Animals. Proceedings of the Third Quadrennial International Conference on Cancer. June 24–29, 1965. Perugia, Italy: University of Perugia, 1966.
170. Crocker, T. T., B. I. Nielsen, and I. Lasnitzki. Carcinogenic hydrocarbons. Effects on suckling rat trachea in organ culture. Arch. Environ. Health 10:240–250, 1965.
171. Crocker, T. T., and L. L. Sanders. Influence of vitamin A and 3,7-dimethyl-2,6-octadienal (citral) on the effect of benzo(a)pyrene on hamster trachea in organ culture. Cancer Res. 30:1312–1318, 1970.
172. Crow, J. F. Chemical risk to future generations. Sci. Citizen 10:113–117, 1968.
173. Cruickshank, C. N. D., and A. Gourevitch. Skin cancer of the hand and forearm. Brit. J. Ind. Med. 9:74–79, 1952.
174. Cruickshank, C. N. D., and J. R. Squire. Skin cancer in the engineering industry from the use of mineral oil. Brit. J. Ind. Med. 7:1–11, 1950.
175. Crummett, W., and R. Hummel. Ultraviolet spectrometry. Anal. Chem. 42:239R–248R, 1970.
176. Cruser, S. A., and A. J. Bard. Electrogenerated chemiluminescence. III.

Intensity-time and concentration-intensity relation and the lifetime of radical cations of aromatic hydrocarbons in *N,N*-dimethylformamide solution. J. Amer. Chem. Soc. 91:267–275, 1969.

177. Curwen, M. P., E. L. Kennaway, and N. M. Kennaway. The incidence of cancer of the lung and larynx in urban and rural districts. Brit. J. Cancer 8:181–198, 1954.
178. Dalhamn, T., M.-L. Edfors, and R. Rylander. Retention of cigarette smoke components in human lungs. Arch. Environ. Health 17:746–748, 1968.
179. Daniel, P. M., O. E. Pratt, and M. M. L. Prichard. Metabolism of labelled carcinogenic hydrocarbons in rats. Nature 215:1142–1146, 1967.
180. Darlington, C. D. The plasmagene theory of the origin of cancer. Brit. J. Cancer 2:118–126, 1948.
181. Daudel, P., and R. Daudel. Chemical Carcinogenesis and Molecular Biology. New York: John Wiley & Sons, Inc., 1966. 158 pp.
182. Dautrebande, L. Physiological and pharmacological characteristics of liquid aerosols. Physiol. Rev. 32:214–275, 1952.
183. Davies, C. N. Size, area, volume and weight of dust particles. Ann. Occup. Hyg. 3:219–225, 1961.
184. Davies, J. H. Correlation spectroscopy. Anal. Chem. 42:101A–105A, 108A, 110A, 112A, May 1970.
185. Davies, W., and J. R. Wilmshurst. Carcinogens formed in the heating of food stuffs. Formation of 3,4-benzopyrene from starch at 370–390°C. Brit. J. Cancer 14:295–299, 1960.
186. Day, T. D. Carcinogenic action of cigarette smoke condensate on mouse skin. An attempt at a quantitative study. Brit. J. Cancer 21:56–81, 1967.
187. Dean, G. Lung cancer among white South Africans. Brit. Med. J. 2:852–857, 1959.
188. Dean, G. Lung cancer and bronchitis in Northern Ireland, 1960–2. Brit. Med. J. 1:1506–1514, 1966.
189. Dean, G. Lung cancer in South Africans and British immigrants. Proc. Roy. Soc. Med. 57:984–987, 1964.
190. Della Porta, G. Some aspects of medical drug testing for carcinogenic activity, pp. 33–47. In R. Truhaut, Ed. Potential Carcinogenic Hazards from Drugs: Evaluation of Risks. (UICC Monograph Series, Vol. 7.) Berlin: Springer-Verlag, 1967.
191. Della Porta, G., P. Shubik, K. Dammert, and B. Terracini. Role of polyoxyethylene sorbitan monostearate in skin carcinogenesis in mice. J. Nat. Cancer Inst. 25:607–625, 1960.
192. DeMaio, L., and M. Corn. Gas chromatographic analysis of polynuclear aromatic hydrocarbons with packed columns. Application to air pollution studies. Anal. Chem. 38:131–133, 1966.
193. DeMaio, L., and M. Corn. Polynuclear aromatic hydrocarbons associated with particulates in Pittsburgh air. J. Air Pollut. Control Assoc. 16:67–71, 1966.
194. Demas, J. N., and G. A. Crosby. Photoluminescence decay curves: An analysis of the effects of flash duration and linear instrumental distortions. Anal. Chem. 42:1010–1017, 1970.
195. Demisch, R. R., and G. F. Wright. The distribution of polynuclear aromatic

hydrocarbons between aqueous and non-aqueous phases. Can. J. Biochem. Physiol. 41:1655–1662, 1963.
196. Dewar, M. J. S. A molecular orbital theory of organic chemistry. VI. Aromatic substitution and addition. J. Amer. Chem. Soc. 74:3357–3363, 1952.
197. Diamond, L. Effect of carcinogenic hydrocarbons on rodent and primate cells *in vitro*. J. Cell. Physiol. 66:183–197, 1965.
198. Diamond, L., and H. V. Gelboin. Alpha-naphthoflavone: An inhibitor of hydrocarbon cytotoxicity and microsomal hydroxylase. Science 166:1023–1025, 1969.
199. Diamond, L., C. Sardet, and G. H. Rothblat. The metabolism of 7,12-dimethylbenz(a)anthracene in cell cultures. Int. J. Cancer 3:838–849, 1968.
200. Dickey, F. H., G. H. Cleland, and C. Lotz. The rôle of organic peroxides in the induction of mutations. Proc. Nat. Acad. Sci. U.S.A. 35:581–586, 1949.
201. DiPaolo, J. A., P. Donovan, and R. Nelson. Quantitative studies of *in vitro* transformation by chemical carcinogens. J. Nat. Cancer Inst. 42:867–874, 1969.
202. DiPaolo, J. A., R. L. Nelson, and P. J. Donovan. Sarcoma-producing cell lines derived from clones transformed *in vitro* by benzo[a]pyrene. Science 165:917–918, 1969.
203. Dipple, A., P. D. Lawley, and P. Brookes. Theory of tumour initiation by chemical carcinogens: Dependence of activity on structure of ultimate carcinogen. Eur. J. Cancer 4:493–506, 1968.
204. Dirksen, E. R., and T. T. Crocker. Ultrastructural alterations produced by polycyclic aromatic hydrocarbons on rat tracheal epithelium in organ culture. Cancer Res. 28:906–923, 1968.
205. Doll, R. The causes of death among gas-workers with special reference to cancer of the lung. Brit. J. Ind. Med. 9:180–187, 1952.
206. Doll, R., R. E. W. Fisher, E. J. Gammon, W. Gunn, G. O. Hughes, F. H. Tyrer, and W. Wilson. Mortality of gasworkers with special reference to cancer of the lung and bladder, chronic bronchitis, and pneumoconiosis. Brit. J. Ind. Med. 22:1–12, 1965.
207. Doll, R., and A. B. Hill. Mortality in relation to smoking: Ten years' observations of British doctors. Brit. Med. J. 1:1399–1410, 1460–1467, 1964.
208. Doll, R., and A. B. Hill. Mortality of British doctors in relation to smoking: Observations on coronary thrombosis. Nat. Cancer Inst. Monogr. 19:205–268, 1966.
209. Doll, R., A. B. Hill, P. G. Gray, and E. A. Parr. Lung cancer mortality and the length of cigarette ends. An international comparison. Brit. Med. J. 5118:322–325, 1959.
210. Domsky, I., W. Lijinsky, K. Spencer, and P. Shubik. Rate of metabolism of 9,10-dimethyl-1,2-benzanthracene in newborn and adult mice. Proc. Soc. Exp. Biol. Med. 113:110–112, 1963.
211. Doniach, I., and J. C. Mottram. On the effect of light upon the incidence of tumours in painted mice. Amer. J. Cancer 39:234–240, 1940.
212. Doniach, I., J. C. Mottram, and F. Weigert. The fluorescence of 3:4 benzpyrene *in vivo*. Part II. The inter-relationship of the derivatives formed in various sites. Brit. J. Exp. Path. 24:9–14, 1943.
213. Dontenwill, W., H. Elmenhorst, G. Reckzeh, H.-P. Harke, and L. Stadler.

Experimentelle Untersuchungen über Aufnahme, Abtransport und Abbau cancerogener Kohlenwasserstoffe im Bereich des Respirationstraktes. Z. Krebsforsch. 71:225–243, 1968.

214. Dorn, H. F. Morbidity and mortality from bladder tumors in North and South America. Acta Un. Int. Cancr. 18:553–559, 1962.

215. Dörr, R. Die Aufnahme von Alkaloiden und Benzpyren durch intakte Pflanzenwurzeln. Naturwissenschaften 52:166, 1965.

216. Druckrey, H. Pharmacological approach to carcinogenesis, pp. 110–130. In G. E. W. Wolstenholme and M. O'Connor, Eds. Ciba Foundation Symposium on Carcinogenesis. Mechanisms of Action. Boston: Little, Brown and Co., 1958.

217. Drushel, H. V., A. L. Sommers, and R. C. Cox. Correction of luminescence spectra and calculation of quantum efficiencies using computer techniques. Anal. Chem. 35:2166–2172, 1963.

218. Dubois, L., A. Corkery, and J. L. Monkman. The chromatography of polycyclic hydrocarbons. Int. J. Air Pollut. 2:236–252, 1960.

219. Duchen, L. W. Bronchogenic carcinoma. Incidence and pathology as seen on the Witwatersrand. S. Afr. J. Med. Sci. 19:65–74, 1954.

220. Duncan, M., P. Brookes, and A. Dipple. Metabolism and binding to cellular macromolecules of a series of hydrocarbons by mouse embryo cells in culture. Int. J. Cancer 4:813–819, 1969.

221. Dunn, J. E., and F. S. Brackett. Photosensitizing properties of some petroleum solvents. Indust. Med. 17:303–308, 1948.

222. Dunn, J. R., W. A. Waters, and I. M. Roitt. The retardation of benzaldehyde autoxidation. Part VI. The actions of a number of polycyclic aromatic hydrocarbons of carcinogenic interest. J. Chem. Soc. 1954:580–586.

223. Eastcott, D. F. The epidemiology of lung cancer in New Zealand. Lancet 1: 37–39, 1956.

224. Eastman, J. W. The dependence of fluorescence on solvent and temperature. Spectrochim. Acta 26A:1545–1557, 1970.

225. Easty, G. C. Organ culture methods. Methods Cancer Res. 5:1–43, 1970.

226. Eberson, L., and K. Nyberg. Studies on electrolyte substitution reactions. II. Anodic acetamidation, a novel anodic substitution. Tetrahedron Lett. 22:2389–2393, 1966.

227. Eckardt, R. E. Industrial Carcinogens. (Modern Monographs in Industrial Medicine, 4.) New York: Grune & Stratton, Inc., 1959. 164 pp.

228. Ehling, U. H. The multiple loci method in chemical mutagenesis, pp. 156–161. In F. Vogel and G. Rohrborn, Eds. Chemical Mutagenesis in Mammals and Man. Berlin: Springer-Verlag, 1970.

229. Ehling, U. H., R. B. Cumming, and H. V. Malling. Induction of dominant lethal mutations by alkylating agents in male mice. Mutat. Res. 5:417–428, 1968.

230. Ejder, E. Methods of representing emission, excitation, and photoconductivity spectra. J. Opt. Soc. Amer. 59:223–224, 1969.

231. Epstein, J. H. Ultraviolet light carcinogenesis, pp. 215–236. In W. Montagna and R. L. Dobson, Eds. Advances in Biology of Skin. Vol. VII. Carcinogenesis. Proceedings of a Symposium on the Biology of Skin held at the University of Oregon Medical School, 1965. Oxford: Pergamon Press, 1966.

232. Epstein, S. S. A "catch-all" toxicological screen. Experientia 25:617, 1969.
233. Epstein, S. S. Photoactivation of polynuclear hydrocarbons. Arch. Environ. Health 10:233–239, 1965.
234. Epstein, S. S., J. Andrea, S. Joshi, and N. Mantel. Hepatocarcinogenicity of griseofulvin following parenteral administration to infant mice. Cancer Res. 27:1900–1906, 1967.
235. Epstein, S. S., E. Arnold, E. Steinberg, D. Mackintosh, H. Shafner, and Y. Bishop. Mutagenic and antifertility effects of TEPA and METEPA in mice. Toxicol. Appl. Pharmacol. 17:23–40, 1970.
236. Epstein, S. S., W. Bass, E. Arnold, and Y. Bishop. Mutagenicity of trimethylphosphate in mice. Science 168:584–586, 1970.
237. Epstein, S. S., W. Bass, E. Arnold, Y. Bishop, S. Joshi, and I. D. Adler. Sterility and semisterility in male progeny of male mice treated with the chemical mutagen TEPA. Toxicol. Appl. Pharmacol. 19:134–146, 1971.
238. Epstein, S. S., and M. Burroughs. Some factors influencing the photodynamic response of *Paramecium caudatum* to 3,4-benzpyrene. Nature 193:337–338, 1962.
239. Epstein, S. S., M. Burroughs, and M. Small. The photodynamic effect of the carcinogen, 3,4-benzpyrene, on *Paramecium caudatum*. Cancer Res. 23:35–44, 1963.
240. Epstein, S. S., N. P. Buu-Hoi, and D. P. Hien. On the association between photodynamic and enzyme-inducing activities on polycyclic compounds. Cancer Res. 31:1087–1094, 1971.
241. Epstein, S. S., S. Joshi, J. Andrea, N. Mantel, E. Sawicki, T. Stanley, and E. C. Tabor. Carcinogenicity of organic particulate pollutants in urban air after administration of trace quantities to neonatal mice. Nature 212:1305–1307, 1966.
242. Epstein, S. S., S. R. Joshi, E. Arnold, E. C. Page, and Y. Bishop. Abnormal zygote development in mice after paternal exposure to a chemical mutagen. Nature 225:1260–1261, 1970.
243. Epstein, S. S., and N. Mantel. Hepatocarcinogenicity of the herbicide maleic hydrazide following parenteral administration to infant Swiss mice. Int. J. Cancer 3:325–335, 1968.
244. Epstein, S. S., N. Mantel, and T. W. Stanley. Photodynamic assay of neutral subfractions of organic extracts of particulate atmospheric pollutants. Environ. Sci. Technol. 2:132–138, 1968.
245. Epstein, S. S., and H. Shafner. Chemical mutagens in the human environment. Nature 219:385–387, 1968.
246. Epstein, S. S., M. Small, H. L. Falk, and N. Mantel. On the association between photodynamic and carcinogenic activities in polycyclic compounds. Cancer Res. 24:855–862, 1964.
247. Epstein, S. S., M. Small, J. Koplan, N. Mantel, H. L. Falk, and E. Sawicki. Photodynamic bioassay of polycyclic air pollutants. Arch. Environ. Health 7:531–537, 1963.
248. Epstein, S. S., M. Small, J. Koplan, N. Mantel, and S. H. Hunter. Photodynamic bioassay of benzo[a]pyrene with *Paramecium caudatum*. J. Nat. Cancer Inst. 31:163–168, 1963.
249. Epstein, S. S., M. Small, E. Sawicki, and H. L. Falk. Photodynamic bioas-

say of polycyclic atmospheric pollutants. J. Air Pollut. Control Assoc. 15:174–176, 1965.

250. Erickson, R. E., P. S. Bailey, and J. C. Davis, Jr. Structure of the mono-özonide of 9,10-dimethylanthracene. A transannular ozone adduct. Tetrahedron 18:388–395, 1962.

251. Eschenroeder, A., and J. R. Martinez. Mathematical Modeling of Photochemical Smog. AIAA Paper 70-116. Presented to the American Institute of Aeronautics and Astronautics 8th Aerospace Sciences Meeting, New York, 1970. 11 pp.

252. Esmen, N. A., and M. Corn. Residence time of particles in urban air. Atmos. Environ. 5:571–578, 1971.

253. Étienne, A., and A. Staehelin. Sur quelques α-azanthrancènes et quelques benzacridines. Bull. Soc. Chim. France 21:748–755, 1954.

254. Evans, H. J. The Effects of External Agents on Differentiation in the Cultivated Mushroom. Ph.D. dissertation, University College of Wales, 1955.

255. Falconer, D. S., B. M. Slizynski, and C. Auerbach. Genetical effects of nitrogen mustard in the house mouse. J. Genet. 51:81–88, 1952.

256. Falk, H. L. Carcinogenesis, mutagenesis, and teratogenesis, pp. 155–164. In Man's Health and the Environment—Some Research Needs. Report of the Task Force on Research Planning in Environmental Health Science. U.S. Department of Health, Education, and Welfare. Washington, D.C.: U.S. Government Printing Office, 1970.

257. Falk, H. L., and P. Kotin. Pesticide synergists and their metabolites: Potential hazards. Ann. N.Y. Acad. Sci. 160:299–313, 1969.

258. Falk, H. L., P. Kotin, and I. Markul. The disappearance of carcinogens from soot in human lungs. Cancer 11:482–489, 1958.

259. Falk, H. L., P. Kotin, and A. Mehler. Polycyclic hydrocarbons as carcinogens for man. Arch. Environ. Health 8:721–730, 1964.

260. Falk, H. L., P. Kotin, and A. Miller. Aromatic polycyclic hydrocarbons in polluted air as indicators of carcinogenic hazards. Int. J. Air Pollut. 2:201–209, 1960.

261. Falk, H. L., P. Kotin, and S. Thompson. Inhibition of carcinogenesis. The effect of polycyclic hydrocarbons and related compounds. Arch. Environ. Health 9:169–179, 1964.

262. Falk, H. L., I. Markul, and P. Kotin. Aromatic hydrocarbons. IV. Their fate following emission into atmosphere and experimental exposure to washed air and synthetic smog. A.M.A. Arch. Ind. Health 13:13–17, 1956.

263. Falk, H. L., A. Miller, and P. Kotin. Elution of 3,4-benzpyrene and related hydrocarbons from soots by plasma proteins. Science 127:474–475, 1958.

264. Faust, W. J., and M. J. Sterba. Minimizing Exhaust Emissions—A Realistic Approach. ASTM Symposium, Toronto, Canada, June 1970. Philadelphia: American Society for Testing Materials, 1971.

265. Federal Radiation Council. Radiation Exposure of Uranium Miners. Washington, D.C.: Federal Radiation Council, 1968. 31 pp.

266. Fell, H. B., and L. M. Rinaldini. The effects of vitamins A and C on cells and tissues in culture, pp. 659–699. In E. N. Willmer, Ed. Cells and Tissues in Culture. Methods, Biology, and Physiology. Vol. 1. London: Academic Press Inc., 1965.

267. Fieser, L. F., and E. B. Hershberg. The orientation of 3,4-benzpyrene in substitution reactions. J. Amer. Chem. Soc. 61:1565–1574, 1939.
268. Fife, J. G. Carcinoma of the skin in machine tool setters. Brit. J. Ind. Med. 19:123–125, 1962.
269. Filley, G. F. Pulmonary Insufficiency and Respiratory Failure. Philadelphia: Lea & Febiger, 1967. 162 pp.
270. Findlay, B. F., and M. S. Hirt. An urban-induced meso-circulation. Atmos. Environ. 3:537–542, 1969.
271. Findlay, F. D., C. J. Fortin, and D. R. Snelling. Deactivation of O_2 ('Δg). Chem. Phys. Lett. 3:204–206, 1969.
272. Findlay, G. M. Ultra-violet light and skin cancer. Lancet 2:1070–1073, 1928.
273. Fisherman, E. W. Does the allergic diathesis influence malignancy? J. Allergy 31:74–78, 1960.
274. Fjelde, A., and J. L. Turk. Induction of an immunological response in local lymph nodes by chemical carcinogens. Nature 205:813–815, 1965.
275. Foote, C. S. Mechanisms of photosensitized oxidation. Science 162:963–970, 1968.
276. Foote, C. S. Photosensitized oxygenations and the role of singlet oxygen. Accounts Chem. Res. 1:104–110, 1968.
277. Foote, C. S., Y. C. Chang, and R. W. Denny. Chemistry of singlet oxygen. X. Carotenoid quenching parallels biological protection. J. Amer. Chem. Soc. 92:5216–5219, 1970.
278. Freeman, A. E., P. J. Price, R. J. Bryan, R. J. Gordon, R. V. Gilden, G. J. Kelloff, and R. J. Huebner. Transformation of rat and hamster embryo cells by extracts of city smog. Proc. Nat. Acad. Sci. U.S.A. 68:445–449, 1971.
279. Fried, J., and D. E. Schumm. One electron transfer oxidation of 7,12-dimethylbenz[a] anthracene, a model for the metabolic activation of carcinogenic hydrocarbons. J. Amer. Chem. Soc. 89:5508–5509, 1967.
280. Friedel, R. A., and M. Orchin. Ultraviolet Spectra of Aromatic Compounds. New York: John Wiley & Sons, Inc., 1961. 52 pp.
281. Friedlander, S. K., and J. H. Seinfeld. A dynamic model of photochemical smog. Environ. Sci. Technol. 3:1175–1181, 1969.
282. Fuks, N. A. The Mechanics of Aerosols. New York: The Macmillan Co., 1964. 408 pp.
283. Furth, J., and E. Lorenz. Carcinogenesis by ionizing radiations, pp. 1145–1201. In A. Hollaender, Ed. Radiation Biology. Vol. 1. Part 2. New York: McGraw-Hill Book Co., Inc., 1954.
284. Gabridge, M. G., A. Denunzio, and M. S. Legator. Cycasin: Detection of associated mutagenic activity *in vivo*. Science 163:689–691, 1969.
285. Gabridge, M. G., A. Denunzio, and M. S. Legator. Microbial mutagenicity of streptozotocin in animal-mediated assays. Nature 221:68–70, 1969.
286. Gabridge, M. G., and M. S. Legator. A host-mediated microbial assay for the detection of mutagenic compounds. Proc. Soc. Exp. Biol. 130:831–834, 1969.
287. Gabridge, M. G., E. J. Oswald, and M. S. Legator. The role of selection in the host-mediated assay for mutagenicity. Mutat. Res. 7:117–119, 1969.
288. Galuškinová, V. 3,4-Benzpyrene determination in the smoky atmosphere of social meeting rooms and restaurants. A contribution to the problem of the noxiousness of so-called passive smoking. Neoplasma 11:465–468, 1964.

289. Gardner, M. B. Biological effects of urban air pollution. III. Lung tumors in mice. Arch. Environ. Health 12:305–313, 1966.
290. Geacintov, N., G. Oster, and T. Cassen. Polymeric matrices for organic phosphors. J. Opt. Soc. Amer. 58:1217–1229, 1968. Erratum: *ibid.* 59:367, 1969.
291. Gelboin, H. V. A microsome-dependent binding of benzo[a]pyrene to DNA. Cancer Res. 29:1272–1276, 1969.
292. Gelboin, H. V. Carcinogens, enzyme induction, and gene action. Adv. Cancer Res. 10:1–81, 1967.
293. Gelboin, H. V., and F. J. Wiebel. Studies on the mechanism of aryl hydrocarbon hydroxylase induction and its role in cytotoxicity and tumorenicity. Ann. N.Y. Acad. Sci. 171:529–549. 1971.
294. Gelboin, H. V., F. Wiebel, and L. Diamond. Dimethylbenzanthracene tumorigenesis and aryl hydrocarbon hydroxylase in mouse skin: Inhibition by 7,8-benzoflavone. Science 170:169–171, 1970.
295. Generoso, W. M. Chemical induction of dominant lethals in female mice. Genetics 61:461–470, 1969.
296. Gilbert, J. A. S., and A. J. Lindsey. Polycyclic hydrocarbons in tobacco smoke: Pipe smoking experiments. Brit. J. Cancer 10:646–648, 1956.
297. Gillette, J. R. Factors that affect the stimulation of the microsomal drug enzymes induced by foreign compounds. Adv. Enzyme Regul. 1:215–223, 1963.
298. Gilliland, M. R., A. J. Howard, and D. Hamer. Polycyclic hydrocarbons in crude peat wax. Chem. Ind. 1960:1357–1358.
299. Gleason, W. S., A. D. Broadbent, E. Whittle, and J. N. Pitts, Jr. Singlet oxygen in the environmental sciences. IV. Kinetics of the reaction of oxygen ($'\Delta g$) with tetramethylethylene and 2,5-dimethylfuran in the gas phase. J. Amer. Chem. Soc. 92:2068–2075, 1970.
300. Goetz, A., H. J. R. Stevenson, and O. Preining. The design and performance of the aerosol spectrometer. J. Air Pollut. Control Assoc. 10:378–383, 414, 416, 1970.
301. Gold, P., M. Gold, and S. O. Freedman. Cellular location of carcinoembryonic antigens of the human digestive system. Cancer Res. 28:1331–1334, 1968.
302. Goldsmith, E. I., and J. Moor-Jankowski. Experimental medicine and surgery in primates. Opening remarks. Ann. N.Y. Acad. Sci. 162:5–6, 1969.
303. Goldsmith, J. R. Effects of air pollution on human health, pp. 547–615. In A. C. Stern, Ed. Air Pollution. Vol. 1. Air Pollution and Its Effects. (2nd ed.) New York: Academic Press Inc., 1968.
304. Gollnick, K., and G. O. Schenck. Oxygen as a dienophile, pp. 255–344. In J. Hamer, Ed. 1,4-Cycloaddition Reactions. The Diels–Alder Reaction in Heterocyclic Syntheses. New York: Academic Press Inc., 1967.
305. Good, R. A., and J. Finstad. Essential relationship between the lymphoid system, immunity, and malignancy. Nat. Cancer Inst. Monogr. 31:41–58, 1969.
306. Gorelova, N. D., and P. P. Dikun. 3,4-Benzopyrene content of sausage and smoked fish prepared using fuel gas and coke. Gig. Sanit. 30(7):120–122, 1965. (in Russian)
307. Goshman, L. M., and C. Heidelberger. Binding of tritium-labeled polycyclic hydrocarbons to DNA of mouse skin. Cancer Res. 27:1678–1688, 1967.
308. Gottschalk, R. G. Quantitative studies on tumor production in mice by benzpyrene. Proc. Soc. Exp. Biol. Med. 50:369–373, 1942.

309. Gräf, W., and H. Diehl. Über den naturbedingten Normalpegel kanzerogener polycyclischer Aromate und seine Ursache. Arch. Hyg. Bakt. 150:49–59, 1966. (summary in English)
310. Graham, S., M. L. Levin, A. M. Lilienfeld, and P. Sheebe. Ethnic derivation as related to cancer at various sites. Cancer 16:13–27, 1963.
311. Green, B. Influence of pH and metal ions on the fluorescence of polycyclic hydrocarbons in aqueous DNA solution. Eur. J. Biochem. 14:567–574, 1970.
312. Grimmer, G. Cancerogene Kohlenwasserstoffe in der Umgebung des Menschen. Erdoel Kohle 19:578–583, 1966.
313. Grimmer, G., and A. Hildebrandt. Content of polycyclic hydrocarbons in crude vegetable oils. Chem. Ind. 1967:2000–2002.
314. Griswold, M. H., C. S. Wilder, S. J. Cutler, and E. S. Pollack. Cancer in Connecticut, 1935–1951. Hartford: Connecticut State Department of Health, 1955. 141 pp.
315. Grobstein, C. Differentiation: Environmental factors, chemical and cellular, pp. 463–488. In E. N. Willmer, Ed. Cells and Tissues in Culture. Methods, Biology, and Physiology. Vol. I. London: Academic Press Inc., 1965.
316. Gross, G. P. First Annual Report on Gasoline Composition and Vehicle Exhaust Gas Polynuclear Aromatic Content. Durham, N.C.: U.S. Department of Health, Education, and Welfare, 1970.
317. Gross, M. A., O. G. Fitzhugh, and N. Mantel. Evaluation of safety for food additives: An illustration involving the influence of methyl salicylate on rat reproduction. Biometrics 26:181–194, 1970.
318. Gross, P., E. A. Pfitzer, J. Watson, R. T. P. deTreville, M. Kaschak, E. B. Tolker, and M. A. Babyak. Experimental carcinogenesis. Bronchial intramural adenocarcinomas in rats from x-ray irradiation of the chest. Cancer 23:1046–1060, 1969.
319. Grovenstein, E., Jr., and A. J. Mosher. Reaction of atomic oxygen with aromatic hydrocarbons. J. Amer. Chem. Soc. 92:3810–3812, 1970.
320. Grover, P. L., and P. Sims. Enzyme-catalysed reactions of polycyclic hydrocarbons with deoxyribonucleic acid and protein *in vitro*. Biochem. J. 110:159–160, 1968.
321. Grover, P. L., and P. Sims. Interactions of K-region epoxides of phenanthrene and dibenz[a,h]anthracene with nucleic acids and histone. Biochem. Pharmacol. 19:2251–2259, 1970.
322. Grover, P. L., P. Sims, E. Huberman, H. Marquardt, T. Kuroki, and C. Heidelberger. *In vitro* transformation of rodent cells by K-region derivatives of polycyclic hydrocarbons. Proc. Nat. Acad. Sci. U.S.A. 68:1098–1101, 1971.
323. Guddal, E. Isolation of polynuclear aromatic hydrocarbons from the roots of *Chrysanthemum vulgare* Bernh. Acta Chem. Scand. 13:834–835, 1959.
324. Gunther, F. A., and F. Buzzetti. Occurrence, isolation, and identification of polynuclear hydrocarbons as residues. Residue Rev. 9:90–113, 1965.
325. Gunther, F. A., F. Buzzetti, and W. E. Westlake. Residue behavior of polynuclear hydrocarbons on and in oranges. Residue Rev. 17:81–104, 1967.
326. Haagensen, C. D. Occupational neoplastic disease. Amer. J. Cancer 15:641–703, 1931.
327. Hadidian, Z., T. N. Fredrickson, E. K. Weisburger, J. H. Weisburger, R. M.

Glass, and N. Mantel. Tests for chemical carcinogens. Report on the activity of derivatives of aromatic amines, nitrosamines, quinolines, nitroalkanes, amides, epoxides, aziridines, and purine antimetabolites. J. Nat. Cancer Inst. 41:985–1036, 1968.

328. Haenni, E. O., J. W. Howard, and F. L. Joe, Jr. Dimethyl sulfoxide: A superior analytical extraction solvent for polynuclear hydrocarbons and for some highly chlorinated hydrocarbons. J. Assoc. Offic. Agr. Chem. 45:67–70, 1962.
329. Haenszel, W. Cancer mortality among the foreign-born in the United States. J. Nat. Cancer Inst. 26:37–132, 1961.
330. Haenszel, W., D. B. Loveland, and M. G. Sirken. Lung-cancer mortality as related to residence and smoking histories. I. White males. J. Nat. Cancer Inst. 28:947–1001, 1962.
331. Haenszel, W., S. C. Marcus, and E. G. Zimmerer. Cancer Morbidity in Urban and Rural Iowa. Public Health Monograph 37; Public Health Service Publication 462. Washington, D.C.: U.S. Government Printing Office, 1956. 85 pp.
332. Haenszel, W., and K. E. Traeuber. Lung-cancer mortality as related to residence and smoking histories. II. White females. J. Nat. Cancer Inst. 32:803–838, 1964.
333. Hagstrom, R. M., H. A. Sprague, and E. Landau. The Nashville air pollution study. VII. Mortality from cancer in relation to air pollution. Arch. Environ. Health 15:237–248, 1967.
334. Hakama, M., and E. A. Saxén. Cereal consumption and gastric cancer. Int. J. Cancer 2:265–268, 1967.
335. Hall, W. K. The formation of cation radicals on the surface of silica–alumina catalysts. J. Catalysis 1:53–61, 1962.
336. Hamburg, F. C. Economically feasible alternatives to open burning in railroad freight car dismantling. J. Air Pollut. Control Assoc. 19:477–483, 1969.
337. Hammond, E. C. Quantitative relationship between cigarette smoking and death rates. Nat. Cancer Inst. Monogr. 28:3–8, 1968.
338. Hammond, E. C. Smoking in relation to mortality and morbidity. Findings in first thirty-four months of follow-up in a prospective study started in 1959. J. Nat. Cancer Inst. 32:1161–1188, 1964.
339. Hammond, E. C. Smoking in relation to the death rates of one million men and women. Nat. Cancer Inst. Monogr. 19:127–204, 1966.
340. Hammond, E. C., O. Auerbach, D. Kirman, and L. Garfinkel. Effects of cigarette smoking on dogs. I. Design of experiment, mortality, and findings in lung parenchyma. Arch. Environ. Health 21:740–753, 1970.
341. Hammond, E. C., and D. Horn. Smoking and death rates—report on 44 months of follow-up of 187,783 men. Part I. Total mortality. Part II. Death rates by cause. J.A.M.A. 166:1159–1172, 1294–1308, 1958.
342. Hangebrauck, R. P., R. P. Lauch, and J. E. Meeker. Emissions of polynuclear hydrocarbons from automobiles and trucks. Amer. Ind. Hyg. Assoc. J. 27:47–56, 1966.
343. Hangebrauck, R. P., D. J. von Lehmden, and J. E. Meeker. Sources of Polynuclear Hydrocarbons in the Atmosphere. Public Health Service Publication 999-AP-33. Cincinnati: U.S. Department of Health, Education, and Welfare, 1967. 48 pp.

344. Harris, R. J. C., and G. Negroni. Production of lung carcinomas in C57BL mice exposed to a cigarette smoke and air mixture. Brit. Med. J. 4:637–641, 1967.
345. Härting, F. H., and W. Hesse. Der Lungenkrebs, die Bergkrankheit in den Schneeberger Gruben. Vierteljahresschr. Gerichtl. Med. Oeff. Sanitaetsw. 31:102–132, 313–337, 1879.
346. Hartwell, J. L., and P. Shubik. Survey of Compounds Which Have Been Treated for Carcinogenic Activity. Public Health Service Publication 149. (2nd ed.) Washington, D.C.: U.S. Government Printing Office, 1951. 583 pp.
347. Hatch, T. F. Significant dimensions of the dose-response relationship. Arch. Environ. Health 16:571–578, 1968.
348. Hecker, E. Cocarcinogenic principles from the seed oil of *Croton tiglium* and from other euphorbiaceae. Cancer Res. 28:2338–2348, 1968.
349. Heidelberger, C. Chemical carcinogenesis, chemotherapy: Cancer's continuing core challenges. G. H. A. Clowes Memorial Lecture. Cancer Res. 30:1549–1569, 1970.
350. Heidelberger, C. Studies on the cellular and molecular mechanisms of hydrocarbon carcinogenesis. Eur. J. Cancer 6:161–172, 1970.
351. Heidelberger, C., M. E. Baumann, L. Griesbach, A. Ghobar, and T. M. Vaughan. The carcinogenic activities of various derivatives of dibenzanthracene. Cancer Res. 22:78–83, 1962.
352. Heidelberger, C., H. I. Hadler, and G. Wolf. The metabolic degradation in the mouse of 1,2,5,6-dibenzanthracene-9,10-C^{14}. III. Some quinone metabolites retaining the intact ring system. J. Amer. Chem. Soc. 75:1303–1308, 1953.
353. Heidelberger, C., and P. T. Iype. Malignant transformation *in vitro* by carcinogenic hydrocarbons. Science 155:214–217, 1967.
354. Heidelberger, C., and H. B. Jones. The distribution of radioactivity in the mouse following administration of dibenzanthracene labeled in the 9 and 10 positions with carbon 14. Cancer 1:252–260, 1948.
355. Heidelberger, C., M. R. Kirk, and M. S. Perkins. The metabolic degradation in the mouse of dibenzanthracene labeled in the 9 and 10 positions with carbon 14. Cancer 1:261–275, 1948.
356. Heidelberger, C., and S. M. Weiss. The distribution of radioactivity in mice following administration of 3,4-benzpyrene-5-C^{14} and 1,2,5,6-dibenzanthracene-9,10-C^{14}. Cancer Res. 11:885–891, 1951.
357. Heller, I. Occupational cancers. J. Ind. Hyg. 12:169–197, 1930.
358. Hendricks, N. V., C. M. Berry, J. G. Lione, and J. J. Thorpe. Cancer of the scrotum in wax pressmen. I. Epidemiology. A.M.A. Arch. Ind. Health 19: 524–529, 1959.
359. Henry, S. A. Cancer of the Scrotum in Relation to Occupation. London: Oxford University Press, 1946. 120 pp.
360. Henry, S. A. Occupational cutaneous cancer attributable to certain chemicals in industry. Brit. Med. Bull. 4:389–401, 1947.
361. Hercules, D. M. Chemiluminescence from electron-transfer reactions. Accounts Chem. Res. 2:301–307, 1969.
362. Hercules, D. M. Electron spectroscopy. Anal. Chem. 42:20A–28A, 30A, 32A, 34A–35A, 38A–40A, Jan. 1970.

363. Hercules, D. M. Fluorescence and Phosphorescence Analysis. Principles and Applications. New York: John Wiley & Sons, Inc., 1966. 258 pp.
364. Hercules, D. M. Organic electro-luminescence. In A. Weissberger and B. W. Rossiter, Eds. Physical Methods of Organic Chemistry. Part II. Electrochemical Methods. (4th ed.) New York: Wiley-Interscience. (to be published)
365. Hermann, T. S. Identification of trace amounts of organophosphorous pesticides by frustrated multiple internal reflectance spectroscopy. Appl. Spectrosc. 19:10–14, 1965.
366. Herrold, K. M., and L. J. Dunham. Induction of carcinoma and papilloma of the tracheobronchial mucosa of the Syrian hamster by intratracheal instillation of benzo[a]pyrene. J. Nat. Cancer Inst. 28:467–491, 1962.
367. Herron, J. T., and R. E. Huie. Reactions of O_2 $'\Delta g$ with olefins and their significance in air pollution. Environ. Sci. Technol. 4:685–686, 1970.
368. Hertwig, P. Vererbbare Semisterilität bei Mäusen nach Röntgenbestrahlung, verursacht durch reziproke Chromosomentranslokationen. Z. Indukt. Abstamm. Verebungsl. 79:1–27, 1940.
369. Hesse, G., I. Daniel, and G. Wohlleben. Aluminiumoxyde für die chromatographische Analyse und Versuche zu ihrer Standardisierung. Angew. Chem. 64:103–107, 1952.
370. Heston, W. E., and M. A. Schneiderman. Analysis of dose-response in relation to mechanism of pulmonary tumor induction in mice. Science 117:109–111, 1953.
371. Hidy, G. M. The dynamics of aerosols in the lower troposphere. In Assessment of Airborne Particles. Proceedings of the 3rd University of Rochester Conference on Environmental Toxicology, June 1, 1970. Springfield, Ill.: Charles C Thomas, 1971.
372. Hidy, G. M., and J. R. Brock. An assessment of the global sources of tropospheric aerosols, p. 113. (abstract) In The 2nd International Clean Air Congress; Proceedings Digest. Washington, D.C., Dec. 6–11, 1970.
373. Hidy, G. M., and J. R. Brock. The Dynamics of Aerocolloidal Systems. New York: Pergamon Press, 1970. 371 pp.
374. Hidy, G. M., and S. K. Friedlander. The nature of the Los Angeles aerosol, p. 60. (abstract) In The 2nd International Clean Air Congress; Proceedings Digest. Washington, D.C., Dec. 6–11, 1970.
375. Hilding, A. C. Ciliary streaming in the bronchial tree and the time element in carcinogenesis. New Eng. J. Med. 256:634–640, 1957.
376. Hilding, A. C. On cigarette smoking, bronchial carcinoma and ciliary action. I. Smoking habits and measurement of smoke intake. New Eng. J. Med. 254: 775–781, 1956.
377. Hinds, W. T. Diffusion over coastal mountains of Southern California. Atmos. Environ. 4:107–124, 1970.
378. Hirayama, K. Handbook of Ultraviolet and Visible Absorption Spectra of Organic Compounds. New York: Plenum Press Data Division, 1967. 642 pp.
379. Hitosugi, M. Epidemiological study of lung cancer with special reference to the effect of air pollution and smoking habits. Inst. Public Health Bull. 17: 237–256, 1968.
380. Hochrainer, D., and P. M. Brown. Sizing of aerosol particles by centrifugation. Environ. Sci. Technol. 3:830–835, 1969.

381. Hodkinson, J. R. The effect of particle shape on measures for the size and concentration of suspended and settled particles. Amer. Ind. Hyg. Assoc. J. 26:64–71, 1965.
382. Hoffman, C. S., Jr., R. L. Willis, G. H. Patterson, and E. S. Jacobs. Polynuclear Aromatic Hydrocarbon Emission from Vehicles. Presented to the 160th National Meeting of the American Chemical Society, Los Angeles, California, March 1971.
383. Hoffman, E. F., and A. G. Gilliam. Lung cancer mortality. Geographic distribution in the United States for 1948–1949. Public Health Rep. 69:1033–1042, 1954.
384. Hoffmann, D., E. Theisz, and E. L. Wynder. Studies on the carcinogenicity of gasoline exhaust. J. Air Pollut. Control Assoc. 15:162–165, 1965.
385. Hoffmann, D., and G. Rathkamp. Quantitative determination of 1-alkylindoles in cigaret smoke. Anal. Chem. 42:366–370, 1970.
386. Hoffmann, D., G. Rathkamp, and S. Nesnow. Quantitative determination of 9-methylcarbazoles in cigarette smoke. Anal. Chem. 41:1256–1259, 1969.
387. Hoffmann, D., and E. L. Wynder. A study of air pollution carcinogenesis. II. The isolation and identification of polynuclear aromatic hydrocarbons from gasoline engine exhaust condensate. Cancer 15:93–102, 1962.
388. Hoffmann, D., and E. L. Wynder. Chemical analysis and carcinogenic bioassays of organic particulate pollutants, pp. 187–247. In A. C. Stern, Ed. Air Pollution. Vol. 2. Analysis, Monitoring, and Surveying. (2nd ed.) New York: Academic Press Inc., 1968.
389. Hoffmann, D., and E. L. Wynder. Short-term determination of carcinogenic aromatic hydrocarbons. Anal. Chem. 32:295–296, 1960.
390. Hoffmann, D., and E. L. Wynder. Studies on gasoline engine exhaust. J. Air Pollut. Control Assoc. 13:322–327, 1963.
391. Holzworth, G. C. Atmospheric contaminants at remote California coastal sites. J. Meteorol. 16:68–79, 1969.
392. Holzworth, G. C. Large scale weather influences in community air pollution potential in the United States. J. Air Pollut. Control Assoc. 19:248–254, 1969.
393. Homburger, F. Chemical carcinogenesis in the Syrian golden hamster. A review. Cancer 23:313–338, 1959.
394. Homburger, F., and S. S. Hsueh. Rapid induction of subcutaneous fibrosarcoma by 7,12-dimethylbenz(a)anthracene in an inbred line of Syrian hamsters. Cancer Res. 30:1449–1452, 1970.
395. Horrocks, D. L. Effect of solvent on excimer fluorescence yields. J. Chem. Phys. 51:5443–5448, 1969.
396. Horton, A. W., D. T. Denman, and R. P. Trosset. Carcinogenesis of the skin. II. The accelerating properties of aliphatic and related hydrocarbons. Cancer Res. 17:758–766, 1957.
397. Howard, J. W., and T. Fazio. A review of polycyclic aromatic hydrocarbons in foods. Ind. Med. Surg. 39:435–440, 1970.
398. Huberman, E., and L. Sachs. Cell susceptibility to transformation and cytotoxicity by the carcinogenic hydrocarbon benzo[a]pyrene. Proc. Nat. Acad. Sci. U.S.A. 56:1123–1129, 1966.

399. Huebner, R. J., and G. J. Todaro. Oncogenes of RNA tumor viruses as determinants of cancer. Proc. Nat. Acad. Sci. U.S.A. 64:1087–1094, 1969.
400. Hueper, W. C. Carcinogens in the human environment. Arch. Path. 71:237–267, 355–380, 1961.
401. Hueper, W. C. Chemically induced skin cancers in man. Nat. Cancer Inst. Monogr. 10:377–391, 1963.
402. Hueper, W. C. Occupational Tumors and Allied Diseases. Springfield, Ill.: Charles C Thomas, 1942. 896 pp.
403. Hueper, W. C., and W. D. Conway. Chemical Carcinogenesis and Cancers. Springfield, Ill.: Charles C Thomas, 1964. 744 pp.
404. Hueper, W. C., P. Kotin, E. C. Tabor, W. W. Payne, H. Falk, and E. Sawicki. Carcinogenic bioassays on air pollutants. Arch. Path. 74:89–116, 1962.
405. Huggins, C., L. Grand, and R. Fukunishi. Aromatic influences on the yields of mammary cancers following administration of 7,12-dimethylbenz(a)-anthracene. Proc. Nat. Acad. Sci. U.S.A. 51:737–742, 1964.
406. Huggins, C., L. C. Grand, and F. P. Brillantes. Mammary cancer induced by a single feeding of polynuclear hydrocarbons, and its suppression. Nature 189:204–207, 1961.
407. Huggins, C., and N. C. Yang. Induction and extinction of mammary cancer. Science 137:257–262, 1962.
408. Hünigen, E., N. Jaskulla, and K. Wettig. Die Herabsetzung Krebsfördernder Schadstoffe in Ottomotoren-Abgasen durch Kraftstoffzusätz und Schmierstoffauswahl, pp. 191–193. Proceedings of the International Clean Air Congress. Part 1. London, 1966.
409. Husar, R., N. Barsic, M. Tomaides, B. Y. H. Liu, and K. T. Whitby. Five spectra and miscellaneous experiments, pp. 3–133. In K. T. Whitby, Ed. Aerosol Measurements in Los Angeles. Part 3. Minnesota Aerosol Analyzing System and Miscellaneous Experiments Particle Laboratory Publication 141. Minneapolis: University of Minnesota, 1970.
410. Imagawa, D. T., M. Yoshimori, and J. M. Adams. The death rate in mice with pulmonary tumors induced with urethan and influenza virus. Proc. Amer. Assoc. Cancer Res. 2:217, 1957. (abstract)
411. Inbar, M., and L. Sachs. Structural difference in sites on the surface membrane of normal and transformed cells. Nature 223:710–712, 1969.
412. Innes, J. R. M., B. M. Ulland, M. G. Valerio, L. Petrucelli, L. Fishbein, E. R. Hart, A. J. Pallota, R. R. Bates, H. L. Falk, J. J. Gart, M. Klein, I. Mitchell, and J. Peters. Bioassay of pesticides and industrial chemicals for tumorigenicity in mice. A preliminary note. J. Nat. Cancer Inst. 42:1101–1114, 1969.
413. Inscoe, M. N. Photochemical changes in thin-layer chromatograms of polycyclic, aromatic hydrocarbons. Anal. Chem. 36:2505–2606, 1964.
414. Isbell, A. F., Jr., and D. T. Sawyer. Gas–solid chromatography with salt-modified porous silica beads. Anal. Chem. 41:1381–1387, 1969.
415. Ishikawa, S., D. H. Bowden, V. Fisher, and J. P. Wyatt. The "emphysema profile" in two midwestern cities in North America. Arch. Environ. Health 18:660–666, 1969.
416. Jerina, D. M., J. W. Daly, B. Witkop, P. Zaltzman-Nirenberg, and S. Uden-

friend. 1,2-Naphthalene oxide as an intermediate in the microsomal hydroxylation of naphthalene. Biochemistry 9:147–155, 1970.
417. Johnson, B. H., and T. Aczel. Analysis of complex mixtures of aromatic compounds by high-resolution mass spectrometry at low-ionizing voltages. Anal. Chem. 39:682–685, 1967.
418. Jones, P. R., and S. Siegel. Temperature effects on the phosphorescence of aromatic hydrocarbons in poly(methylmethacrylate). J. Chem. Phys. 50: 1134–1140, 1969.
419. Jull, J. W. The induction of tumours of the bladder epithelium in mice by the direct application of a carcinogen. Brit. J. Cancer 5:328–330, 1951.
420. Junge, C. E. Atmospheric radioactivity, pp. 209–288. In Air Chemistry and Radioactivity. International Geophysics Series. Vol. 4. New York: Academic Press Inc., 1963.
421. Junge, C. E. Comments on "Concentration and size distribution measurements of atmospheric aerosols and a test of the theory of self-preserving size distributions." J. Atmos. Sci. 26:603–608, 1969.
422. Kahn, H. A. The Dorn study of smoking and mortality among U.S. Veterans: Report on eight and one-half years of observation. Nat. Cancer Inst. Monogr. 19:1–125, 1966.
423. Kao, F-T., and T. T. Puck. Genetics of somatic mammalian cells. VII. Induction and isolation of nutritional mutants in Chinese hamster cells. Proc. Nat. Acad. Sci. U.S.A. 60:1275–1281, 1968.
424. Kao, F-T., and T. T. Puck. Genetics of somatic mammalian cells. IX. Quantitation of mutagenesis by physical and chemical agents. J. Cell Physiol. 74:245–258, 1969.
425. Kao, S.-K, and D. Henderson. Large-scale dispersion of clusters of particles in various flow patterns. J. Geophys. Res. 75:3104–3113, 1970.
426. Kato, R. Chromosome breakage induced by a carcinogenic hydrocarbon in Chinese hamster cells and human leukocytes *in vitro*. Hereditas 59:120–141, 1968.
427. Kato, R., M. Bruze, and Y. Tegner. Chromosome breakage induced *in vivo* by a carcinogenic hydrocarbon in bone marrow cells of the Chinese hamster. Hereditas 61:1–8, 1969.
428. Katz, M. Measurement of Air Pollutants. Guide to the Selection of Methods. Geneva: World Health Organization, 1969. 123 pp.
429. Kawai, M., H. Amamoto, and K. Harada. Epidemiologic study of occupational lung cancer. Arch. Environ. Health 14:859–864, 1967.
430. Kawai, M., T. Matsuyama, H. Amamoto, and M. Nakamura. A study of occupational lung cancers of the generator gas plant workers in Yawata Iron and Steel Works, Japan. J. Labour Hyg. Iron Steel Ind. 10:5–9, 1961. (in Japanese)
431. Kearns, D. R., A. U. Khan, C. K. Duncan, and A. H. Maki. Detection of the naphthalene-photosensitized generation of singlet ($'\Delta$ g) oxygen by paramagnetic resonance spectroscopy. J. Amer. Chem. Soc. 91:1039–1040, 1969.
432. Kelly, M. G., and R. W. O'Gara. Induction of tumors in newborn mice with dibenz[a,h]anthracene and 3-methylcholanthrene. J. Nat. Cancer Inst. 26: 651–679, 1961.
433. Kelly, M. G., R. W. O'Gara, R. H. Adamson, K. Gadekar, C. C. Botkin, W. H.

Reese, Jr., and W. T. Kerber. Induction of hepatic cell carcinomas in monkeys with *N*-nitrosodiethylamine. J. Nat. Cancer Inst. 36:323–351, 1966.
434. Kennaway, E. L., and N. M. Kennaway. A further study of the incidence of cancer of the lung and larynx. Brit. J. Cancer 1:260–298, 1947.
435. Kent, S. P. Spontaneous and induced malignant neoplasms in monkeys. Ann. N.Y. Acad. Sci. 85:819–827, 1960.
436. Kirkland, J. J. High-speed liquid chromatography with controlled surface porosity support. J. Chromatogr. Sci. 7:7–12, 1969.
437. Klein, G. Experimental studies in tumor immunology. Fed. Proc. 28:1739–1753, 1969.
438. Klein, G. Tumor-specific transplantation antigens. G. H. A. Clowes Memorial Lecture. Cancer Res. 28:625–635, 1968.
439. Klein, M. Development of hepatomas in inbred albino mice following treatment with 20-methylcholanthrene. Cancer Res. 19:1109–1113, 1959.
440. Klein, M. Induction of skin tumors in the mouse with minute doses of 9, 10-dimethyl-1,2-benzanthracene alone or with croton oil. Cancer Res. 16:123–127, 1956.
441. Klein, M. Influence of low dose of 2-acetylaminofluorene on liver tumorigenesis in mice. Proc. Soc. Exp. Biol. Med. 101:637–638, 1959.
442. Kohn-Speyer, A. C. Effect of ultra-violet radiation on the incidence of tar cancer in mice. Lancet 2:1305–1306, 1929.
443. Koller, P. C. Segmental interchange in mice. Genetics 29:247–263, 1944.
444. Koller, P. C., and C. A. Auerbach. Chromosome breakage and sterility in the mouse. Nature 148:501–502, 1941. (letter to the editor)
445. Kooyman, E. C., and E. Farenhorst. The relative reactivities of polycyclic aromatics towards trichloromethyl radicals. Trans. Faraday Soc. 49:58–67, 1953.
446. Kortum, G., and W. Braun. Photochemische Reaktionen des Anthracens in adsorbiertem Zustand. Justus Liebigs Ann. Chem. 632:104–115, 1960.
447. Kotin, P. The influence of pathogenic viruses on cancers induced by inhalation. Can. Cancer Conf. 6:475–498, 1966.
448. Kotin, P. The role of atmospheric pollution in the pathogenesis of pulmonary cancer. A review. Cancer Res. 16:375–393, 1956.
449. Kotin, P., and H. L. Falk. Atmospheric factors in pathogenesis of lung cancer. Adv. Cancer Res. 7:475–514, 1963.
450. Kotin, P., and H. L. Falk. The role and action of environmental agents in the pathogenesis of lung cancer. I. Air pollutants. Cancer 12:147–163, 1959.
451. Kotin, P., and H. L. Falk. The role and action of environmental agents in the pathogenesis of lung cancer. II. Cigaret smoke. Cancer 13:250–262, 1960.
452. Kotin, P., H. L. Falk, and R. Busser. Distribution, retention, and elimination of C^{14}-3,4-benzpyrene after administration to mice and rats. J. Nat. Cancer Inst. 23:541–555, 1959.
453. Kotin, P., H. L. Falk, P. Mader, and M. Thomas. Aromatic hydrocarbons. I. Presence in the Los Angeles atmosphere and the carcinogenicity of exhaust extracts. A.M.A. Arch. Ind. Hyg. Occup. Med. 9:153–163, 1964.
454. Kotin, P., H. L. Falk, and C. J. McCammon. The experimental induction of pulmonary tumors and changes in the respiratory epithelium in C57BL mice

following their exposure to an atmosphere of ozonized gasoline. Cancer 11:473-481, 1958.

455. Kotin, P., H. L. Falk, and M. Thomas. Aromatic hydrocarbons. II. Presence in the particulate phase of gasoline-engine exhausts and the carcinogenicity of exhaust extracts. A.M.A. Arch. Ind. Hyg. Occup. Med. 9:164-177, 1954.
456. Kotin, P., and D. V. Wiseley. Production of lung cancer in mice by inhalation exposure to influenza virus and aerosols of hydrocarbons. Progr. Exp. Tumor Res. 3:186-215, 1963.
457. Kotrappa, P. Shape factors for aerosols of coal, UO_2 and Th_2 in the respirable size range. In Assessment of Airborne Particles. Proceedings of the 3rd University of Rochester Conference on Environmental Toxicology, June 1970. Springfield, Ill.: Charles C Thomas, 1971.
458. Kracht, H. J., U. E. Klein, and M. Baghirzade. Erfahrungen mit dem Talgdrüsenschwundtest. Verh. Deutsch. Ges. Path. 45:170-174, 1961.
459. Kreyberg, L. 3:4-Benzopyrene in industrial air pollution. Some reflexions. Brit. J. Cancer 13:618-622, 1959.
460. Kreyberg, L. Occurrence and aetiology of lung cancer in Norway in the light of pathological anatomy. Brit. J. Prev. Soc. Med. 10:145-158, 1956.
461. Kreyberg, L. The occurrence of lung cancer in Norway. Brit. J. Cancer 8:209-214, 1954.
462. Kummler, R. H., and M. H. Bortner. Production of O_2 ($'\Delta$ g) by energy transfer from excited benzaldehyde. Environ. Sci. Technol. 3:944-946, 1969.
463. Kuratsune, M. Benzo[a]pyrene content of certain pyrogenic materials. J. Nat. Cancer Inst. 16:1485-1496, 1956.
464. Kuratsune, M., and T. Hirohata. Decomposition of polycyclic aromatic hydrocarbons under laboratory illuminations. Nat. Cancer Inst. Monogr. 9:117-125, 1962.
465. Kuratsune, M., and W. C. Hueper. Polycyclic aromatic hydrocarbons in roasted coffee. J. Nat. Cancer Inst. 24:463-469, 1960.
466. Kuroda, S. Occupational pulmonary cancer of generator gas workers. Ind. Med. Surg. 6:304-306, 1937.
467. Kuroki, T., and H. Sato. Transformation and neoplastic development *in vitro* of hamster embryonic cells by 4-nitroquinoline-1-oxide and its derivatives. J. Nat. Cancer Inst. 41:53-71, 1968.
468. Kuschner, M. The causes of lung cancer. The J. Burns Amberson Lecture. Amer. Rev. Resp. Dis. 98:573-590, 1968.
469. Kutscher, W., R. Tomingas, and H. P. Weisfeld. Untersuchungen über die Schädlichkeit von Russen unter besonderer Berücksichtigung ihrer cancerogenen Wirkung. 5. Mitteilung: Über die Ablösbarkeit von 3,4-Benzpyren durch Blutserum und einege Eiweissfaktoren des Serums. Arch. Hyg. Bakt. 151:646-655, 1967. (summary in English)
470. Lamb, R. G., and M. Neiburger. An interim version of a generalized urban air pollution model. Atmos. Environ. 5:239-264, 1971.
471. Lane, W. R., and B. R. D. Stone. Structure and density of particulate aggregates, pp. 417-426. In Proceedings of the International Conference on Mechanisms of Corrosion by Fuel Impurities. Marchwood, near Southampton, Hampshire, England. London: Butterworth & Co., 1963.

472. Lane-Petter, W., Ed. Animals for Research. Principles of Breeding and Management. New York: Academic Press Inc., 1963. 531 pp.
473. Láng, L., Ed. Absorption Spectra in the Ultraviolet and Visible Region. New York: Academic Press Inc., 1961–.
474. Laskin, S., M. Kuschner, and R. T. Drew. Studies in pulmonary carcinogenesis, pp. 321–350. In M. G. Hanna, Jr., P. Nettesheim, and J. R. Gilbert, Eds. Inhalation Carcinogenesis. AEC Symposium Series, No. 18. Washington, D.C.: U.S. Atomic Energy Commission, 1970.
475. Lasnitzki, I. Growth pattern of the mouse prostate gland in organ culture and its response to sex hormones, vitamin A, and 3-methylcholanthrene. Nat. Cancer Inst. Monogr. 12:381–403, 1963.
476. Lasnitzki, I. Observations on effects of condensates from cigarette smoke on human foetal lung *in vitro.* Brit. J. Cancer 12:547–552, 1958.
477. Lasnitzki, I. The action of hormones on cell and organ cultures, pp. 591–658. In E. N. Willmer, Ed. Cells and Tissues in Culture. Methods, Biology, and Physiology. Vol. 1. London: Academic Press Inc., 1965.
478. Lasnitzki, I. The effect of a hydrocarbon-enriched fraction of cigarette smoke condensate on human fetal lung grown *in vitro.* Cancer Res. 28:510–516, 1968.
479. Lasnitzki, I. The effect of a hydrocarbon-enriched fraction from cigarette smoke on mouse tracheas grown *in vitro.* Brit. J. Cancer 22:105–109, 1968.
480. Lasnitzki, I. The effect of 3:4-benzopyrene on human foetal lung grown *in vitro.* Brit. J. Cancer 10:510–516, 1956.
481. Lasnitzki, I. The effects of actinomycin D and methylcholanthrene on the cytology and RNA and protein synthesis in prostatic epithelium grown *in vitro.* Cancer Res. 29:318–326, 1969.
482. Lasnitzki, I. The influence of A hypervitaminosis on the effect of 20-methylcholanthrene on mouse prostate glands grown *in vitro.* Brit. J. Cancer 9:434–441, 1955.
483. Lawther, P. J., B. T. Commins, and R. E. Waller. A study of the concentrations of polycyclic aromatic hydrocarbons in gas works retort houses. Brit. J. Ind. Med. 22:13–20, 1965.
484. Ledford, C. J., C. P. Morie, and C. A. Glover. Separation of polynuclear aromatic hydrocarbons in cigaret smoke high-resolution liquid chromatography. Tobacco Sci. 14:158–160, 1970.
485. Leighton, P. A. Photochemistry of Air Pollution. New York: Academic Press Inc., 1961. 300 pp.
486. Leitch, A. Mule-spinners' cancer and mineral oils. Brit. Med. J. 2:941–943, 1924.
487. Leiter, J., and M. J. Shear. Production of tumors in mice with tars from city air dusts. J. Nat. Cancer Inst. 3:167–174, 1942.
488. Leiter, J., M. B. Shimkin, and M. J. Shear. Production of subcutaneous sarcomas in mice with tars extracted from atmospheric dusts. J. Nat. Cancer Inst. 3:155–165, 1942.
489. Lemke, E. E., C. Thomas, and W. E. Zwaicker. Profile of Air Pollution in Los Angeles County. Los Angeles County Air Pollution Control District, 1969.

490. Leubner, I. H. Comment. (Observed phosphorescence lifetimes and glass relaxation at 77°K.) J. Phys. Chem. 73:2088–2090, 1969.
491. Leuchtenberger, C., and R. Leuchtenberger. In L. Severi, Ed. Lung Tumours in Animals. Proceedings of the Third Quadrennial International Conference on Cancer. June 24–29, 1965. Perugia, Italy: University of Perugia, 1966.
492. Levin, M. L., W. Haenszel, B. E. Carroll, P. R. Gerhardt, V. H. Handy, and S. C. Ingraham, II. Cancer incidence in urban and rural areas of New York State. J. Nat. Cancer Inst. 24:1243–1257, 1960.
493. Levy, B. M. Induction of fibrosarcoma in the primate *Tamarinus nigricollis.* Nature 200:182–183, 1963.
494. Lew, E. A. Cancer of the respiratory tract. Recent trends in mortality. J. Int. Coll. Surg. 24:12–27, 1955.
495. Leymann. Steinkohlenteer- oder Steinkohlenteer-Pechkrätze und -Krebs. Zentralbl. Gewerbehyg. 5:2–7, 35–40, 51–55, 170–174, 1917.
496. Lijinsky, W., and P. Shubik. The detection of polycyclic aromatic hydrocarbons in liquid smoke and some foods. Toxicol. Appl. Pharmacol. 7:337–343, 1965.
497. Lione, J. G., and J. S. Denholm. Cancer of the scrotum in wax pressmen. II. Clinical observations. A.M.A. Arch. Ind. Health 19:530–539, 1959.
498. Lipsett, F. R. Energy transfer in polyacene solid solutions. VIII. A bibliography for 1968. Molec. Cryst. Liquid Cryst. 6:175–204, 1969.
499. Lipsett, F. R., G. Bechthold, F. D. Blair, F. V. Cairns, and D. H. O'Hara. Apparatus for measurement of luminescence spectra with a digital recording system. Appl. Opt. 9:1312–1318, 1970.
500. Little, J. B., B. N. Grossman, and W. F. O'Toole. Respiratory carcinogenesis in hamsters induced by polonium-210 alpha radiation and benzo[a]pyrene, pp. 383–392. In P. Nettesheim, M. G. Hanna, Jr., and J. W. Deatherage, Jr., Eds. Morphology of Experimental Respiratory Carcinogenesis. AEC Symposium Series, No. 21. Springfield, Va.: National Technical Information Service, Department of Commerce, 1970.
501. Littlewood, A. B. The coupling of gas chromatography with methods of identification. I. Mass spectrometry. Chromatographia 1968:37–42.
502. Lloyd, J. W. Long-term mortality study of steelworkers. V. Respiratory cancer in coke plant workers. J. Occup. Med. 13:53–68, 1971.
503. Ludwig, J. H., G. B. Morgan, and T. B. McMullen. Trends in urban air quality. Trans. Amer. Geophys. Union 51:468–475, 1970.
504. Luther, M. Cancer in Subhuman Primates. Public Health Service Publication 1138. Bibliography Series, No. 44. Washington, D.C.: U.S. Government Printing Office, 1962. 61 pp.
505. Lyons, M. J. Assay of possible carcinogenic hydrocarbons from cigarette smoke. Nature 177:630–631, 1956.
506. MacMahon, R. The ethnic distribution of cancer mortality in New York City, 1955. Acta Un. Int. Cancr. 16:1716–1724, 1960.
507. Maher, V. M., E. C. Miller, J. A. Miller, and W. Szybalski. Mutations and decreases in density of transforming DNA produced by derivatives of the carcinogens 2-acetyl-aminofluorene and *N*-methyl-4-aminoazobenzene. Molec. Pharmacol. 4:411–426, 1968.

508. Maisin, J., and A. De Jonghe. Au sujet de l'action de la lumière et de l'ozone sur certains corps cancérigènes. C. R. Soc. Biol. 117:111–114, 1934.
509. Maisin, J., and E. Picard. Production experimentale d'un épithélioma épidermoïde de la vessie chez le rat blanc. C. R. Soc. Biol. 91:799–801, 1924.
510. Mallet, L. Recherche des hydrocarbures polybenzéniques du type benzo-3.4-pyrène dans la faune des milieux marins (Manche, Atlantique et Méditerranée). C. R. Acad. Sci. (Paris) 253:168–170, 1961.
511. Mallet, L., and M. Héros. Pollution des terres végétales par les hydrocarbures polybenzéniques du type benzo-3.4-pyrène. C. R. Acad. Sci. (Paris) 254:958–960, 1962.
512. Mancuso, T. F., and E. J. Coulter. Cancer mortality among native white, foreign-born white, and nonwhite male residents of Ohio: Cancer of the lung, larynx, bladder, and central nervous system. J. Nat. Cancer Inst. 20:79–105, 1958.
513. Mancuso, T. F., E. M. MacFarlane, and J. D. Porterfield. Distribution of cancer mortality in Ohio. Amer. J. Public Health 45:58–70, 1955.
514. Mann, C. K., and K. K. Barnes. Electrochemical Reactions in Nonaqueous Systems. New York: Marcel Dekker, 1970. 560 pp.
515. Manos, N. E., and G. F. Fisher. An index of air pollution and its relation to health. J. Air Pollut. Control Assoc. 9:5–11, 1959.
516. Mantel, N. Some statistical viewpoints in the study of carcinogenesis. Progr. Exp. Tumor Res. 11:431–443, 1969.
517. Mantel, N., and W. R. Bryan. "Safety" testing of carcinogenic agents. J. Nat. Cancer Inst. 27:455–470, 1961.
518. Mantel, N., W. E. Heston, and J. M. Gurian. Thresholds in linear dose-response models for carcinogenesis. J. Nat. Cancer Inst. 27:203–215, 1961.
519. Marchesani, V. J., T. Towers, and H. C. Wohlers. Minor sources of air pollutant emissions. J. Air Pollut. Control Assoc. 20:19–22, 1970.
520. Marchetti, A. P., and D. R. Kearns. Investigation of singlet–triplet transitions by the phosphorescence excitation method. IV. The singlet–triplet absorption spectra of aromatic hydrocarbons. J. Amer. Chem. Soc. 89:768–777, 1967.
521. Mastromatteo, E. Cutting oils and squamous-cell carcinoma. Part I. Incidence in a plant with a report of six cases. Brit. J. Ind. Med. 12:240–243, 1955.
522. Masuda, Y., and M. Kuratsune. Photochemical oxidation of benzo[a]pyrene. Air Water Pollut. 10:805–811, 1966.
523. McCarthy, W. J., and J. D. Winefordner. Phosphorimetry as a means of chemical analysis, pp. 371–442. In G. G. Guilbault, Ed. Fluorescence: Theory, Instrumentation, and Practice. New York: Marcel Dekker, Inc., 1967.
524. McDonald, S., Jr., and D. L. Woodhouse. On the nature of mouse lung adenomata, with special reference to the effects of atmospheric dust on the incidence of these tumours. J. Path. Bact. 54:1–12, 1942.
525. McFadden, W. H. Introduction of gas-chromatographic samples to a mass spectrometer. Separation Sci. 1:723–746, 1966.
526. McFadden, W. H. Mass-spectrometric analysis of gas-chromatographic effluents, pp. 265–332. In J. C. Giddings and R. A. Keller, Eds. Advances in Chromatography. Vol. 4. New York: Marcel Dekker, Inc., 1967.
527. McIntire, K. R., and G. L. Princler. Prolonged adjuvant stimulation in germ-

free BALB/C mice: Development of plasma cell neoplasia. Immunology 17: 481–487, 1969.
528. McKee, H. C., and W. A. McMahon. Polynuclear Aromatic Content of Vehicle Emissions. Project 21-2139, TR-1. San Antonio, Texas: Southwest Research Institute, 1967. 37 pp.
529. Meranze, D. R., M. Gruenstein, and M. B. Shimkin. Effect of age and sex on the development of neoplasms in Wistar rats receiving a single intragastric instillation of 7,12-dimethylbenz(a)anthracene. Int. J. Cancer 4:480–486, 1969.
530. Meyer, E. C., and Liebow, A. A. Relationship of interstitial pneumonia, honeycombing and atypical epithelial proliferation to cancer of the lung. Cancer 18:322–351, 1965.
531. Mihalyi, E. Estimation of stray-light and fluorescence effects in differential spectroscopy. Arch. Biochem. 110:325–330, 1965.
532. Miller, J. A. Carcinogenesis by chemicals: An overview. G. H. A. Clowes Memorial Lecture. Cancer Res. 30:559–576, 1970.
533. Miller, J. A., and E. C. Miller. The carcinogenic aminoaza dyes. Adv. Cancer Res. 1:340–396, 1953.
534. Miller, L., W. E. Smith, and S. W. Berliner. Tests for effect of asbestos on benzo[a]pyrene carcinogenesis in the respiratory tract. Ann. N.Y. Acad. Sci. 132:489–500, 1965.
535. Mirvish, S., G. Cividalli, and I. Berenblum. Slow elimination of urethan in relation to its high carcinogenicity in newborn mice. Proc. Soc. Exp. Biol. Med. 116:265–268, 1964.
536. Mold, J. D., T. B. Walker, and L. G. Veasey. Selective separation of polycyclic aromatic compounds by countercurrent distribution with a solvent system containing tetramethyluric acid. Anal. Chem. 35:2071–2074, 1963.
537. Mondal, S., and C. Heidelberger. *In vitro* malignant transformation by methylcholanthrene of the progeny of single cells derived from C3H mouse prostate. Proc. Nat. Acad. Sci. U.S.A. 65:219–225, 1970.
538. Mondal, S., P. T. Iype, L. M. Griesbach, and C. Heidelberger. Antigenicity of cells derived from mouse prostate cells after malignant transformation *in vitro* by carcinogenic hydrocarbons. Cancer Res. 30:1593–1597, 1970.
539. Moore, G. E., and F. G. Bock. A summary of research techniques for investigating the cigarette smoking–lung cancer problem. Surgery 39: 120–130, 1956.
540. Moriconi, E. J., B. Rakoczy, and W. F. O'Connor. Ozonolysis of polycyclic aromatics. VIII. Benzo[a]pyrene. J. Amer. Chem. Soc. 83:4618–4623, 1961.
541. Moriconi, E. J., and L. Salce. Ozonation of polycyclic aromatics. XV. Carcinogenicity and K- and/or L-region additivity towards ozone. Adv. Chem. Ser. 77:65–73, 1968.
542. Moriconi, E. J., and F. A. Spano. Heteropolar ozonization of aza-aromatics and their N-oxides. J. Amer. Chem. Soc. 86:38–46, 1964.
543. Moriconi, E. J., and L. B. Taranko. Ozonolysis of polycyclic aromatics. X. 7,12-Dimethylbenz[a]anthracene. J. Org. Chem. 28:1831–1834, 1963.
544. Moriconi, E. J., and L. B. Taranko. Ozonolysis of polycyclic aromatics. XI. 3-Methylcholanthrene. J. Org. Chem. 28:2526–2533, 1963.

545. Morton, J. J., E. M. Luce-Clausen, and E. B. Mahoney. The effect of visible light on the development of tumors induced by benzpyrene in the skin of mice. Amer. J. Roentgen. 43:896–898, 1940.
546. Morton, J. J., E. M. Luce-Clausen, and E. B. Mahoney. Visible light and skin tumors induced with benzpyrene in mice. Cancer Res. 2:256–260, 1942.
547. Moscona, A., O. A. Trowell, and E. N. Willmer. Methods, pp. 19–98. In E. N. Willmer, Ed. Cells and Tissues in Culture. Methods, Biology, and Physiology. Vol. 1. London: Academic Press Inc., 1965.
548. Mueller, P. K., H. L. Helwig, A. E. Alcocer, W. K. Gong, and E. E. Jones. Concentration of fine particles and lead in car exhaust, pp. 60–77. In Symposium on Air-Pollution Measurement Methods. ASTM Special Technical Publication 352. Philadelphia: American Society for Testing Materials, 1964.
549. Mueller, P. K., R. W. Mosley, and L. B. Pierce. Carbonate and Non-carbonate Carbon in Atmospheric Particles. Air Industrial Laboratory Report 72. Berkeley: California State Department of Health, 1970. 16 pp.
550. Muhich, A. J., A. J. Klee, and P. W. Britton. 1968 National Survey of Community Solid Waste Practices. Public Health Service Publication 1866. Cincinnati: U.S. Department of Health, Education, and Welfare, 1968.
551. Mukai, F., and W. Troll. The mutagenicity and initiating activity of some aromatic amine metabolites. Ann. N.Y. Acad. Sci. 163:828–836, 1969.
552. Murray, R. W., and M. L. Kaplan. Gas-phase reactions of singlet oxygen from a chemical source. J. Amer. Chem. Soc. 90:4161–4162, 1968.
553. Murray, R. W., W. C. Lumma, Jr., and J. W.-P. Lin. Singlet oxygen sources in ozone chemistry. Decomposition of oxygen-rich intermediates. J. Amer. Chem. Soc. 92:3205–3207, 1970.
554. Nakamizo, M., and Y. Kanda. Fluorescence spectra of organic compounds in solution—I. On the positions of the O,O-bands of the fluorescence spectra. Spectrochim. Acta 19:1235–1248, 1963.
555. Nebert, D. W., and H. V. Gelboin. Substrate-inducible microsomal arylhydroxylase in mammalian cell culture. I. Assay and properties of induced enzyme. J. Biol. Chem. 243:6242–6249, 1968.
556. Nebert, D. W., and H. V. Gelboin. Substrate-inducible microsomal aryl hydroxylase in mammalian cell culture. II. Cellular responses during enzyme induction. J. Biol. Chem. 243:6250–6261, 1968.
557. Nebert, D. W., and H. V. Gelboin. The *in vivo* and *in vitro* induction of aryl hydrocarbon hydroxylase in mammalian cells of different species, tissues, strains, and developmental and hormonal states. Arch. Biochem. Biophys. 134:76–89, 1969.
558. Nebert, D. W., and H. V. Gelboin. The role of ribonucleic acid and protein synthesis in microsomal aryl hydrocarbon hydroxylase induction in cell culture. The independence of transcription and translation. J. Biol. Chem. 245:160–168, 1970.
559. Nebert, D. W., J. Winker, and H. V. Gelboin. Aryl hydrocarbon hydroxylase activity in human placenta from cigarette smoking and nonsmoking women. Cancer Res. 29:1763–1769, 1969.
560. Nettesheim, P., M. G. Hanna, Jr., D. H. Doherty, R. F. Newell, and A. Hellman. Effects of chronic exposure to artificial smog and chromium oxide dust on the incidence of lung tumors in mice, pp. 305–320. In M. G. Hanna,

Jr., P. Nettesheim, and J. R. Gilbert, Eds. Inhalation Carcinogenesis. AEC Symposium Series, No. 18. Washington, D.C.: U. S. Atomic Energy Commission, 1970.

561. Neuberger, H., C. L. Hosier, and W. C. Kocmond. Vegetation as aerosol filter, pp. 693–702. In S. W. Tromp and W. H. Weihe, Eds. Biometeorology. Vol. 2. Proceedings of the Third International Biometeorological Congress held at Pau, S. France, 1–7 September 1963, organized by The International Society of Biometeorology. Oxford: Pergamon Press Ltd., 1967.

562. Neve, E. F. Causation of cancer. Practitioner 122:355–359, 1929.

563. Newman, M. S., and S. Blum. A new cyclization reaction leading to epoxides of aromatic hydrocarbons. J. Amer. Chem. Soc. 86:5598–5600, 1964.

564. Noonan, F. M., and B. Linsky. Internal reflection spectroscopy applied to air pollution. Atmos. Environ. 4:125–128, 1970.

565. Noyes, W. F. Carcinogen induced neoplasia with metastasis in a South American primate, *Saguinus oedipus*. Proc. Soc. Exp. Biol. Med. 131:223–225, 1969.

566. Noyes, W. F. Carcinogen-induced sarcoma in the primitive primate, *Tupaia glis*. Proc. Soc. Exp. Biol. Med. 127:594–596, 1968.

567. Nurmukhametov, R. N. Electronic absorption and luminescence spectra of aromatic polycyclic hydrocarbons. Russ. Chem. Rev. 35:469–486, 1966.

568. O'Donnell, H., T. L. Montgomery, and M. Corn. Routine assessment of the particle size–weight distribution of urban aerosols. Atmos. Environ. 4:1–7, 1970.

569. O'Donovan, W. J. Epitheliomatous ulceration among tar workers. Brit. J. Derm. 32:215–228, 245–252, 1920.

570. Offen, H. W., and D. E. Hein. Environmental effects on phosphorescence. VI. Matrix site effects for triphenylene. J. Chem. Phys. 50:5274–5278, 1969.

Old, L. J., E. A. Boyse, D. A. Clarke, and E. A. Carswell. Antigenic properties of chemically induced tumors. Part II. Antigens of tumor cells. Ann. N.Y. Acad. Sci. 101:80–106, 1962.

572. Olsen, D., and J. L. Haynes. Preliminary Air Pollution Survey of Organic Carcinogens. A Literature Review. National Air Pollution Control Administration Publication APTD 69–43. Raleigh, N.C.: U.S. Department of Health, Education, and Welfare, 1969. 117 pp.

573. Oró, J., and J. Han. Application of combined chromatography–mass spectrometry to the analysis of aromatic hydrocarbons formed by pyrolysis of methane. J. Gas Chromatogr. 5:480–485, 1967.

574. Palekar, L., M. Kuschner, and S. Laskin. The effect of 3-methylcholanthrene on rat trachea in organ culture. Cancer Res. 28:2098–2104, 1968.

575. Panteleev, V. V., M. L. Petukh, O. I. Putrenko, T. A. Yankovskaya, and A. A. Yankovskii. The sensitivity of emission spectral analysis when using a laser. Zh. Prikl. Spektrosk. 12:1106–1108, 1970.

576. Paoletti, J., and J.-B. Le Pecq. Corrections for instrumental errors in measurement of fluorescence and polarization of fluorescence. Anal. Biochem. 31:33–41, 1969.

577. Parker, C. A. Photoluminescence of Solutions. Amsterdam: Elsevier Publishing Company, 1968. 544 pp.

578. Parker, C. A., and C. G. Hatchard. Photoreaction of benz[a]pyrene in solutions containing polymer. Photochem. Photobiol. 5:699–703, 1966.
579. Parker, C. A., and W. T. Rees. Determination of 3:4-Benzpyrene in the Atmosphere of a Submarine. AML Report A/68 (M). Poole, England: Admiralty Materials Laboratory, 1966. 9 pp.
580. Parmenter, C. S., and J. D. Rau. Fluorescence quenching in aromatic hydrocarbons by oxygen. J. Chem. Phys. 51:2242–2246, 1969.
581. Pasceri, R., and S. K. Friedlander. Measurements of the particle size distribution of the atmospheric aerosol: II. Experimental results and discussion. J. Atmos. Sci. 22:577–584, 1965.
582. Passey, R. D. Experimental soot cancer. Brit. Med. J. 2:1112–1113, 1922.
583. Passwater, R. A. Guide to Fluorescence Literature. 2 vols. New York: Plenum Press Data Division, 1967. 736 pp.
584. Pasternak, G., A. Graffi, F. Hoffman, and K.-H. Horn. Resistance against carcinomas of the skin induced by dimethylbenzanthracene (DMBA) in mice of the strain XVII/Bln. Nature 203:307–308, 1964.
585. Pathak, M. A., F. Daniels, Jr., and T. B. Fitzpatrick. The presently known distribution of furocoumarins (psoralens) in plants. J. Invest. Derm. 39: 225–239, 1962.
586. Patterson, A. M., L. T. Capell, and D. F. Walker. The Ring Index. A List of Ring Systems Used in Organic Chemistry. (2nd ed.) Washington, D.C.: American Chemical Society, 1960. 1425 pp.
587. Payne, W. W. Methods of sampling, separation, and analysis of air pollutants: Special requirements for carcinogenic bioassay. Nat. Cancer Inst. Monogr. 9:75–80, 1962.
588. Peacock, P. R. Evidence regarding the mechanism of elimination of 1:2-benzpyrene, 1:2:5:6–dibenzanthracene, and anthracene from the bloodsteam of injected animals. Brit. J. Exp. Path. 17:164–172, 1936.
589. Peover, M. E. Electrochemistry of aromatic hydrocarbons and related substances, pp. 1–51. In A. J. Baird, Ed. Electroanalytical Chemistry. Vol. 2. New York: Marcel Dekker, Inc., 1967.
590. Perkampus, H. H., and L. Pohl. Über die Fluoreszenzspektren dünner Filme aromatischer Kohlenwasserstoffe. Z. Phys. Chem. 40:162–188, 1964.
591. Peterson, C. M., H. J. Paulus, and C. H. Foley. The number size distribution of atmospheric particles during temperature inversions. J. Air Pollut. Control Assoc. 19:795–801, 1969.
592. Petrov, N. N. Results of experiments in carcinogenesis in monkeys over a 20-year period (1939–1960). Probl. Oncol. 6:1709–1715, 1960.
593. Pfeiffer, C. A., and E. Allen. Attempts to produce cancer in rhesus monkeys with carcinogenic hydrocarbons and estrogens. Cancer Res. 8:97–127, 1948.
594. Pierce, G. B. Differentiation of normal and malignant cells. Fed. Proc. 29: 1248–1254, 1970.
595. Pietrzyk, D. J. Organic polarography. Anal. Chem. 42:139R–152R, 1970.
596. Pitot, H. C., and C. Heidelberger. Metabolic regulatory circuits and carcinogenesis. Cancer Res. 23:1694–1700, 1963.
597. Poglazova, M. N., G. E. Fedoseeva, A. J. Khesina, M. N. Meissel, and L. M.

Shabad. Destruction of benzo(a)pyrene by soil bacteria. Life Sci. 6:1053–1062, 1967.
598. Pooler, F., Jr. Airflow over a city in terrain of moderate relief. J. Appl. Meteor. 2:446–456, 1963.
599. Porath, J. Molecular sieving and adsorption. Nature 218:834–838, 1968.
600. Pott, P. Chirurgical Observations Relative to the Cataract, the Polypus of the Nose, the Cancer of the Scrotum, the Different Kinds of Ruptures, and the Mortification of the Toes and Feet. London: L. Hawes, W. Clarke, and R. Collins, 1775. 208 pp.
601. Potter, V. R. Biochemical perspectives in cancer research. Cancer Res. 24: 1085–1098, 1964.
602. Prehn, R. T. A clonal selection theory of chemical carcinogenesis. J. Nat. Cancer Inst. 32:1–17, 1964.
603. Prehn, R. T. Specific isoantigenicities among chemically induced tumors. Ann. N.Y. Acad. Sci. 101:107–113, 1962.
604. Prehn, R. T. Tumor-specific antigens of putatively nonviral tumors. Cancer Res. 28:1326–1330, 1968.
605. Prindle, R. A. Some considerations in the interpretation of air pollution health effects data. J. Air Pollut. Control Assoc. 9:12–19, 1959.
606. Pringsheim, P. Fluorescence and Phosphorescence. New York: Interscience Publishers, Inc., 1949. 794 pp.
607. Puck, T. T., and P. I. Marcus. A rapid method for viable cell titration and clone production with HeLa cells in tissue culture: The use of x-irradiated cells to supply conditioning factors. Proc. Nat. Acad. Sci. U.S.A. 41:432–437, 1955.
608. Puck, T. T., P. I. Marcus, and S. J. Cieciura. Clonal growth of mammalian cells *in vitro*. Growth characteristics of colonies from single HeLa cells with and without a "feeder" layer. J. Exp. Med. 103:273–284, 1956.
609. Pullinger, B. D. The first effects on mouse skin of some polycyclic hydrocarbons. J. Path. Bact. 50:463–471, 1940.
610. Pullman, A., and B. Pullman. A quantum chemist's approach to the mechanism of chemical carcinogenesis, pp. 9–24. In Jerusalem Symposia on Quantum Chemistry and Biochemistry. Vol. 1. 1969.
611. Pullman, A., and B. Pullman. Electronic structure and carcinogenic activity of aromatic molecules. New developments. Adv. Cancer Res. 3:117–169, 1955.
612. Pybus, F. C. Cancer and atmospheric pollution. Newcastle Med. J. 28:31–66, 1963.
613. Pylev, L. N. Effect of the dispersion of soot in deposition of 3,4-benzpyrene in lung tissue of rats. Hyg. Sanit. 32:174–179, Apr.-June, 1967.
614. Pylev, L. N. Experimental induction of lung cancer in rats by intratracheal administration of 9,10-dimethyl-1,2-benzanthracene. Bull. Exp. Biol. Med. (U.S.S.R.) 52:1316–1319, 1961.
615. Pylev, L. N. Experimental induction of lung cancer in rats with 3,4-benzpyrene. Vestn. Akad. Med. Nauk S.S.S.R. 19(11):41–45, 1964. (In Russian)
616. Pylev, L. N. Induction of lung cancer in rats by intratracheal insufflation of cancerogenic hydrocarbons. Acta Un. Int. Cancr. 19:688–691, 1962.
617. Pylev, L. N. Late appearance of tumors in rats after the administration of

9,10-dimethyl-1,2-benzanthracene into the lungs. Vopr. Onkol. 10(8):53-60, 1964. (in Russian)
618. Pylev, L. N., F. J. C. Roe, and G. P. Warwick. Elimination of radioactivity after intratracheal instillation of tritiated 3,4-benzopyrene in hamsters. Brit. J. Cancer 23:103-115, 1969.
619. Pysh, E. S., and N. C. Yang. Polarographic oxidation potentials of aromatic compounds. J. Amer. Chem. Soc. 85:2124-2130, 1963.
620. Rall, D. P. Difficulties in the extrapolation of the results of toxicity studies in laboratory animals to man, pp. 62-73. In Physiological Characterization of Health Hazards in Man's Environment. Washington, D.C.: National Academy of Sciences, 1967.
621. Rand, R. N. Practical spectrophotometric standards. Clin. Chem. 15:839-863, 1969.
622. Rao, C. N. R. Chemical Applications of Infrared Spectroscopy, pp. 163-166. New York: Academic Press Inc., 1963.
623. Rapp, F., and J. L. Melnick. Cell, tissue and organ cultures in virus research, pp. 263-316. In E. N. Willmer, Ed. Cells and Tissues in Culture. Methods, Biology, and Physiology. Vol. 3. London: Academic Press Inc., 1966.
624. Rathkamp, G., and D. Hoffmann. Fluorenes and Fluoranthenes in Cigarette Smoke. Presented to the 24th Tobacco Chemists' Research Conference, Montreal, Canada, 1970.
625. Raven, R. W., and F. J. C. Roe, Eds. The Prevention of Cancer. London: Butterworth & Co., 1967. 397 pp.
626. Reckner, L. R., W. E. Scott, and W. F. Biller. The composition and odor of diesel exhaust. Proc. Amer. Petrol. Inst. 45 (Sect. 3):133-147, 1965.
627. Refining survey. Oil Gas J. 67:115-137. 1969.
628. Reid, D. C., J. Cornfield, R. E. Markush, D. Seigel, E. Pedersen, and W. Haenszel. Studies of disease among migrants and native populations in Great Britain, Norway, and the United States. III. Prevalence of cardiorespiratory symptoms among migrants and native-born in the United States. Nat. Cancer Inst. Monogr. 19:321-346, 1966.
629. Reid, L., Ed. The Pathology of Emphysema. Chicago: Year Book Medical Publishers, 1967. 372 pp.
630. Reiquam, H. An atmospheric transport and accumulation model for airsheds. Atmos. Environ. 4:233-247, 1970.
631. Riddick, J. A., and W. B. Bunger. Techniques of Chemistry. Vol. 2. Organic Solvents: Physical Properties and Methods of Purification. (3rd ed.) New York: John Wiley & Sons, Inc., 1970. 1072 pp.
632. Roe, F. J. C. Comparison of carcinogenicity of tobacco smoke condensate and particulate air pollutants and a demonstration that their effects may be additive, pp. 110-111. In K. H. Weber, Ed. Alkylierend Wirkende Verbindungen. Hamburg: Forschungsinstitut im Berband der Cigaretten-industrie, 1968.
633. Roe, F. J. C. The induction of skin cancer in mice by combinations of cigarette smoke condensate and particulate matter from London air, p. 759. In Ninth International Cancer Congress. Abstracts of Papers. Tokyo, Japan, Oct. 23-29, 1966.

634. Roe, F. J. C., and G. A. Grant. Inhibition by germ-free status of development of liver and lung tumours in mice exposed neonatally to 7,12-dimethylbenz(a)anthracene: Implications in relation to tests for carcinogenicity. Int. J. Cancer 6:133–144, 1970.
635. Roe, F. J. C., and W. E. H. Peirce. Further studies on the tumour-promoting action of citrus oils, pp. 212–213. In British Empire Cancer Campaign. Thirty-eighth Annual Report, 1960. Part II. London: British Empire Cancer Campaign, 1960.
636. Röhrborn, G. Mutagenicity tests in mice. I. The dominant lethal method and the control problem. Humangenetik 6:345–361, 1968.
637. Roitt, I. M., and W. A. Waters. Action of benzoyl peroxide on polycyclic aromatic hydrocarbons. J. Chem. Soc. 1952:2695–2705.
638. Röller, M. R., and C. Heidelberger. Attempts to produce carcinogenesis in organ cultures of mouse prostate with polycyclic hydrocarbons. Int. J. Cancer 2:509–520, 1967.
639. Rondia, D. Sur la volatilite des hydrocarbures polycycliques. Air Water Pollut. 9:113–121, 1965.
640. Rook, A. J., D. S. Wilkinson, and F. J. Ebling, Eds. Textbook of Dermatology. 2 vols. Oxford: Blackwell Scientific Publications, 1968. 2980 pp.
641. Rooney, J. J., and R. C. Pink. Formation and stability of hydrocarbon radical-ions on a silica-alumina surface. Trans. Faraday Soc. 58:1632–1641, 1962.
642. Rosinski, J., and C. T. Nagamoto. Particle deposition on and reentrainment from coniferous trees. Part I. Experiments with trees. Kolloid-Z. 204: 78–84, 1966.
643. Rounds, D. E. Environmental influences on living cells. Arch. Environ. Health 12:78–84, 1966.
644. Rounds, D. E., A. Awa, and C. M. Pomerat. Effect of automobile exhaust on cell growth *in vitro*. Arch. Environ. Health 5:319–324, 1962.
645. Royal College of Physicians, London. Committee on Smoking and Atmospheric Pollution. Air Pollution and Health. Summary and Report on Air Pollution and Its Effect on Health. London: Pitman Medical and Scientific Publishing Co., 1970. 80 pp.
646. Rusch, H. P., B. E. Kline, and C. A. Baumann. The nonadditive effect of ultraviolet light and other carcinogenic procedures. Cancer Res. 2:183–188, 1942.
647. Russell, L. B., and W. L. Russell. Pathways of radiation effects in the mother and the embryo, pp. 50–59. In Cold Spring Harbor Symposia on Quantitative Biology. Vol. XIX. The Mammalian Fetus: Physiological Aspects of Development. Long Island, N.Y.: The Biological Laboratory, 1954.
648. Saffiotti, U. Experimental respiratory tract carcinogenesis and its relation to inhalation exposures, pp. 27–54. In M. G. Hanna, Jr., P. Nettesheim, and J. R. Gilbert, Eds. Inhalation Carcinogenesis. Proceedings of a Biology Division, Oak Ridge National Laboratory, conference held in Gatlinburg, Tenn., Oct. 8–11, 1969. AEC Symposium Series, No. 18. Oak Ridge, Tenn.: U.S. Atomic Energy Commission, Division of Technical Information, 1970.
649. Saffiotti, U., S. A. Borg, M. I. Grote, and D. B. Karp. Retention rates of

particulate carcinogens in the lungs. Studies in an experimental model for lung cancer induction. Chicago Med. School Q. 24:10–17, 1964.

650. Saffiotti, U., F. Cefis, and L. H. Kolb. A method for the experimental induction of bronchogenic carcinoma. Cancer Res. 28:104–124, 1968.
651. Saffiotti, U., F. Cefis, and L. H. Kolb. Bronchiogenic carcinoma induction by particulate carcinogens. Proc. Amer. Assoc. Cancer Res. 5:55, 1964. (abstract)
652. Saffiotti, U., F. Cefis, L. H. Kolb, and M. I. Grote. Intratracheal injection of particulate carcinogens into hamster lungs. Proc. Amer. Assoc. Cancer Res. 4:59, 1963. (abstract)
653. Saffiotti, U., F. Cefis, L. H. Kolb, and P. Shubik. Experimental studies of the conditions of exposure to carcinogens for lung cancer induction. J. Air Pollut. Control Assoc. 15:23–25, 1965.
654. Saffiotti, U., F. Cefis, and P. Shubik. Histopathology and histogenesis of lung cancer induced in hamsters by carcinogens carried by dust particles, pp. 537–546. In L. Severi, Ed. Lung Tumours in Animals. Proceedings of the Third Quadrennial International Conference on Cancer. June 24–29, 1965. Perugia, Italy: University of Perugia, 1966.
655. Saffiotti, U., R. Montesano, A. R. Sellakumar, and S. A. Borg. Experimental cancer of the lung. Inhibition by vitamin A of the induction of tracheobronchial squamous metaplasia and squamous cell tumors. Cancer 20: 857–864, 1967.
656. Saffiotti, U., R. Montesano, and N. Tompkins. Benzo(a)pyrene retention in hamster lungs: Studies on particle size and on total dust load. Proc. Amer. Assoc. Cancer Res. 8:57, 1967. (abstract)
657. Salaman, M. H., and F. J. C. Roe. Cocarcinogenesis. Brit. Med. Bull. 20: 139–144, 1964.
658. Sanders, C. L., Jr., R. C. Thompson, and W. J. Bair. Lung cancer: Dose response studies with radionuclides, pp. 258–303. In M. G. Hanna, Jr., P. Nettesheim, and J. R. Gilbert, Eds. Inhalation Carcinogenesis. Proceedings of a Biology Division, Oak Ridge National Laboratory, conference held in Gatlinburg, Tenn., Oct. 8–11, 1969. AEC Symposium Series, No. 18. Oak Ridge, Tenn.: U.S. Atomic Energy Commission, Division of Technical Information, 1970.
659. Sanford, K. K. "Spontaneous" neoplastic transformation of cells *in vitro:* Some facts and theories. Nat. Cancer Inst. Monogr. 26:387–408, 1965.
660. Santamaria, L., and G. G. Giordano. Effects of long-wave ultraviolet hydrocarbon carcinogenesis, pp. 569–580. In F. Urbach, Ed. The Biologic Effects of Ultraviolet Radiation. Oxford: Pergamon Press, 1969.
661. Sawicki, C. R., and E. Sawicki. Thin-layer chromatography in air pollution research, pp. 233–293. In A. Niederwieser and G. Pataki, Eds. Progress in Thin-layer Chromatography and Related Methods. Vol. 3. Ann Arbor, Mich.: Ann Arbor Science Publishers, Inc., 1972.
662. Sawicki, E. Airborne carcinogens and allied compounds. Arch. Environ. Health 14:46–53, 1967.
663. Sawicki, E. Fluorescence analysis in air pollution research. Talanta 16:1231–1266, 1969.

664. Sawicki, E. The separation and analysis of polynuclear aromatic hydrocarbons present in the human environment. Chem.-Anal. 53:24–26, 28–30, 56–62, 88–91, 1964.
665. Sawicki, E., R. C. Corey, A. E. Dooley, J. B. Gisclard, J. L. Monkman, R. E. Neligan, and L. A. Ripperton. Tentative method of routine analysis for polynuclear aromatic hydrocarbon content of atmospheric particulate matter. Health Lab. Sci. 7:45–55, 1970.
666. Sawicki, E., R. C. Corey, A. E. Dooley, J. B. Gisclard, J. L. Monkman, R. E. Neligan, and L. A. Ripperton. Tentative method of spectrophotometric analysis for benzo[a]pyrene in atmospheric particulate matter. Health Lab. Sci. 7(suppl.):68–71, 1970.
667. Sawicki, E., W. C. Elbert, T. R. Hauser, F. T. Fox, and T. W. Stanley. Benzo(a)pyrene content of the air of American communities. Amer. Ind. Hyg. Assoc. J. 21:443–451, 1960.
668. Sawicki, E., W. Elbert, T. W. Stanley, T. R. Hauser, and F. T. Fox. Separation and characterization of polynuclear aromatic hydrocarbons in urban airborne particulates. Anal. Chem. 32:810–815, 1960.
669. Sawicki, E., M. Guyer, and C. R. Engel. Paper and thin-layer electrophoretic separations of polynuclear aza heterocyclic compounds. J. Chromatogr. 30: 522–527, 1967.
670. Sawicki, E., T. R. Hauser, W. C. Elbert, F. T. Fox, and J. E. Meeker. Polynuclear aromatic hydrocarbon composition of the atmosphere in some large American cities. Amer. Ind. Hyg. Assoc. J. 23:137–144, 1962.
671. Sawicki, E., T. R. Hauser, and T. W. Stanley. Ultraviolet, visible and fluorescence spectral analysis of polynuclear hydrocarbons. Int. J. Air Pollut. 2:253–272, 1960.
672. Sawicki, E., and H. Johnson. Characterization of aromatic compounds by low-temperature fluorescence and phosphorescence: Application to air pollution studies. Microchem. J. 8:85–101, 1964.
673. Sawicki, E., S. P. McPherson, T. W. Stanley, J. Meeker, and W. C. Elbert. Quantitative composition of the urban atmosphere in terms of polynuclear aza heterocyclic compounds and aliphatic and polynuclear aromatic hydrocarbons. Int. J. Air Water Pollut. 9:515–524, 1965.
674. Sawicki, E., J. E. Meeker, and M. J. Morgan. The quantitative composition of air pollution source effluents in terms of aza heterocyclic compounds and polynuclear aromatic hydrocarbons. Int. J. Air Water Pollut. 9:291–298, 1965.
675. Sawicki, E., T. W. Stanley, and H. Johnson. Quenchofluorometric analysis for polynuclear compounds. Mikrochim. Ichnoanal. Acta 1965:178–192.
676. Sawicki, E., T. W. Stanley, S. McPherson, and M. Morgan. Use of gas–liquid and thin-layer chromatography in characterizing air pollutants by fluorometry. Talanta 13:619–629, 1966.
677. Schiffman, R., and E. Landau. Use of indexes of air pollution potential in mortality studies. J. Air Pollut. Control Assoc. 2:384–386, 1961.
678. Schmillen, A., and R. Legler. Landolt-Boernstein Numerical Data and Functional Relationships in Science and Technology, Group II: Atomic and Molecular Physics. Vol. 3. Luminescence of Organic Substances. New York: Springer-Verlag, 1967.

679. Schubert, C. C., and R. N. Pease. Reaction of paraffin hydrocarbons with ozonized oxygen: Possible role of ozone in normal combustion. J. Chem. Phys. 24:919–920, 1956.
680. Schubert, C. C., and R. N. Pease. The oxidation of lower paraffin hydrocarbons. I. Room temperature reaction of methane, propane, *n*-butane and isobutane with ozonized oxygen. J. Amer. Chem. Soc. 78:2044–2048, 1956.
681. Schubert, C. C., and R. N. Pease. The oxidation of lower paraffin hydrocarbons. II. Observations on the role of ozone in the slow combustion of isobutane. J. Amer. Chem. Soc. 78:5553–5556, 1956.
682. Schulman, S. G., and J. D. Winefordner. Influence of pH in fluorescence and phosphorescence spectrometric analysis. Talanta 17:607–616, 1970.
683. Schwartz, L., L. Tulipan, and D. J. Birmingham. Occupational Diseases of the Skin. (3rd ed.) Philadelphia: Lea & Febiger, 1957. 981 pp.
684. Scott, A. Cancers in mineral oil refineries, pp. 275–279. In Report of the International Conference on Cancer, London, 17th–20th July, 1928. New York: William Wood and Company, 1928.
685. Searl, T. D., F. J. Cassidy, W. H. King, and R. A. Brown. An analytical method for polynuclear aromatic compounds in coke oven effluents by combined use of gas chromatography and ultraviolet absorption spectrometry. Anal. Chem. 42:954–958, 1970.
686. Seelig, M. G., and E. L. Benignus. Coal smoke soot and tumors of the lung in mice. Amer. J. Cancer 28:96–111, 1936.
687. Seelig, M. G., and E. L. Benignus. The production of experimental cancer of the lung in mice. Amer. J. Cancer 33:549–554, 1938.
688. Seelig, M. G., and Z. K. Cooper. Light and tar cancer. An experimental study, with a critical review of the literature on light as a carcinogenic factor. Surg. Gynec. Obstet. 56:752–761,1933.
689. Segi, M., M. Kurihara, and T. Matsuyama. Cancer Mortality for Selected Sites in Twenty-four Countries. No. 5 (1964–1965). Dept. of Public Health, Tohoku University School of Medicine, Sendai, Japan, 1969.
690. Selkirk, J. K., E. Huberman, and C. Heidelberger. An epoxide is an intermediate in the microsomal metabolism of the chemical carcinogen, dibenz(a,h)anthracene. Biochem. Biophys. Res. Commun. 43:1010–1016, 1971.
691. Shabad, L. M. Experimental cancer of the lung. J. Nat. Cancer Inst. 28: 1305–1332, 1962.
692. Shabad, L. M. Experimental cancer of the lungs. Fed. Proc. (Transl. Suppl.) 22:T331–336, 1963.
693. Shabad, L. M. Studies in the U.S.S.R. on the distribution, circulation, and fate of carcinogenic hydrocarbons in the human environment and the role of their deposition in tissues in carcinogenesis: A review. Cancer Res. 27:1132–1137, 1967.
694. Shabad, L. M., L. N. Pylev, and T. S. Kolesnichenko. Importance of the deposition of carcinogens for cancer induction in lung tissue. J. Nat. Cancer Inst. 33:135–141, 1964.
695. Sharkey, A. G., Jr., J. L. Shultz, T. Kessler, and R. A. Friedel. High-resolution mass spectrometry has advantages in determining organic contaminants in air and water. Res./Develop. 20:30–32, 1969.

696. Shay, H., E. A. Aegerter, M. Gruenstein, and S. A. Komarov. Development of adenocarcinoma of the breast in the Wistar rat following the gastric instillation of methylcholanthrene. J. Nat. Cancer Inst. 10:255–266, 1949.
697. Shellabarger, C. J. Effect of 3-methylcholanthrene and X irradiation, given singly or combined, on rat mammary carcinogenesis. J. Nat. Cancer Inst. 38:73–77, 1967.
698. Shigorin, V. D., and G. P. Shipulo. Laser excitation of organic molecules phosphorescence. Zh. Prikl. Spektrosk. 12:331–333, 1970.
699. Shubik, P., and J. L. Hartwell. Survey of Compounds Which Have Been Tested for Carcinogenic Activity. Supplement 1. Public Health Service Publication 149-1. Washington, D.C.: U.S. Government Printing Office, 1957. 388 pp.
700. Shubik, P., and J. L. Hartwell. Survey of Compounds Which Have Been Tested for Carcinogenic Activity. Supplement 2. Public Health Service Publication 149-2. Washington, D.C.: U.S. Government Printing Office, 1969. 655 pp.
701. Shultz, J. L., R. A. Friedel, and A. G. Sharkey, Jr. Analyses of coal-tar pitch by mass spectrometry. Fuel 44:55–61, 1965.
702. Simmers, M. H. Petroleum asphalt inhalation by mice. Effects of aerosols and smoke on the tracheobronchial tree and lungs. Arch. Environ. Health 9:727–734, 1964.
703. Simmers, M. H., E. Podolak, and R. Kinosita. Carcinogenic effects of petroleum asphalt. Proc. Soc. Exp. Biol. Med. 101:266–268, 1959.
704. Simpson, W. L., C. Carruthers, and W. Cramer. Loss of carcinogenic activity when methylcholanthrene is dissolved in anhydrous lanolin. Cancer Res. 5:1–4, 1945.
705. Simpson, W. L., and W. Cramer. Sebaceous glands and experimental skin carcinogenesis in mice. Cancer Res. 3:515–518, 1943.
706. Sims, P. Qualitative and quantitative studies on the metabolism of a series of aromatic hydrocarbons by rat-liver preparations. Biochem. Pharmacol. 19:795–818, 1970.
707. Sims, P. The metabolism of some aromatic hydrocarbons by mouse embryo cell cultures. Biochem. Pharmacol. 19:285–297, 1970.
708. Slizynski, B. M. Pachytene analysis of Snell's T(5:8)a translocation in the mouse. J. Genet. 50:507–510, 1952.
709. Smith, W. E., N. S. Cooper, and E. L. Wynder. Biological tests of fractions derived from cigarette smoke condensates. Proc. Amer. Assoc. Cancer Res. 1:45, 1954. (abstract)
710. Smith, W. E., D. A. Sunderland, and K. Sigiura. Experimental analysis of the carcinogenic activity of certain petroleum products. A.M.A. Arch. Ind. Hyg. Occup. Med. 4:299–314, 1951.
711. Smith, W. M. Evaluation of Coke Oven Emissions. Presented to the 78th General Meeting of the American Iron and Steel Institute, New York City, May 28–29, 1970.
712. Snell, G. D., E. Bodemann, and W. Hollander. A translocation in the house mouse and its effect on development. J. Exp. Zool. 67:93–104, 1934.
713. Snell, G. D., and D. I. Picken. Abnormal development in the mouse caused by chromosome unbalance. J. Genet. 31:213–235, 1935.
714. Southam, A. H. Mule-spinners cancer, pp. 280–283. In Report of the Inter-

national Conference on Cancer, London, 17th–20th July, 1928. New York: William Wood and Company, 1928.

715. Southam, A. H., and S. R. Wilson. Cancer of the scrotum: The etiology, clinical features, and treatment of the disease. Brit. Med. J. 2:971–973, 1922.

716. Southern, P. F., and W. A. Waters. The thermal decomposition of some *meso*-substituted anthracene photo-oxides. J. Chem. Soc. 1960:4340–4346.

717. Spikes, J. D., and R. Straight. Sensitized photochemical processes in biological systems. Ann. Rev. Phys. Chem. 18:409–436, 1967.

718. Spotswood, T. M. Chromatography of polycyclic aromatic hydrocarbons on acetylated paper. J. Chromatogr. 2:90–94, 1959.

719. Staemmler, M. I. Referat über Beruf und Krebs. Verh. Deutsch. Ges. Path. 30:188–238, 1937.

720. Steer, R. P., K. R. Darnall, and J. N. Pitts, Jr. Base-induced decomposition of peroxyacetyl nitrate. Tetrahedron Lett. 43:3765–3767, 1969.

721. Steer, R. P., J. L. Sprung, and J. N. Pitts, Jr. Singlet oxygen in the environmental sciences. Evidence for the production of O_2 ($'\Delta$ g) by energy transfer in the gas phase. Environ. Sci. Technol. 3:946–947, 1969.

722. Stein, F., N. A. Esmen, and M. Corn. The shape of atmospheric particles in Pittsburgh air. Atmos. Environ. 3:443–453, 1969.

723. Steiner, P. E., and C. G. Loosli. The effect of human influenza virus (type A) on incidence of lung tumors in mice. Cancer Res. 10:385–392, 1950.

724. Stevens, B., and E. Hutton. Delayed fluorescence from micro crystalline aromatic hydrocarbons. Proc. Phys. Soc. 81:893–897, 1963.

725. Stöber, W., and H. Flachsbart. Size-separating precipitation of aerosols in a spinning spiral duct. Environ. Sci. Tech. 3:1280–1296, 1969.

726. Stocks, P. Air pollution and cancer mortality in Liverpool Hospital region and North Wales. Int. J. Air Pollut. 1:1–13, 1958.

727. Stocks, P. Cancer and bronchitis mortality in relation to atmospheric deposit and smoke. Brit. Med. J. 1:74–79, 1959.

728. Stocks, P. Epidemiology of cancer of the lung in England and Wales. Brit. J. Cancer 6:99–111, 1952.

729. Stocks, P. Lung cancer and bronchitis in relation to cigarette smoking and fuel consumption in twenty countries. Brit. J. Prev. Soc. Med. 21:181–185, 1967.

730. Stocks, P. On the relations between atmospheric pollution in urban and rural localities and mortality from cancer, bronchitis and pneumonia, with particular reference to 3:4 benzopyrene, beryllium, molybdenum, vanadium and arsenic. Brit. J. Cancer 14:397–418, 1960.

731. Stocks, P. Recent epidemiological studies of lung cancer mortality, cigarette smoking and air pollution, with discussion of a new hypothesis of causation. Brit. J. Cancer 20:595–623, 1966.

732. Stocks, P., and J. M. Campbell. Lung cancer death rates among non-smokers and pipe and cigarette smokers. An evaluation in relation to air pollution by benzpyrene and other substances. Brit. Med. J. 2:923–939, 1955.

733. Stocks, P., B. T. Commins, and K. V. Aubrey. A study of polycyclic hydrocarbons and trace elements in smoke in Merseyside and other northern localities. Int. J. Air Water Pollut. 4:141–153, 1961.

734. Stokinger, H. E., and D. L. Coffin. Biologic effects of air pollutants, pp.

445–546. In A. C. Stern, Ed. Air Pollution. Vol. 1. Air Pollution and Its Effects. (2nd ed.) New York: Academic Press Inc., 1968.
735. Sugiura, K., W. E. Smith, and D. A. Sunderland. Experimental production of carcinoma in rhesus monkeys. Cancer Res. 16:951–955, 1956.
736. Sulman, E., and F. Sulman. The carcinogenicity of wood soot from the chimney of a smoked sausage factory. Cancer Res. 6:366–367, 1946.
737. Suntzeff, V., A. B. Croninger, E. L. Wynder, E. V. Cowdry, and E. A. Graham. Use of sebaceous-gland test of primary cigarette-tar fractions and of certain noncarcinogenic polycyclic hydrocarbons. Cancer 10:250–254, 1957.
738. Suskind, R. R. Acne: Occupational, pp. 563–573. In P. D. Cantor, Ed. Traumatic Medicine and Surgery for the Attorney. Vol. 6. Psychiatry. Skin and its Appendages. Washington, D.C.: Butterworth Inc., 1962.
739. Suskind, R. R., and A. W. Horton. Etiologic factors and the pathogenesis of premalignant and malignant lesions of the skin, pp. 171–192. In S. Rothman, Ed. The Human Integument: Normal and Abnormal. Washington, D.C.: American Association for the Advancement of Science, 1959.
740. Tanimura, H. Benzo(a)pyrene in an iron and steel works. Arch. Environ. Health 17:172–177, 1968.
741. Tannenbaum, A. Nutrition and cancer, pp. 517–562. In F. Homburger, Ed. The Physiopathology of Cancer. (2nd ed.) New York: Paul B. Hoeber, Inc., 1959.
742. Tannenbaum, A., and H. Silverstone. Nutrition in relation to cancer. Adv. Cancer Res. 1:451–501, 1953.
743. Tarbell, D. S., E. G. Brooker, A. Vanterpool, W. Conway, C. J. Claus, and T. J. Hall. A system for paper chromatography of 3,4-benzpyrene, some derivatives and other polycyclic aromatic hydrocarbons. J. Amer. Chem. Soc. 77:767–768, 1955.
744. Tasseron, J. G., H. Diringer, H. Frohwirth, S. S. Mirvish, and C. Heidelberger. Partial purification of the soluble protein from mouse skin to which carcinogenic hydrocarbons are specifically bound. Biochemistry 9:1636–1644, 1970.
745. Tebbens, B. D., J. F. Thomas, and M. Mukai. Fate of arenes incorporated with airborne soot. Amer. Ind. Hyg. Assoc. J. 27:415–422, 1966.
746. Tentative method of analysis for polynuclear aromatic hydrocarbon content of atmospheric particulate matter. Health Lab. Sci. 7 (Suppl.):31–44, 1970.
747. Tentative method of analysis for suspended particulate matter in the atmosphere: (high-volume method). Health Lab. Sci. 7:279–286, 1970.
748. Tentative method of chromatographic analysis for benzo[a]pyrene and benzo[k]fluoranthene in atmospheric particulate matter. Health Lab. Sci. 7 (Suppl.):60–67, 1970.
749. Tentative method of microanalysis for benzo[a]pyrene in airborne particulates and source effluents. Health Lab. Sci. 7 (Suppl.):56–59, 1970.
750. Tentative method of routine analysis for polynuclear aromatic hydrocarbon content of atmospheric particulate matter. Health Lab. Sci. 7 (Suppl.): 45–55, 1970.
751. Thomas, J. F., M. Mukai, and B. D. Tebbens. Fate of airborne benzo[a]pyrene. Environ. Sci. Tech. 2:33–39, 1968.
752. Thomas, J. F., B. D. Tebbens, E. N. Sanborn, and J. M. Cripps. Fluorescent

spectra of aromatic hydrocarbons found in polluted atmosphere. Int. J. Air Pollut. 2:210–220, 1960.

753. Tipson, R. S. Review of Oxidation of Polycyclic, Aromatic Hydrocarbons. National Bureau of Standards Report 8363. Washington, D.C.: U.S. Government Printing Office, 1964. 89 pp.

754. Toth, B. A critical review of experiments in chemical carcinogenesis using newborn animals. Cancer Res. 28:727–738, 1968.

755. Toth, B., P. N. Magee, and P. Shubik. Carcinogenesis study with dimethylnitrosamine adminstered orally to adult and subcutaneously to newborn BALB/c mice. Cancer Res. 24:1712–1719, 1964.

756. Toth, L. Spektralfluorometrische *in situ*-analyse polycyclisher aromaten nach trennung auf acetylierten cellulosechichten. I. Mit qualitative und quantitative auswertung. J. Chromatogr. 50:72–82, 1970. (summary in English)

757. Tsai, S. C., and G. W. Robinson. Phosphorescence and the true lifetime of triplet states in fluid solutions. J. Chem. Phys. 49:3148–3191, 1968.

758. Ts'o, P. O. P., S. A. Lesco, and R. S. Umans. The physical binding and the chemical linkage of benzpyrene to nucleotides, nucleic acids, and nucleohistones, pp. 106–135. In Jerusalem Symposia on Quantum Chemistry and Biochemistry. Vol. 1, 1969.

759. Twort, C. C., and J. M. Twort. The carcinogenic potency of mineral oils. J. Ind. Hyg. 13:204–226, 1931.

760. Tye, R., and K. L. Stemmer. Experimental carcinogenesis of the lung. II. Influence of phenols in the production of carcinoma. J. Nat. Cancer Inst. 39:175–186, 1967.

761. Ultraviolet Atlas of Organic Compounds. 5 vols. New York: Plenum Press, 1967–1968.

762. Ungnade, H. E., Ed. Organic Electronic Spectral Data. New York: Interscience Publishers, Inc., 1960–.

763. United Nations. Report of the United Nations Scientific Committee on the Effects of Atomic Radiation, p. 99. General Assembly; Official Records: Twenty-first Session. Supplement 14 (A/6314). New York: United Nations, 1966.

764. U.S. Department of Agriculture, Forest Service, Division of Cooperative Forest Fire Control. 1968 Wildfire Statistics. Washington, D.C.: U.S. Department of Agriculture, 1969. 48 pp.

765. U.S. Department of Commerce, Census Bureau. Estimates of Population of Selected Standard Metropolitan Statistical Areas. July 1, 1962; July 1, 1963. Washington, D.C.: U.S. Department of Commerce, 1964. 8 pp.; 6 pp.

766. U.S. Department of Health, Education, and Welfare. Mutagenicity of pesticides, pp. 565–653. In Report of the Secretary's Commission on Pesticides and Their Relationship to Environmental Health. Parts I and II. Washington, D.C.: U.S. Government Printing Office, 1969.

767. U.S. Department of Health, Education, and Welfare. Nationwide Inventory of Air Pollutant Emissions. 1968. National Air Pollution Control Administration Publication AP-73. Washington, D.C.: U.S. Government Printing Office, 1970. 36 pp.

768. U.S. Department of Health, Education, and Welfare. Preliminary Air Pollution Survey of Organic Carcinogens. A Literature Review. National Air

Pollution Control Administration Publication APTD 69-43. Raleigh, N.C.: U.S. Department of Health, Education, and Welfare, 1969. 117 pp.
769. U.S. Department of Health, Education, and Welfare. Smoking and Health. Report of the Advisory Committee to the Surgeon General of the Public Health Service. Public Health Service Publication 1103. Washington, D.C.: U.S. Government Printing Office, 1964. 387 pp.
770. U.S. Department of Health, Education, and Welfare. Teratogenicity of pesticides, pp. 655–677. In Report of the Secretary's Commission on Pesticides and Their Relationship to Environmental Health. Parts I and II. Washington, D.C.: U.S. Government Printing Office, 1969.
771. U.S. Department of Health, Education, and Welfare. The Health Consequences of Smoking: 1969 Supplement to the 1967 Public Health Service Review. Washington, D.C.: U.S. Government Printing Office, 1969. 98 pp.
772. U.S. Department of Health, Education, and Welfare. The Health Consequences of Smoking, a Public Health Service Review: 1967. Public Health Service Publication 1696, Revised 1968. Washington, D.C.: U.S. Government Printing Office, 1968. 227 pp.
773. U.S. Department of Health, Education, and Welfare, National Air Pollution Control Administration. Air Quality Data from the National Air Sampling Networks and Contributing State and Local Networks. 1966 Edition. NAPCA Publication APTD 68-9. Durham, N.C.: U.S. Department of Health, Education, and Welfare, 1968. 157 pp.
774. U.S. Department of Health, Education, and Welfare, Public Health Service, Consumer Protection and Environmental Health Service, Environmental Control Adminstration, Bureau of Solid Waste Management. Technical–Economic Study of Solid Waste Disposal Needs and Practices. Public Health Service Publication 1886. Washington, D.C.: U.S. Department of Health, Education, and Welfare, 1969. 4 pp.
775. U.S. Department of Health, Education, and Welfare, Public Health Service, National Air Pollution Control Administration. Air Quality Criteria for Particulate Matter, NAPCA Publication AP-49. Washington, D.C.: U.S. Government Printing Office, 1969. 211 pp.
776. U.S. Department of Labor, Labor Statistics Bureau. Current Population Survey. Washington, D.C.: U.S. Government Printing Office, 1964. 18 pp.
777. U.S. Department of the Interior, Bureau of Mines. Automobile Disposal, a National Problem. Washington, D.C.: U.S. Government Printing Office, 1967. 569 pp.
778. Uytdenhoef. Quelques considérations sur les dermatoses professionnelles. Arch. Med. Soc. Hyg. 2:830–847, 1939.
779. Vadova, A. V., and V. I. Gel'shtein. Spontaneous tumours in catarrhine monkeys according to the data obtained in the monkey colony of the Sukhumi medico-biological station, pp. 137–158. In I. A. Utkin, Ed. Theoretical and Practical Problems of Medicine and Biology in Experiments on Monkeys. Translated by Ruth Schacter. New York: Pergamon Press Inc., 1960.
780. Van Duuren, B. L. Carcinogenic epoxides, lactones, and halo-ethers and their mode of action. Ann. N.Y. Acad. Sci. 163:633–651, 1969.
781. Van Duuren, B. L. Effects of the environment on the fluorescence of aromatic compounds in solution. Chem. Rev. 63:325–354, 1963.

782. Van Duuren, B. L. The fluorescence spectra of aromatic hydrocarbons and heterocyclic aromatic compounds. Anal. Chem. 32:1436–1442, 1960.
783. Van Duuren, B. L. The polynuclear aromatic hydrocarbons in cigarette-smoke condensate. II. J. Nat. Cancer Inst. 21:623–630, 1958.
784. Van Duuren, B. L. Tumor-promoting agents in two-stage carcinogenesis. Prog. Exp. Tumor Res. 11:31–68, 1969.
785. Van Duuren, B. L., I. Bekersky, and M. Lefar. The peracid oxidation of dibenz[a,h]anthracene. J. Org. Chem. 29:686–689, 1964.
786. Van Duuren, B. L., J. A. Bilbao, and C. A. Joseph. The carcinogenic nitrogen heterocyclics in cigarette-smoke condensate. J. Nat. Cancer Inst. 25:53–61, 1960.
787. Van Duuren, B. L., and S. Melchionne. Inhibition of tumorigenesis. Prog. Exp. Tumor Res. 12:55–94, 1969.
788. Van Duuren, B. L., and L. Orris. The tumor-enhancing principles of *Croton tiglium* L. Cancer Res. 25:1871–1875, 1965.
789. Van Duuren, B. L., A. Sivak, B. M. Goldschmidt, C. Katz, and S. Melchionne. Initiating activity of aromatic hydrocarbons in two-stage carcinogenesis. J. Nat. Cancer Inst. 44:1167–1173, 1970.
790. Van Duuren, B. L., A. Sivak, and L. Langseth. The tumor-promoting activity of tobacco leaf extract and whole cigarette tar. Brit. J. Cancer 21:460–463, 1967.
791. Van Duuren, B. L., A. Sivak, L. Langseth, B. M. Goldschmidt, and A. Segal. Initiators and promoters in tobacco carcinogenesis. Nat. Cancer Inst. Monogr. 28:173–180, 1968.
792. Van Duuren, B. L., A. Sivak, A. Segal, L. Orris, and L. Langseth. The tumor-promoting agents of tobacco leaf and tobacco smoke condensate. J. Nat. Cancer Inst. 37:519–526, 1966.
793. Varma, P. S., and J. L. Das Cupta. Comparative study of the preparation of anthraquinone. Q. J. Indian Chem. Soc. 4:297–298, 1927.
794. Vlès, F., A. de Coulon, and A. Ugo. Recherches sur les propriétés physico-chimiques des tissus en relation avec l'état normal ou pathologique de l'organisme. XXI. Influence de l'obscurité et de la luminère sur la cancèrisation par le goudron. Arch. Phys. Biol. 12:255–277, 1935.
795. Vollmann, H., H. Becker, M. Corell, and H. Streeck. Beiträge zur Kenntnis des Pyrens und seiner Derivate. Justus Liebigs Ann. Chem. 531:1–159, 1937.
796. Vorwald, A. J. Medical aspects of beryllium disease, pp. 167–200. In H. E. Stokinger, Ed. Beryllium. Its Industrial Hygiene Aspects. New York: Academic Press Inc., 1966.
797. Voyatzakis, E., D. Jannakoudakis, T. Dorfmüller, C. Sipitanos, and G. Stalidis. Action de la lumière ultraviolette sur les hydrocarbures polybenzénique adsorbés. Anthracène, naphthalène. C. R. Acad. Sci. (Paris) 251:2696–2707, 1960.
798. Wadleigh, C. H. Wastes in Relation to Agriculture and Forestry. Miscellaneous Publication 1065. Washington, D.C.: U.S. Department of Agriculture, 1968. 112 pp.
799. Walburg, H. E., Jr., G. E. Cosgrove, and A. C. Upton. Influence of microbial environment on development of myeloid leukemia in x-irradiated RFM mice. Int. J. Cancer 3:150–154, 1968.

800. Waller, R. E., and B. T. Commins. Studies of the smoke and polycyclic aromatic hydrocarbon content of the air in large urban areas. Environ. Res. 1:295–306, 1967.
801. Waltz, P., and M. Häusermann. Sur un traitment du tabac en vue de diminuer la teneur en hydrocarbures polycycliques de la fumée de cigarettes. A. Praeventivmed. 8:111–124, 1963.
802. Wanta, R. C. Meteorology and air pollution, pp. 187–226. In A. C. Stern, Ed. Air Pollution. Vol. 1. Air Pollution and Its Effects. (2nd ed.) New York: Academic Press Inc., 1968.
803. Wasserman, E., V. J. Kuck, W. M. Delavan, and W. A. Yager. Electron paramagnetic resonance of $'\Delta$ oxygen produced by gas-phase photosensitization with naphthalene derivatives. J. Amer. Chem. Soc. 91:1040–1041, 1969.
804. Wasserman, E., R. W. Murray, M. L. Kaplan, and W. A. Yager. Electron paramagnetic resonance of $'\Delta$ oxygen from a phosphite–ozone complex. J. Amer. Chem. Soc. 90:4160–4161, 1968.
805. Wasserman, H. H., and J. R. Scheffer. Singlet oxygen reactions from photoperoxides. J. Amer. Chem. Soc. 89:3073–3075, 1967.
806. Watson, J. T. Gas chromatography and mass spectroscopy, pp. 145–225. In L. S. Ettre and W. H. McFadden, Eds. Ancillary Techniques of Gas Chromatography. New York: John Wiley & Sons, Inc., 1969.
807. Wattenberg, L. W., and J. L. Leong. Inhibition of the carcinogenic action of benzo(a)pyrene by flavones. Cancer Res. 30:1922–1925, 1970.
808. Wattenberg, L. W., and J. L. Leong. Inhibition of the carcinogenic action of 7,12-dimethylbenz(a)anthracene by beta-naphthoflavone. Proc. Soc. Exp. Biol. Med. 128:940–943, 1968.
809. Weber, G. Enumeration of components in complex systems by fluorescence spectrophotometry. Nature 190:27–29, 1961.
810. Wehry, E. L. Structural and environmental factors in fluorescence, pp. 37–132. In G. G. Guilbault, Ed. Fluorescence: Theory, Instrumentation, and Practice. New York: Marcel Dekker, Inc., 1967.
811. Weinberg, N. L., and H. R. Weinberg. Electrochemical oxidation of organic compounds. Chem. Rev. 68:449–523, 1968.
812. Weisburger, J. H., and E. K. Weisburger. Tests for chemical carcinogens. Methods Cancer Res. 1:307–398, 1967.
813. Welch, R. M., Y. E. Harrison, A. H. Conney, P. J. Poppers, and M. Finster. Cigarette smoking: Stimulatory effect on metabolism of 3,4-benzpyrene by enzymes in human placenta. Science 160:541–542, 1968.
814. Went, F. W. Organic matter in the atmosphere, and its possible relation to petroleum formation. Proc. Nat. Acad. Sci. U.S.A. 46:212–221, 1960.
815. Whitby, K. T., R. S. Husar, and B. Y. H. Liu. The Aerosol Spectra of Los Angeles Smog. Presented to the American Chemical Society Kendall Award Symposium, Los Angeles, March 29, 1971.
816. Whiting, M. C., A. J. N. Bolt, and J. H. Parish. The reaction between ozone and saturated compounds. Adv. Chem. Ser. 77:4–14, 1968.
817. Wilk, M., W. Bez, and J. Rochlitz. Neue Reaktion der Carcinogen Kohlenwasserstoffe 3,4-Benzpyren, 9,10-Dimethyl-1,2-Benzanthracen und 20-Methylcholanthren. Tetrahedron 22: 2599–2608, 1966. (abstract in English)

818. Wilk, M., and W. Girke. Radical cations of carcinogenic alternant hydrocarbons, amines, and azo dyes, and their reactions with nucleobases, pp. 91–105. In E. D. Bergman and B. Pullman, Eds. Physico-chemical Mechanisms of Carcinogenesis. Proceedings of an International Symposium Held in Jerusalem, 21–25 October, 1968. Jerusalem: Israel Academy of Sciences and Humanities, 1969.

819. Wilk, M., J. Rochlitz, and H. Bende. Säulenchromatographie von polycyclischen aromatischen Kohlenwasserstoffen an lipophilem Sephadex LH-20. J. Chromatogr. 24:414–416, 1966.

820. Willmer, E. N., Ed. Cells and Tissues in Culture: Methods, Biology, and Physiology. 3 vols. London: Academic Press Inc., 1965–1966. 2423 pp.

821. Wilmshurst, J. R. Gas chromatographic analysis of polynuclear arenes. J. Chromatogr. 17:50–59, 1965.

822. Windsor, M. W. Luminescence and energy transfer, pp. 345–431. In D. Fox, M. M. Labes, and A. Weissberger, Eds. Physics and Chemistry of the Organic Solid State. Vol. 2. New York: Interscience Publishers, Inc., 1965.

823. Winefordner, J. D. Time-resolved phosphorimetry. Accounts Chem. Res. 2:361–367, 1969.

824. Winkelstein, W., Jr., and S. Kantor. Prostatic cancer: Relationship to suspended particulate air pollution. Amer. J. Public Health 59:1134–1138, 1969.

825. Winkelstein, W., and S. Kantor. Stomach cancer. Positive association with suspended particulate air pollution. Arch. Environ. Health 18:544–547, 1960.

825a. Winkelstein, W., S. Kantor, E. W. Davis, C. S. Maneri, and W. E. Mosher. The relationship of air pollution and economic status to total mortality and selected respiratory system mortality in men. I. Suspended particulates. Arch. Environ. Health 14:162–171, 1962.

826. Wolff, E. Embryogenesis *in vitro*, pp. 531–589. In E. N. Willmer, Ed. Cells and Tissues in Culture: Methods, Biology, and Physiology. Vol. 1. London: Academic Press Inc., 1965.

827. Wynder, E. L., and D. Hoffmann. Present status of laboratory studies on tobacco carcinogenesis. Acta Path. Microbiol. Scand. 52:119–132, 1961.

828. Wynder, E. L., and D. Hoffmann. Selected laboratory methods in tobacco carcinogenesis. Methods Cancer Res. 4:3–52, 1968.

829. Wynder, E. L., and D. Hoffmann. Some laboratory and epidemiological aspects of air pollution carcinogenesis. J. Air Pollut. Control Assoc. 15: 155–159, 1965.

830. Wynder, E. L., and D. Hoffmann. The epidermis and the respiratory tract as bioassay systems in tobacco carcinogenesis. Brit. J. Cancer 24: 574–582, 1970.

831. Wynder, E. L., and D. Hoffmann, Eds. Tobacco and Tobacco Smoke. Studies in Experimental Carcinogenesis. New York: Academic Press Inc., 1968. 730 pp.

832. Wynder, E. L., J. Kmet, N. Dungal, and M. Segi. An epidemiological investigation of gastric cancer. Cancer 16:1461–1496, 1963.

833. Yasuhira, K. Experimental induction of lung cancer in rat and mouse with 20-methylcholanthrene in Freund's adjuvant. Acta Path. Jap. 17:475–493, 1967.

834. Yuile, C. L., H. L. Berke, and T. Hull. Lung cancer following polonium 210 inhalation in rats. Radiat. Res. 31:760–774, 1967.
835. Zander, M. Phosphorimetry. The Application of Phosphorescence to the Analysis of Organic Compounds. Translated by T. H. Goodwin. New York: Academic Press Inc., 1968. 206 pp.
836. Zweig, A. Electron transfer luminescence in solution. Adv. Photochem. 6: 425–451, 1968.

Index